AF806499

METHODS IN MOLECULAR BIOLOGY

Series Editor
John M. Walker
School of Life and Medical Sciences
University of Hertfordshire
Hatfield, Hertfordshire, AL10 9AB, UK

For further volumes:
http://www.springer.com/series/7651

Thyroid Hormone Nuclear Receptor

Methods and Protocols

Edited by

Michelina Plateroti

Centre Léon Bérard, Cancer Research Center of Lyon, Lyon, France

Jacques Samarut

École normale supérieure de Lyon , Lyon, France

Humana Press

Editors
Michelina Plateroti
Centre Léon Bérard
Cancer Research Center of Lyon
Lyon, France

Jacques Samarut
École normale supérieure de Lyon
Lyon, France

ISSN 1064-3745 ISSN 1940-6029 (electronic)
Methods in Molecular Biology
ISBN 978-1-4939-7901-1 ISBN 978-1-4939-7902-8 (eBook)
https://doi.org/10.1007/978-1-4939-7902-8

Library of Congress Control Number: 2018943137

© Springer Science+Business Media, LLC, part of Springer Nature 2018
This work is subject to copyright. All rights are reserved by the Publisher, whether the whole or part of the material is concerned, specifically the rights of translation, reprinting, reuse of illustrations, recitation, broadcasting, reproduction on microfilms or in any other physical way, and transmission or information storage and retrieval, electronic adaptation, computer software, or by similar or dissimilar methodology now known or hereafter developed.
The use of general descriptive names, registered names, trademarks, service marks, etc. in this publication does not imply, even in the absence of a specific statement, that such names are exempt from the relevant protective laws and regulations and therefore free for general use.
The publisher, the authors and the editors are safe to assume that the advice and information in this book are believed to be true and accurate at the date of publication. Neither the publisher nor the authors or the editors give a warranty, express or implied, with respect to the material contained herein or for any errors or omissions that may have been made. The publisher remains neutral with regard to jurisdictional claims in published maps and institutional affiliations.

Printed on acid-free paper

This Humana Press imprint is published by the registered company Springer Science+Business Media, LLC part of Springer Nature.
The registered company address is: 233 Spring Street, New York, NY 10013, U.S.A.

Preface

Thyroid hormones (THs), namely T3 and T4, play a fundamental role in the development, growth, and metabolic homeostasis in all vertebrates by affecting the expression of multiple genes. As a consequence, an altered thyroid status affects many organs and systems. TH's action is mediated mainly by the T3 receptors (TRs), which belong to the nuclear hormone receptor superfamily of transcription factors. Interestingly, the THs and the TRs are also present in nonvertebrates, opening to important questions concerning their function in simpler animals as well as their role along the evolution.

As for other nuclear receptors, studies on TRs have broadened the field of interest to transcription processes, epigenetics, intracellular signaling, structural biology, and molecular epidemiology. Given the growing amount of data on the pleiotropic functions of the THs and TRs, the aim of this volume is to summarize the advances in this field both from a conceptual and methodological point of view. In particular, by summarizing key topics each chapter addresses a comprehensive review of the most common methodologies. Indeed, this includes classical approaches of animal or cell engineering, biochemistry, cellular and molecular biology. However, new technologies are also illustrated that are very promising due to their high efficiency. This is the case of the genome editing approaches to modify TRs in animals or in cell lines or the large-scale deep sequencing techniques to search for TR mutations, TR binding sites, or to analyze the transcriptome in TR's modified cells or animals. Finally, methods to quantify the THs and their derivatives are now extremely refined and permit the detection of very low levels of molecules without the use of radioactive iodine.

The different topics covered in the volume span from a general introduction, with an historical perspective, about the THs and TRs, through the lessons learned from TR's engineered mice and the molecular mechanisms of TR's action. Moreover, human genetics of TRs is described in relation to the resistance to thyroid hormone (RTH) syndrome linked to mutations of TRα or TRβ genes. Also, how the hypothalamus-pituitary-thyroid (HPT) axis is functioning and the relevance of the local THs production and metabolism are largely presented and discussed. Last but not least, nonmammalian models are also treated that gives an evo-devo perspective to the volume. Throughout all these chapters the roles and mechanisms of action of THs and TRs are covered in the field of physiology, development, oncogenesis, genetics, and molecular dynamics.

Taken together, this book should then be of wide interest to scientists of fundamental fields as well as for physicians to medical applications.

Lyon, France

Michelina Plateroti
Jacques Samarut

Contents

Contributors

J. H. DUNCAN BASSETT · *Department of Medicine, Molecular Endocrinology Laboratory, Imperial College London, London, UK*

JUAN BERNAL · *Instituto de Investigaciones Biomedicas, Consejo Superior de Investigaciones Científicas (CSIC), Universidad Autónoma de Madrid (UAM), Madrid, Spain; Center for Biomedical Research on Rare Diseases (CIBERER), Instituto de Salud Carlos III, Madrid, Spain*

ANTONIO C. BIANCO · *Division of Endocrinology, Diabetes and Metabolism, Rush University Medical Center, Chicago, IL, USA*

NATACHA BOUHOURS-NOUET · *Centre de référence des maladies rares de la thyroïde et des Récepteurs hormonaux, service EDN, CHU d'Angers, Institut MITOVASC, Université d'Angers, Angers, France*

GREGORY A. BRENT · *Department of Medicine, David Geffen School of Medicine at UCLA and VA Greater Los Angles Healthcare System, Los Angeles, CA, USA; Department of Physiology, David Geffen School of Medicine at UCLA and VA Greater Los Angles Healthcare System, Los Angeles, CA, USA*

C. BRIET · *Centre de référence des maladies rares de la thyroïde et des Récepteurs hormonaux, service EDN, CHU d'Angers, Institut MITOVASC, Université d'Angers, Angers, France*

NICOLAS BUISINE · *Function and Mechanism of Action of Thyroid Hormone Receptor group, UMR 7221 CNRS and Muséum National d'Histoire Naturelle, Sorbonne Universités, Paris, France*

GÉRARD CABELLO · *DMEM, Univ Montpellier, INRA, Montpellier, France*

FRANÇOIS CASAS · *DMEM, Univ Montpellier, INRA, Montpellier, France*

SHEUE-YANN CHENG · *Laboratory of Molecular Biology, Center for Cancer Research, National Cancer Institute, National Institutes of Health, Bethesda, MD, USA*

PAUL J. DAVIS · *Pharmaceutical Research Institute, Albany College of Pharmacy and Health Sciences, Rensselaer, NY, USA; Albany Medical College, Albany, NY, USA*

RODRIGO R. DA CONCEIÇÃO · *Division of Endocrinology, Diabetes and Metabolism, Rush University Medical Center, Chicago, IL, USA; Programa de Pos-Graduacao em Medicina, Endocrinologia Clinica, Universidade Federal de Sao Paulo, Sao Paulo, SP, Brazil*

BARBARA A. DEMENEIX · *Evolution des Régulations Endocriniennes, Département "Adaptation du Vivant", UMR 7221 Muséum National d'Histoire Naturelle /CNRS, Paris, France*

JORDAN DEVEREAUX · *Program in Chemical Biology, Department of Physiology and Pharmacology, Oregon Health and Science University, Portland, OR, USA*

SKYLAR J. FERRARA · *Program in Chemical Biology, Department of Physiology and Pharmacology, Oregon Health and Science University, Portland, OR, USA*

JEAN-BAPTISTE FINI · *Evolution des Régulations Endocriniennes, Département "Adaptation du Vivant", UMR 7221 Muséum National d'Histoire Naturelle /CNRS, Paris, France*

FRÉDÉRIC FLAMANT · *Institut de Génomique Fonctionnelle de Lyon, Ecole Normale Supérieure de Lyon, INRA USC 1370, CNRS UMR 5242, Université de Lyon, Université Lyon 1, Lyon Cedex 07, France*

LIEZHEN FU · *Section on Molecular Morphogenesis, Eunice Kennedy Shriver National Institute of Child Health and Human Development (NICHD), National Institutes of Health (NIH), Bethesda, MD, USA*

KARINE GAUTHIER · *Institut de Génomique Fonctionnelle de Lyon, Université de Lyon, Université Lyon 1, CNRS UMR 5242, Ecole Normale Supérieure de Lyon, Lyon cedex 07, France*

WENYUE GUAN · *Institut de Génomique Fonctionnelle de Lyon, Ecole Normale Supérieure de Lyon, INRA USC 1370, CNRS UMR 5242, Université de Lyon, Université Lyon 1, Lyon Cedex 07, France*

ROMAIN GUYOT · *Institut de Génomique Fonctionnelle de Lyon, Ecole Normale Supérieure de Lyon, INRA USC 1370, CNRS UMR 5242, Université de Lyon, Université Lyon 1, Lyon Cedex 07, France*

F. ILLOUZ · *Centre de référence des maladies rares de la thyroïde et des Récepteurs hormonaux, service EDN, CHU d'Angers, Institut MITOVASC, Université d'Angers, Angers, France*

GWENNEG KERDIVEL · *Function and Mechanism of Action of Thyroid Hormone Receptor group, UMR 7221 CNRS and Muséum National d'Histoire Naturelle, Sorbonne Universités, Paris, France*

JOSEF KÖHRLE · *Charité–Universitätsmedizin Berlin, corporate member of Freie Universität Berlin, Humboldt-Universität zuBerlin, and Berlin Institute of Health, Institut für Experimentelle Endokrinologie, Berlin, Germany*

RONNY LESMANA · *Department of Physiology, Universitas Padjadjaran, Bandung, Indonesia*

HUNG-YUN LIN · *Taipei Cancer Center, Taipei Medical University, Taipei, Taiwan; Ph.D. Program for Cancer Biology and Drug Discovery, College of Medical Science and Technology, Taipei Medical University, Taipei, Taiwan*

YAN-YUN LIU · *Department of Medicine, David Geffen School of Medicine at UCLA and VA Greater Los Angles Healthcare System, Los Angeles, CA, USA; Department of Physiology, David Geffen School of Medicine at UCLA and VA Greater Los Angles Healthcare System, Los Angeles, CA, USA*

FEDERICA MARELLI · *Department of Clinical Sciences and Community Health, University of Milan, Milan, Italy; Lab of Endocrine and Metabolic Research, IRCCS Istituto Auxologico Italiano, Milan, Italy*

BEATRIZ MORTE · *Center for Biomedical Research on Rare Diseases (CIBERER), Instituto de Salud Carlos III, Madrid, Spain*

SHAKER A. MOUSA · *Pharmaceutical Research Institute, Albany College of Pharmacy and Health Sciences, Rensselaer, NY, USA*

BILAL B. MUGHAL · *Evolution des Régulations Endocriniennes, Département "Adaptation du Vivant", UMR 7221 Muséum National d'Histoire Naturelle /CNRS, Paris, France*

THEODORA PAPPA · *Department of Medicine, The University of Chicago, Chicago, IL, USA*

Luca Persani · *Department of Clinical Sciences and Community Health, University of Milan, Milan, Italy; Lab of Endocrine and Metabolic Research, IRCCS Istituto Auxologico Italiano, Milan, Italy*
Michelina Plateroti · *Département de la Recherche, Centre Léon Bérard, Centre de Recherche en Cancerologie de Lyon (CRCL), INSERM U1052, CNRS UMR5286, Université de Lyon, Université Lyon 1, Lyon, France*
D. Prunier-Mirebeau · *Centre de référence des maladies rares de la thyroïde et des Récepteurs hormonaux, service EDN, CHU d'Angers, Institut MITOVASC, Université d'Angers, Angers, France*
Samuel Refetoff · *Department of Medicine, The University of Chicago, Chicago, IL, USA; Department of Pediatrics, The University of Chicago, Chicago, IL, USA; Committee on Genetics, The University of Chicago, Chicago, IL, USA*
P. Rodien · *Centre de référence des maladies rares de la thyroïde et des Récepteurs hormonaux, service EDN, CHU d'Angers, Institut MITOVASC, Université d'Angers, Angers, France*
Laurent M. Sachs · *Function and Mechanism of Action of Thyroid Hormone Receptor group, UMR 7221 CNRS and Muséum National d'Histoire Naturelle, Sorbonne Universités, Paris, France*
Jacques Samarut · *Institut de Génomique Fonctionnelle de Lyon, Ecole Normale Supérieure de Lyon, Université de Lyon, Hospices Civils de Lyon, Lyon, France*
Thomas S. Scanlan · *Program in Chemical Biology, Department of Physiology and Pharmacology, Oregon Health and Science University, Portland, OR, USA*
Yun-Bo Shi · *Section on Molecular Morphogenesis, Eunice Kennedy Shriver National Institute of Child Health and Human Development (NICHD), National Institutes of Health (NIH), Bethesda, MD, USA*
Rohit A. Sinha · *Program of Cardiovascular and Metabolic Disorders, Duke-NUS Medical School Singapore, Singapore, Singapore; Department of Endocrinology, SGPGIMS, Lucknow, India*
Maria Sirakov · *Dipartimento di Scienze e Tecnologie Ambientali Biologiche e Farmaceutiche (DiSTABiF), Università degli Studi della Campania "Luigi Vanvitelli", Caserta, Italy; Département de la Recherche, Centre Léon Bérard, Centre de Recherche en Cancerologie de Lyon (CRCL), INSERM U1052, CNRS UMR5286, Université de Lyon, Université Lyon 1, Lyon, France*
Luan Wen · *Section on Molecular Morphogenesis, Eunice Kennedy Shriver National Institute of Child Health and Human Development (NICHD), National Institutes of Health (NIH), Bethesda, MD, USA*
Graham R. Williams · *Department of Medicine, Molecular Endocrinology Laboratory, Imperial College London, London, UK*
F. E. Wondisford · *Department of Medicine, Rutgers-Robert Wood Johnson Medical School, Henry Rutgers Term Chair and Chancellor Scholar Rutgers, The State University of New Jersey, New Brunswick, NJ, USA*
Chantal Wrutniak-Cabello · *DMEM, Univ Montpellier, INRA, Montpellier, France*
Winifred W. Y. Yau · *Program of Cardiovascular and Metabolic Disorders, Duke-NUS Medical School Singapore, Singapore, Singapore*

PAUL M. YEN · *Program of Cardiovascular and Metabolic Disorders, Duke-NUS Medical School Singapore, Singapore, Singapore; Department of Medicine, Duke Molecular Physiology Institute, Duke University Medical Center, Durham, NC, USA; Department of Pharmacology and Cancer Biology, Duke Molecular Physiology Institute, Duke University Medical Center, Durham, NC, USA*

JIN ZHOU · *Program of Cardiovascular and Metabolic Disorders, Duke-NUS Medical School Singapore, Singapore, Singapore*

XUGUANG ZHU · *Laboratory of Molecular Biology, Center for Cancer Research, National Cancer Institute, National Institutes of Health, Bethesda, MD, USA*

Chapter 1

Thyroid Hormone Receptors: Several Players for One Hormone and Multiple Functions

Jacques Samarut and Michelina Plateroti

Abstract

Thyroid hormone receptors (TRs) were cloned based on their homology with the retroviral oncogene v-ERBA. In Vertebrates two genes, *THRA* and *THRB*, encode respectively many isotypes and isoforms of receptors TRα and TRβ, resulting from alternative splicing and/or internal transcription start sites. We present here a wide overview of this diversity and of their mechanisms of action as transcription regulators, as well as alternative actions through cytoplasmic signaling.

Key words Thyroid hormone receptors (TRs), *THRA*, *THRB*, Cytoplasmic signaling

1 Introduction

The aim of this introductory chapter is to give a general overview on TRs for the newcomers in the field. As the literature is now quite large, in most cases we have referred to review papers.

Thyroid hormones (TH), namely L-thyroxine or T4 and 3,5,3′-L-triiodothyronine or T3, have been discovered in the early twentieth century from their effects in controlling growth and development of newborn Vertebrates, and for their role in activating basal metabolic rate. The function of both hormones, but mostly T3, at the cellular level, led quickly to the hypothesis that they were acting through binding to specific proteins within the cells. Indeed, this hypothesis was definitively supported by experimental arguments, and the pioneering work of Tata in the 1960s [1] paved the way for successive investigations that demonstrated a direct effect of the THs on transcription, after the identification of specific receptors for T3 as transcriptional regulators. This historical overview has been reviewed by Tata [2].

Michelina Plateroti and Jacques Samarut (eds.), *Thyroid Hormone Nuclear Receptor: Methods and Protocols*, Methods in Molecular Biology, vol. 1801, https://doi.org/10.1007/978-1-4939-7902-8_1,
© Springer Science+Business Media, LLC, part of Springer Nature 2018

2 Identification of Nuclear Receptors for Thyroid Hormones

Right at the beginning of the 1980s a bunch of receptors for steroid hormones were cloned almost simultaneously by several groups [3]. Interestingly, within this context, thyroid hormone receptors were cloned through two different ways. The group of Vennström was interested in cloning and characterizing the normal cellular counterpart of the v-erbA oncogene, one of the two oncogenes encoded by the Avian Erythroleukemia Virus (AEV), an acute avian leukemia virus inducing erythroleukemia. The structure of the c-ErbA protein showed a strong homology with nuclear hormone receptor proteins and was demonstrated to specifically bind the hormone T3 [4]. Almost simultaneously, the group of Evans cloned and characterized another thyroid hormone receptor based on sequence homology with the v-erbA oncogene and steroid hormone receptors [5]. These two thyroid hormone nuclear receptors turned out to show minor sequence differences between each other, which generated the concept of two isotypes, respectively TRα and TRβ in Vertebrates. It has successively been shown that these two isotypes are encoded by two independent genes, respectively, *THRA* and *THRB*, located on chromosomes 17 and 3 respectively in humans. No further evidence has been found so far for any additional TR gene encoding a third isotype, as it is the case for RARs and RXRs [3].

As for other nuclear hormone receptors, the main characteristic of the TRs is the presence of a DNA- and a hormone binding domain, known as DBD and HBD, respectively. The amino-acid sequence of these domains is highly conserved across species [3]. The DBD contains two zinc finger structures that interact with DNA [6], while the HBD made of 12 alpha helices allows for hormone binding and for interaction with co-regulators [3]. Activating Function 1 and Activating Function 2 domains, which are located at the amino-terminal and carboxy-terminal ends respectively, are responsible for the hormone-independent or hormone-dependent transactivation properties of the receptors [6]. The amino-terminal end also contains a sequence for nuclear localization (NLS) [7].

From the study of TR transcripts in various murine tissues it resulted clearly that each of the *THR* genes encodes multiple isoforms generated by alternative promoter usage and alternative splicing [8].

2.1 TRα Isoforms

In human and mouse the *THRA/Thra* locus encodes several proteins but only TRα1 is a true nuclear receptor harboring the typical structure that includes a DBD and a HBD [8]. This isoform has been clearly demonstrated to be a bona fide T3-modulated transcription factor. The TRα2 protein shares 100% homology with the first 370 amino acids of TRα1, but an alternative-splicing generates a protein highly divergent from TRα1 in the region encoding the HBD, resulting in its inability to bind T3 [9]. In both human and mouse the TRΔα1 and TRΔα2 truncated isotypes are TRα isoforms transcribed from an internal promoter located in intron 7 of the *THRA/Thra* gene, and consequently lack both the DBD and HDB [10]. These last three isoforms may behave as antagonists of TRα1 or TRβ receptors [11]. Finally, the p43 isoform translated from an internal initiation codon lacks the extreme amino-terminus but possesses both the DBD and HBD, therefore displaying all the characteristics of a functional nuclear receptor for T3. The p43 protein is localized in the mitochondria and has been described as a regulator of mitochondrial genome transcription [12], (*see* Chapter 14 by Wrutniak-Cabello et al., this book).

2.2 TRβ Isoforms

The *THRB* locus encodes two receptor isoforms, TRβ1 and TRβ2, which differ from each other by the length of their amino-termini [8]. All these isoforms are conserved between mouse and human as suggested by direct observation or comparison of respective genomic sequences. The rat *THRB* locus encodes in addition the TRβ3 and TRΔβ3 isoforms that can be translated separately or from a single transcript containing an internal translation start site. TRβ3 is a bona fide T3 nuclear receptor, while TRΔβ3 retains T3 binding ability but lacks the DBD and exerts a cell- and TRE-specific inhibitory function on the TRβ1, TRβ3, and TRα1 receptors [13, 14].

3 TRs as Regulators of Transcription

For TRs, as for other nuclear hormone receptors, there is a large literature describing their major features as transcription factors. As such, they can interact with several cofactors, including co-activators and co-repressors, able to remodel the chromatin. Indeed, TRs modulate the transcription of target genes by binding to specific DNA sequences called thyroid hormone response elements (TREs). TREs organized as direct repeat of a five-bases motif separated by four bases (DR4) are more common than other types of TR-consensus sequences [6, 15]. On DNA, TRs usually form heterodimers with the Retinoid X Receptors (RXRs), which are also part of the nuclear hormone receptor superfamily [16]. From a molecular point of view, the binding of T3 to TRs, in the case of positively regulated genes, induces structural changes of the receptors causing the dissociation of co-repressors (CoRs) and their recruitment of co-activator complexes (CoAs), which finally result in stimulation of transcription [17]. Few genes are negatively regulated by T3, such as *TRH* and *TSH,* and in both cases the presence of TREs has been reported in their promoter. So far, the detailed mechanisms underlying the active repression by liganded TRs have been poorly defined, although CoRs and CoAs may also be involved in the control of negatively regulated genes [18].

Notably, the TRs may play a regulatory function independently of their ligand. The unliganded TRs, also called apo-receptors, have the ability to bind to TREs in the chromatin and to recruit CoRs including histone deacetylases, thus generating a repressive transcriptional complex. In the presence of the ligand, the apo-receptors transform into holo-receptors and a conformational change induces the release of the repressive complexes and the recruitment of activating complexes, including histone actetyl transferases [19]. Then TRs can be considered as change-over transcriptional switches depending on their specific conformation and T3 binding. This property is rather unique among nuclear hormone receptors and has been documented so far only for TRs and RARs [20].

The fact that TRs have a conformation-specific function and that they might control gene expression in an "all or none" fashion can have some biological significance. Indeed, it might allow different tissues to respond differently depending on the local concentration of THs. In fact, it is known that local tissue and cell concentration of T3 are tightly controlled by TH-transporters [21] and TH-metabolizing enzymes such as the deiodinases [22], which in turn indirectly control T3/TR-dependent gene expression. This local control of TH levels can also ensure a proper time-scale regulation of TR target genes during development. In support of these assumptions, some studies demonstrated the role of apo-TRs in vivo. In congenital hypothyroid Pax8 knockout mice, TRα but not TRβ apo-receptors are responsible for the repression of T3-regulated genes and for delayed tissues/organs development, leading to the death of pre-weaned animals. Indeed, the suppression of TRα but not of TRβ in these animals reverts these phenotypes and finally obviates precocious death, demonstrating that in hypothyroid conditions, TRα1 apo-receptors block essential physiological functions [23]. Another example of TRα1 apo-receptor function in vivo is represented by the early foetal period of development in the mouse embryo when circulating T3 levels are almost undetectable because the thyroid gland develops at the end of gestation. Consequently, TRs are essentially in apo-receptor configuration during this period. It has been demonstrated that in the mouse foetal heart the TRα1 apo-receptors block the transcription of their target genes, thereby slowing down the heart rate. After birth, when TH levels increase, liganded-TRα1 reverses this condition and activates gene expression leading to an increase in the heart rate [24]. Apo-TRα receptor also regulates metamorphosis timing in tadpoles [23].

Interestingly, the physiopathological relevance of the observations conducted in mouse models has fully been demonstrated in patients affected by the syndromes of Resistance to Thyroid

Hormones induced by mutated TRα (RTHα). In fact, these patients express TRα with poor or null binding of T3 due to mutations in their HBD, resulting in a more or less severe incapacity of the receptor to switch from the apo- to the holo-conformation [25], (*see* the Chapter 9 by Briet et al. in this book).

4 Tissue Distribution and Roles of TRs

Mouse *THRA* and *THRB* genes are widely expressed within different organs and tissues [8]. TRα1 and TRα2 are found almost ubiquitously, and in particular TRα1 is expressed early during development in both amphibians and mouse [8, 26]. In the mouse, the TRΔα transcripts are found mostly in the small intestinal epithelium, lung, brain and early embryo. TRβ1 is also widely expressed and it represents the most abundant TR isoform transcripts in the liver. TRβ2 expression is restricted to the pituitary, hypothalamic TRH neurons, the developing inner ear and retina. The TRβ3 isoform transcripts have been described in rat liver, kidney, and lung and the TRΔβ3 in rat skeletal muscle, heart, spleen, and brain [14]. TR proteins are generally expressed at low level and their study in tissue or cells by immunolabeling approaches is rendered difficult due to poorly efficient available antibodies.

The roles of the different TR isotypes and isoforms in biological functions in vivo have been largely addressed by experimental genetics in the mouse through the generation of specific mutant animals [27]. In fact, a large series of TR-knockout animals and of animals carrying TR-point mutations have been constructed with the aim to abrogate all or specific isoforms encoded by the *THRA* or *THRB* loci. A few studies in the frog by specific knocking down approaches also addressed the function of TRs in vivo and helped to define specific functions for each TR isotype [28], (*see* the Chapter 20 by Fu et al. in this book).

In all these mutants, however, TR mutations are ubiquitous and constitutive possibly influencing the interpretation of results, as direct and indirect mechanisms may be responsible for the phenotypic alterations observed. The generation of conditional, tissue- and time-specific, mouse models based upon the CRE/Lox strategy has strongly helped in better defining the functions of the TRs in specific tissues both during development and in adulthood [27].

From all these studies it clearly appears that TRα and TRβ are controlling distinct functions. TRα1 is a major actor in development and growth and controls the morphogenesis of several tissues and organs including bone, intestine, and brain. Major adverse consequences of TRα1 depletion in mouse are linked to retarded postnatal development at weaning, which appears to be a period of strong requirement of TH signaling. From these observations and in parallel to the TH-dependent amphibian metamorphosis, it has been proposed that TRα1 might control a metamorphosis-like process in newborn mammals similar to that observed in developing tadpoles [27]. Interestingly, the specific knockout of the p43 TRα isoform alters mitochondrial activities, supporting the specificity of this isoform in mitochondrial function [29], (*see* the Chapter 14 by Wrutniak-Cabello et al. in this book).

The TRβ1 isoform plays a major role in adult liver, heart, thyroid, and pituitary and appears strongly involved in physiological regulations. The best example is illustrated by the function of TRβ1, elucidated from TRβ knockout mice, in the regulation of TH production by the thyroid gland. In fact, this regulation depends on a complex negative feedback loop controlling TSH and TRH expression respectively in pituitary and hypothalamus [30, 31]. Finally, one other important lesson learned from TRβ knockout mice is that TRβ isoforms control the development of sensory functions, including color vision and audition [32].

5 Other Mechanisms of Thyroid Hormone Signaling

Several years after their discovery as transcription factors binding directly to DNA, TRs have been found to play alternative roles [33]. In fact, in some cases the receptors are tethering transcription complexes on the chromatin of some genes, working rather like transcription modulators. Accordingly, those genes do not contain TREs. More interestingly, THs can also modulate cell physiology by signaling at the cell membrane [34] (*see* the Chapter 7 by Lin et al. in this book) or inside the cytoplasm [35, 36]. These actions have been referred to as non-genomic action of thyroid hormones.

Based on the existence of these multiple mechanisms, recently a new classification has been proposed for these various modes of action [37]. In short, we can distinguish TR-dependent signaling of thyroid hormone with direct or indirect binding to DNA, TR-dependent signaling of thyroid hormone in the cytoplasm without binding to chromatin and TR-independent thyroid hormone signaling.

6 Conclusions

TRs appear then as major mediators of developmental and physiological functions in Vertebrates. Although they were first identified as direct transcription activators, their mechanisms of action turned out to be quite diversified. Indeed, due to their multiple mechanisms of action, they can behave either as transcription activators or repressors, co-factors within transcription complexes or components of cytoplasmic signaling. There is no doubt that this diversity of functions, also linked to the diversity of isotypes and isoforms, accounts for the multiple physiological effects of thyroid hormones in the organism all along the life.

References

1. Pitt-Rivers R, Tata JR (1959) The thyroid hormones. Pergamon Press, London
2. Tata JR (2013) The road to nuclear receptors of thyroid hormone. Biochim Biophys Acta 1830(7):3860–3866. https://doi.org/10.1016/j.bbagen.2012.02.017
3. Mangelsdorf DJ, Thummel C, Beato M et al (1995) The nuclear receptor superfamily: the second decade. Cell 83(6):835–839
4. Sap J, Muñoz A, Damm K et al (1986) The c-erbA protein is a high affinity receptor for thyroid hormone. Nature 324(6098):635–640
5. Weinberger C, Thompson CC, Ong ES et al (1986) The c-erb-A gene encodes a thyroid hormone receptor. Nature 324(6098):641–646
6. Yen PM (2001) Physiological and molecular basis of thyroid hormone action. Physiol Rev 81(3):1097–1142
7. Andersson ML, Vennström B (1997) Chicken thyroid hormone receptor alpha requires the N-terminal amino acids for exclusive nuclear localization. FEBS Lett 416(3):291–296
8. Flamant F, Samarut J (2003) Thyroid hormone receptors : lessons from knockout and knock-in mutant mice. Trends Endocrinol Metab 14:85–90
9. Mitsuhashi T, Tennyson GE, Nikodem VM (1988) Alternative splicing generates messages encoding rat c-erbA proteins that do not bind thyroid hormone. Nat Rev Endocrinol 12(2):111–121. https://doi.org/10.1038/nrendo.2015.205.
10. Chassande O, Fraichard A, Gauthier K et al (1997) Identification of transcripts initiated from an internal promoter in the c-erbAα locus encode inhibitors of RARalpha and T3R activities. Mol Endoc 11:1278–1290
11. Plateroti M, Gauthier K, Domon-Dell C et al (2001) Functional interference between thyroid hormone receptor TRα and natural truncated isoforms TRΔα in the control of intestine development. Mol Cell Biol 21:4761–4772
12. Wrutniak-Cabello C, Casas F, Cabello G (2017) Mitochondrial T3 receptor and targets. Mol Cell Endocrinol 458:112–120. https://doi.org/10.1016/j.mce.2017.01.054
13. Harvey CB, Bassett JH, Maruvada P et al (2007) The rat thyroid hormone receptor (TR) Deltabeta3 displays cell-, TR isoform-, and thyroid hormone response element-specific actions. Endocrinology 148(4):1764–1773
14. Williams GR (2000) Cloning and characterization of two novel thyroid hormone receptor beta isoforms. Mol Cell Biol 20(22):8329–8342
15. Chatonnet F, Guyot R, Benoît G, Flamant F (2013) Genome-wide analysis of thyroid hormone receptors shared and specific functions in neural cells. Proc Natl Acad Sci U S A 110(8):E766–E775. https://doi.org/10.1073/pnas.1210626110
16. Laudet V (1997) Evolution of the nuclear receptor superfamily: early diversification from an ancestral orphan receptor. J Mol Endocrinol 19(3):207–226
17. Rosenfeld MG, Glass CK (2001) Coregulator codes of transcriptional regulation by nuclear receptors. J Biol Chem 276(40):36865–36868
18. Steinfelder HJ, Wondisford FE (1997) Thyrotropin (TSH) beta-subunit gene expression--an example for the complex regulation of pituitary hormone genes. Exp Clin Endocrinol Diabetes 105(4):196–203
19. Perissi V, Rosenfeld MG (2005) Controlling nuclear receptors: the circular logic of cofactor cycles. Nat Rev Mol Cell Biol 6(7):542–554
20. Hörlein AJ, Näär AM, Heinzel T et al (1995) Ligand-independent repression by the thyroid hormone receptor mediated by a nuclear receptor co-repressor. Nature 377(6548):397–404
21. van der Deure WM, Hansen PS, Peeters RP et al (2008) Thyroid hormone transport and metabolism by organic anion transporter 1C1 and consequences of genetic variation. Endocrinology 149(10):5307–5314. https://doi.org/10.1210/en.2008-0430
22. Bianco AC, Kim BW (2006) Deiodinases: implications of the local control of thyroid hormone action. J Clin Invest 116(10):2571–2579
23. Flamant F, Poguet A-L, Plateroti M et al (2002) Congenital hypothyroid Pax8$^{-/-}$ mutant mice can be rescued by inactivating the thyroid hormone receptor TRα gene. Mol. Endocrinology 16:24–32
24. Mai W, Janier MF, Allioli N et al (2004) Thyroid hormone receptor α is a molecular switch of cardiac function between fetal and post-natal life. Proc Nat Acad Sci 101:10332–10337

25. van Gucht ALM, Moran C, Meima ME et al (2017) Resistance to thyroid hormone due to heterozygous mutations in thyroid hormone receptor alpha. Curr Top Dev Biol 125:337–355. https://doi.org/10.1016/bs.ctdb.2017.02.001
26. Wen L, Shi YB (2016) Regulation of growth rate and developmental timing by Xenopus thyroid hormone receptor α. Develop Growth Differ 58(1):106–115. https://doi.org/10.1111/dgd.12231
27. Flamant F, Gauthier K (2013) Thyroid hormone receptors: the challenge of elucidating isotype-specific functions and cell-specific response. Biochim Biophys Acta 1830(7):3900–3907. https://doi.org/10.1016/j.bbagen.2012.06.003
28. Shi YB (2013) Unliganded thyroid hormone receptor regulates metamorphic timing via the recruitment of histone deacetylase complexes. Curr Top Dev Biol 105:275–297. https://doi.org/10.1016/B978-0-12-396968-2.00010-5
29. Pessemesse L, Schlernitzauer A, Sar C et al (2012) Depletion of the p43 mitochondrial T3 receptor in mice affects skeletal muscle development and activity. FASEB J 26(2):748–756. https://doi.org/10.1096/fj.11-195933
30. Gauthier K, Chassande O, Plateroti M et al (1999) Different functions for the thyroid hormone receptors TRα and TRβ in the control of thyroid hormone production and post-natal development. EMBO J 18:623–631
31. Göthe S, Wang Z, Ng L et al (1999) Mice devoid of all known thyroid hormone receptors are viable but exhibit disorders of the pituitary-thyroid axis, growth and bone maturation. Genes Dev 13(10):1329–1341
32. Jones I, Srinivas M, Ng L, Forrest D (2003) The thyroid hormone receptor beta gene: structure and functions in the brain and sensory systems. Thyroid 13(11):1057–1068
33. Gauthier K, Flamant F (2014) Nongenomic, TRβ-dependent, thyroid hormone response gets genetic support. Endocrinology 155(9):3206–3209. https://doi.org/10.1210/en.2014-1597
34. Davis PJ, Goglia F, Leonard JL (2016) Nongenomic actions of thyroid hormone. Nat Rev Endocrinol 12(2):111–121. https://doi.org/10.1038/nrendo.2015.205
35. Kalyanaraman H et al (2014) Nongenomic thyroid hormone signaling occurs through a plasma membrane-localized receptor. Sci Signal 7(326):ra48
36. Martin NP, Marron Fernandez de Velasco E, Mizuno F et al (2014) A rapid cytoplasmic mechanism for PI3 kinase regulation by the nuclear thyroid hormone receptor, TRβ, and genetic evidence for its role in the maturation of mouse hippocampal synapses in vivo. Endocrinology 155(9):3713–3724. https://doi.org/10.1210/en.2013-2058.
37. Flamant F, Cheng SY, Hollenberg AN et al (2017) Thyroid hormone signaling pathways. Time for a more precise nomenclature. Endocrinology 158(7):2052–2057. https://doi.org/10.1210/en.2017-00250

Chapter 2

Two Protocols to Study the Interactions of Thyroid Hormone Receptors with Other Proteins and Chromatin

Wenyue Guan, Romain Guyot, and Frédéric Flamant

Abstract

Understanding the transcriptional function of thyroid hormone receptors implies a precise analysis of their interactions with chromatin and other protein components of the cells. We present here two protocols that are routinely used in our laboratory. The first co-immunoprecipitation procedure allows addressing the capacity of proteins to form stable multiprotein complexes with TRs in cells. The chromatin affinity purification enables us to define the sites occupied by TRs on chromatin. In this case the lack of high quality antibodies is circumvented by introducing an N-terminal tag in TR, with unspecific affinity for immunoglobulins.

Key words Thyroid hormone receptors, Co-immunoprecipitation, Chromatin affinity purification

1 Introduction

Triiodo-L-thyronine (T3) is the most active form or thyroid hormone. It regulates gene expression by binding to the TRα1, TRβ1, and TRβ2 (TRs) nuclear receptors. A model, common to other nuclear receptors, describes this mode of action [1]. Briefly TRs possess an N-terminal DNA binding domain linked by a hinge to a C-terminal ligand binding domain. This ligand binding domain is necessary not only for T3 binding, but also for generating a dimerization interface, which enables the formation of heterodimers with RXR, another nuclear receptor. TR/RXR heterodimers bind DNA mainly at the so-called DR4 response element, made of a tandem of two half-sites (5'AGGTCA3') separated by 4 nucleotides. Unliganded DNA-bound TRs recruit transcription corepressor complexes to chromatin. The binding of T3 in the ligand binding pocket of the C-terminal domain results in a conformational change in which the C-terminal helix (helix 12) of TR repositions. This destabilizes the interaction with corepressors, and instead favors the recruitment of transcription coactivators. This model does not take into account all complications brought by recent investigations. First, a plethora of interactions have been

Michelina Plateroti and Jacques Samarut (eds.), *Thyroid Hormone Nuclear Receptor: Methods and Protocols*, Methods in Molecular Biology, vol. 1801, https://doi.org/10.1007/978-1-4939-7902-8_2,
© Springer Science+Business Media, LLC, part of Springer Nature 2018

reported between TRs and both cytoplasmic and nuclear proteins, which do not necessarily belong to the coactivator/corepressor categories. Second, recent genome wide studies outlined the difficulty to establish a firm link between chromatin occupancy by TRs, and the eventual regulation by T3 of proximal genes: more than 70,000 DR4 consensus sites are present in the mammalian genomes, among which on several thousand are occupied in a given cell-type. Importantly, only a small fraction of the genes proximal to these genomic sites occupied by TRs are regulated by T3. For these reasons we propose here two protocols, which can be of great help to explore these difficult issues.

2 Protocol for Proteins Co-Immunoprecipitation

In vitro screening can be performed at high throughput and often identifies a large number of proteins able to interact with purified TRs. Co-immunoprecipitation (Co-IP) is a necessary complement, as it addresses the actual capacity of proteins to form stable multiprotein complexes with TRs in cells. This method does not inform on possible direct interaction between the two proteins, but provides key information on the relevance of the in vitro identified interactions. The protocol can be used to address the interaction between endogenous proteins, provided that good quality antibodies are available. It can also be performed on transfected cells, to increase protein concentrations, or on cell transfected with tagged protein, if suitable antibodies are not available.

2.1 Reagents and Buffers

Phosphate buffer saline (PBS) pH 7.4: 1.47 mM KH_2PO_4; 137 mM NaCl; 8.1 mM Na_2HPO_4-$7H_2O$; 2.7 mM KCl.

IP lysis buffer: 50 mM Tris–HCl pH 7.5: 150 mM KCl, 1% Triton X-100, 1 mM EDTA, 8% glycerol.

IP dilution buffer: 50 mM Tris–HCl pH 7.5: 150 mM KCl, 0,1% Triton X-100, 1 mM EDTA, 8% glycerol.

IP washing buffer: 50 mM Tris–HCl PH7.5: 150 mM KCl, 0.1% Triton X-100, 1 mM EDTA.

Protease inhibitor Cocktail (PIC): cOmplete™ EDTA-free Protease Inhibitor Cocktail Tablets, Roche Inc. Add 1 tablet in 500 μL H_2O as 100× stock buffer and keep in −20 °C. Avoid freeze/thaw cycle.

Bradford protein assay (BCA Protein Assays Thermofisher).

Bovine Serum Albumin (BSA, Roche; fraction V, fatty acid free, molecular biology grade). A stock solution (10 mg/mL) can be aliquoted and stored at −20 °C.

2.2 Procedure

1. Scrap adherent cells in prechilled PBS (volume: 5 mL).
2. Pellet cells in Eppendorf tube (200 × *g*, 5 min, 4 °C) and resuspend in ice-cold IP lysis buffer + PIC (1% of stock solution). Use 300 μL of buffer for 10^7 cells and incubate on ice for 30–50 min with occasional vortexing.
3. Clear lysate by centrifugation (30,000 × *g*, 4 °C, 10 min) and discard pellet.
4. Quantify the total proteins of cell lysis with BCA assay. Keep 50 μg of protein as input fraction, and use 500 μg of protein for immunoprecipitation.
5. Wash magnetic Protein A/G beads (Invitrogen) with IP binding buffer + BSA (2 μg/μL).
6. Prepare bead slurry by mixing one volume of Protein A/G beads and one volume of IP binding buffer + BSA (2 μg/μL).
7. Dilute the cell lysis (500 μg protein) 1:4 in IP dilution buffer + PIC (1×).
8. Incubate the diluted cell lysate with 20 μL Protein A/G slurry at 4 °C for 1 h on vertical wheel with gentle shaking.
9. Collect the beads with magnetic stand for 2 min. Keep the supernatant.
10. Incubate the precleaned cell lysis with antibody at 4 °C for 3–4 h or overnight. The quantity of antibody to be used depends on the abundance of target protein and the titer of the antibody and should be empirically optimized.
11. Add 20–50 μL prepared Protein A/G slurry into the lysate and incubate for another 2 h at 4 °C.
12. Collect the beads with magnetic stand for 2 min, wash the beads with 1 mL of prechilled IP washing buffer + PIC (1×) at 4 °C for 10 min.
13. Repeat the washing **step 4** times.
14. Collect the beads with magnetic stand for 2 min, remove as much supernatant as possible.
15. Add SDS loading buffer to the beads (and also to 50 μg input samples) and boil the samples at 100 °C for 10 min.
16. Proceed to SDS-PAGE and western blot analysis.

2.3 General Considerations

Choice between Protein A or Protein G

Protein A and G beads have varying affinities to different IgG subtypes (*see* below the chart showing the affinity of proteins A and G for various species immunoglobulin subtypes). In general, Rabbit IgG binds equally well to both Protein A and Protein G while mouse IgG has a preference for Protein G.

Species	Immunoglobulin	Affinity to Protein A	Affinity to Protein G
Human	IgG (normal)	++++	++++
	IgG1	++++	++++
	IgG2	++++	++++
	IgG3	–	++++
	IgG4	++++	++++
	IgM	–	–
	IgA	–	–
	IgE	–	–
Mouse	IgG1	+	++++
	IgG2a	++++	++++
	IgG2b	+++	+++
	IgG3	++	+++
Rat	IgG1	–	+
	IgG2a	–	++++
	IgG2b	–	++
	IgG2c	+	++
Goat	IgG	+/–	++
Rabbit	IgG	++++	+++
Sheep	IgG	+/–	++

Protein A/G beads blocking and Pre-cleaning step

The Protein A/G beads blocking (**step 5**) is often performed with IP binding buffer + 2 μg/μL of BSA for 10–30 min at RT. Both the Protein A/G beads blocking and the precleaning (**step 8**) are designed to prevent unspecific proteins binding to the beads.

Controls for Co-IP reactions

As immune reagents are not always as specific as they are sold to be, it is highly recommended to perform several negative controls in Co-IP assays. A simple way to monitor the specificity of the desired interaction is to use other antibodies, directed against irrelevant proteins, in the same experiments, as negative controls. A reverse experiment, where the antibodies used for IP and Western blotting are switched is suitable. Whenever available, pre-immune serum should be used.

Acrylamide concentration in SDS-PAGE should be optimized to favor the resolution in the range of molecular weight to separate (*see* table below).

% Acrylamide	MW Range (KDa)
8	50–500
10	20–300
12	10–200
15	3–100

2.4 Troubleshooting

2.4.1 No/Weak Signals of Proteins Detected After Co-IP

Make sure that proteins are detected by Western blotting on cell lysates.

Ensure fresh protease inhibitors are added when the sample is lysed.

Try more stringent lysis buffer to ensure complete cell lysis. The least stringent lysis buffer that allows detection of interaction between two proteins is the optimal lysis buffer.

Increase the amount of antibody and/or sample used for Co-IP. However, high concentration of cell lysate will increase background signal.

Increase the incubation time for antibody and target proteins and for Protein A/G beads and antibodies.

Use less stringent washing buffer, or decrease the time for each washing step or cut down the number of washing steps.

2.4.2 High Background or Unspecific Bands Detected

Ascertain complete removal of protein aggregates before IP (**step 3**).

Change buffers for **steps 5** and **8**.

Decrease the amount of antibodies used and the sample used for IP, or decrease the incubation time for **step 10**.

Choose an affinity purified antibody.

Adjust washing stringency and steps. The more stringent washing buffer that will not disrupt the anticipated interactions is the optimal washing buffer. Stringency can be changed by using different detergent than Triton ×100.

3 Chromatin Affinity Purification

This protocol is a modified version of a protocol from Upstate Biotechnologies (www.upstate.com) optimized for the protein G-Streptavidin Binding Protein "GS" tag [2]. It has been used to address the association of tagged protein to chromatin in adherent C17.2 neural cells transfected in a stable manner to express TRs with a N-terminal GS tag at genome wide scale [3].

3.1 Reagents and Buffers

Phosphate buffer saline (PBS) pH 7.4: 1.47 mM KH_2PO_4; 137 mM NaCl; 8.1 mM Na_2HPO_4-$7H_2O$; 2.7 mM KCl.

Protease inhibitor Cocktail (PIC): cOmplete™ EDTA-free Protease Inhibitor Cocktail Tablets, Roche Inc. Add 1 tablet in 500 μL H_2O as 100× stock buffer and keep in −20 °C. Avoid freeze/thaw cycle.

Lysis Buffer: 1% SDS, 10 mM EDTA, 50 mM Tris (pH 8.1).

Dilution Buffer: 0.01% SDS, 1% Triton ×100, 2 mM EDTA, 20 mM Tris–HCl (pH 8.1), 150 mM NaCl.

Low Salt Washing Buffer: 0.1% SDS, 1% Triton ×100, 2 mM EDTA, 20 mM Tris–HCl (pH 8.1), 150 mM NaCl.

High Salt Washing Buffer: 0.1% SDS, 1% Triton ×100, 2 mM EDTA, 20 mM Tris–HCl (pH 8.1), 500 mM NaCl.

LiCl Washing Buffer: 0.25 M LiCl, 1% IGEPAL-CA630, 1% deoxycholic acid (sodium salt), 1 mM EDTA, 10 mM Tris–HCl (pH 8.1).

TE Buffer: 10 mM Tris–HCl (pH 8.1), 1 mM EDTA.

Elution Buffer: 0.1 M $NaHCO_3$, 1% SDS. Used freshly prepared and kept up to 2 weeks at room temperature.

Bovine Serum Albumin (BSA, Sigma; fraction V, Protease free, molecular biology grade). 2 mg/mL stock solution kept at −20 °C.

Herring sperm DNA. Stock solution is 10 mg/mL in 5 mM Tris–HCl, 0.1 mM EDTA; pH 7.5. Autoclave to shear and denature DNA. Store at −20 °C.

IgG-loaded magnetic beads: Dynabeads™ M-280 Sheep anti-Rabbit IgG (Invitrogen), stored at 4 °C.

3.2 Procedure

3.2.1 Cross-Linking and Harvesting Cells

1. Rinse adherent cells with PBS twice (5 mL per 100 mm dish of confluent cells).
2. Incubate for 10 min in PBS 1× containing 1% formaldehyde at 37 °C (2.5 mL).
3. Quench formaldehyde by adding 500 μL of 1 M glycine solution, incubate for 5 min at room temperature.
4. Wash twice with ice-cold PBS 1X + PIC.
5. Scrape and collect fixed cells in cold PBS 1X + PIC.
6. Prepare aliquots (~8.10^6 cells/tube) centrifuge (2000 × *g*, 5 min at 4 °C) and store at −80 °C.

3.2.2 Cells Lysis and DNA Shearing

7. Resuspend cell pellet (~8.10^6 cells) in 300 μL of Lysis Buffer + PIC by pipetting up and down. Sonicate to reduce DNA to ~200–700 bp fragments with Bioruptor® (Diagenode) (High; 30 s on 30 s off during 18 min at 4 °C). Covaris® can also be used to sonicate DNA. In this case, time and voltage should be optimized to reduce DNA to ~200–700 bp fragments.

Centrifuge 5 min at 15,000 × *g* at 4 °C and keep the supernatant.

8. Use 50 μL of this lysate supernatant as control input sample (store at −20 °C and process it directly to **step 20**).

3.2.3 Precipitation

9. Add dilution Buffer + PIC + BSA (2 μg/mL) + Herring sperm DNA (2 μg/mL) to 250 μL of lysate supernatant from **step 7** to reach 2.5 mL in 15 mL Polypropylene conical tube with screw cap.
10. Incubate for 1 h on a vertical wheel in a cold room.
11. During this time prepare IgG loaded beads. Homogenize bead slurry with vortex and take 150 μL. Wash twice with 500 μL dilution Buffer + PIC + BSA (2 μg/mL) + herring sperm DNA (2 μg/mL), using a magnet for rapid sedimentation of beads.
12. Incubate 150 μL washed beads with diluted cell lysate, from **step 9**, 24 h on a wheel at 4 °C.

3.2.4 Washes

13. Split into two 1.5 mL tubes and let the beads collect for 3 min or until the supernatant is completely clear (discard it) and then between each wash by using a magnet.
14. Wash the beads 10 min with 750 μL of ice-cold Low Salt Wash Buffer at 4 °C.
15. Wash 15 min with 750 μL of ice-cold High Salt Wash Buffer at 4 °C.
16. Wash 30 min with 750 μL of ice-cold LiCl Wash Buffer at 4 °C.
17. Wash twice (5 min each) with 750 μL of ice-cold TE 1× buffer at 4 °C. Pool the beads from the two tubes during second wash. (These steps can also be done at room temperature).

*3.2.5 Elution (**Steps 21** and **22** are Performed at Room Temperature)*

18. Let the beads collect for 3 min or until the supernatant is completely clear (discard it) and then between each elution step.
19. Elute DNA-protein complexes by incubating (15 min) with 250 μL Elution Buffer. Keep the supernatant and repeat this step. Pool the two 250 μL supernatants.

3.2.6 Reverse Cross-Link and DNA Purification

20. Incubate the eluate (from **step 19**) and the Input (from **step 8**) at 65 °C overnight with 200 mM NaCl.
21. Purify DNA with Minelute® PCR Purification kit (Qiagen) according to the manufacturer's recommendations. Elute the column with 10 μL of 5mM Tris 0.1mM EDTA. Repeat this step a second time for a final elution of 20 μL. Store at −20 °C. Samples can be repeatedly thawed and frozen without loss of PCR signal.

3.3 Trouble Shooting

Both cross-linking time and formaldehyde concentration can affect the efficiency of chromatin shearing and the efficiency of precipitation. Shorter cross-linking times (5–10 min), lower formaldehyde concentrations (1%, wt./vol.) can be tried to improve shearing efficiency; however, for some proteins, especially those that do not directly bind DNA, this might reduce the efficiency of cross-linking and thus the yield of precipitated chromatin.

Cell pellets can be stored at −80 °C for at least 1 year (from **step 6**). Keep material to perform positive controls in later experiments.

Sonication is a critical step. Conditions must be determined empirically for each cell or tissue type, and sonicator model. As heating must be avoided, use sonicator with refrigerate water bath (Bioruptor, Covaris). Unsufficient shearing will increase background. Excessive shearing will prevent later PCR amplification. Optimum conditions should provide a smear of DNA (run on 1% agarose gel stain with EtBr) with an average size of 200–700 bp.

When ChAP efficiency is measured by qPCR, background should be estimated by amplifying controls 5′ et 3′ sequences located next to the tested site (around 1 kb away). For most transcription factors, expect to precipitate at least 1% of the input fraction at specific sites. Genome wide mapping is usually considered only if signal to noise ratio exceeds 5, preferably 10.

Herring sperm DNA is required to prevent unspecific binding of DNA to beads. Do not use any excess, as it will contaminate samples and might introduce artifact in genome wide studies.

References

1. Yen PM, Ando S, Feng X, Liu Y, Maruvada P, Xia X (2006) Thyroid hormone action at the cellular, genomic and target gene levels. Mol Cell Endocrinol 246(1–2):121–127
2. Burckstummer T, Bennett KL, Preradovic A, Schutze G, Hantschel O, Superti-Furga G, Bauch A (2006) An efficient tandem affinity purification procedure for interaction proteomics in mammalian cells. Nat Methods 3(12):1013–1019
3. Chatonnet F, Guyot R, Benoit G, Flamant F (2013) Genome-wide analysis of thyroid hormone receptors shared and specific functions in neural cells. Proc Natl Acad Sci U S A 110(8): E766–E775

Chapter 3

Expression Analysis of Genes Regulated by Thyroid Hormone in Neural Cells

Juan Bernal and Beatriz Morte

Abstract

The actions of thyroid hormones on brain development and function are due primarily to regulation of gene expression. Identification of direct transcriptional responses requires cell culture approaches given the difficulty of in vivo studies. Here, we describe the use of primary cells in culture obtained from embryonic mouse cerebral cortex, to identify the set of genes regulated directly and indirectly by T3 using RNA-Seq.

Key words Cerebral cortex, Cycloheximide, Neurons, Primary culture, RNA-Seq, Thyroid hormones

1 Introduction

As in other tissues, the main actions of thyroid hormones on brain development and function are exerted by T3 through regulation of gene expression by binding to the nuclear receptors. Understanding the actions of thyroid hormones in the brain in molecular terms requires therefore the identification of the gene network regulated by T3 in nerve tissues [1]. This is a formidable task given the complexity of brain structure and the fact that the actions of T3 on gene expression in brain depend on the particular brain structure and developmental time. The changes in gene expression induced by T3 should contain primary and secondary responses. The primary responses are those set in motion directly by the T3-receptor complex interacting with the regulatory regions of the target genes, and with the nuclear coregulators. These interactions result in positive or negative responses of gene expression at the transcriptional level. Additionally, some primary responses have been described as a result of direct receptor interaction with extra nuclear components of some signaling pathways. Secondary responses are those taking place as a result of these primary events and may be extremely complex, and may involve interactions between different

Michelina Plateroti and Jacques Samarut (eds.), *Thyroid Hormone Nuclear Receptor: Methods and Protocols*, Methods in Molecular Biology, vol. 1801, https://doi.org/10.1007/978-1-4939-7902-8_3,
© Springer Science+Business Media, LLC, part of Springer Nature 2018

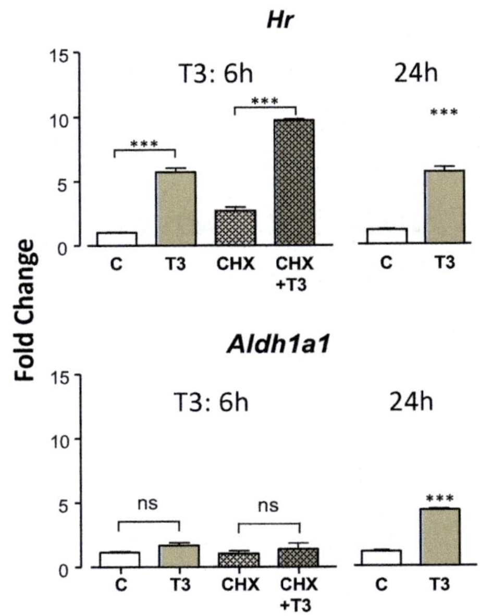

Fig. 1 Effect of T3 on gene expression in the presence and absence of cycloheximide. Cerebrocortical cells were incubated with T3 for 6 or 24 h. The effect of T3 at 6 h was evaluated by incubating the cells in the presence or absence of cycloheximide to inhibit protein synthesis. Here, we illustrate the results with two genes, ***Hr*** and ***Aldh1a1***. ***Hr*** was induced by T3 in the presence and absence of cycloheximide and therefore represents a direct effect of T3. ***Aldh1a1*** is not induced by T3 at 6 h, and likely represents an indirect response. From [2]: Gil-Ibanez P, Bernal J, Morte B (2014) Thyroid hormone regulation of gene expression in primary cerebrocortical cells: role of thyroid hormone receptor subtypes and interactions with retinoic acid and glucocorticoids. PLoS One 9 (3):e91692. doi:https://doi.org/10.1371/journal.pone.0091692

cell types. Dissection of these different pathways in vivo is extremely difficult. Our approach has been to identify primary and secondary transcriptional responses in primary cells derived from the embryonic mouse brain using RNA-Seq to compare gene expression in the presence and absence of T3. To identify directly regulated genes we used the protein inhibitor cycloheximide [2, 3] (Fig. 1). It is assumed that genes up or down regulated by T3 in the presence of cycloheximide are the result of changes in the rate of transcription. Once the primary and secondary responses to T3 are identified, the data can be compared with genomic databases from specific cells to identify cell-specific responses to T3.

We describe the methodology we used in our laboratory for culturing primary cells from the embryonic mouse brain, and the procedures for analyzing the RNA-Seq data.

2 Materials

2.1 Buffers, Culture Media, and Plates

1. NB medium: Neurobasal Medium containing 1× Glutamax, 10 U/mL Penicillin, and 10 U/mL Streptomycin.
2. NB27 medium: NB medium containing 2% B27 supplement (Gibco).
3. Plating solution: NB27 medium (1/2) + Horse serum (1/2) + laminin (1/1000). Prepare fresh.
4. Dissecting solution: PBS containing 1% BSA and 0.1% glucose.
5. Papain solution: Heat papain at 37 °C and dilute 1/25 in the dissecting solution. Filter through a 0.2 μm Millipore® filter. Prepare fresh.
6. DNAse I solution: Dilute DNAse I at a final concentration ~1.5 mg/mL in the dissecting solution (*see* **Note 1**).
7. Fetal calf serum (FCS) deprived of thyroid hormones, following the protocol by Samuels et al. [4].
8. T3 solution: 1.48 mM stock solution of 1 mg/mL in 0.05 M NaOH. Working dilutions are prepared in medium containing 0.1% BSA or serum deprived of thyroid hormones.

2.2 Immunofluorescence

1. Cell permeabilizing buffer: PBS containing 0.5% triton X-100.
2. Blocking buffer: PBS containing 1% BSA, 2% cold water fish skin gelatin, 0.05% Triton X-100.
3. Primary Antibodies: monoclonal against mouse glial fibrillary acidic protein (Clone G-A-5, Sigma cat#G3893) for astrocytes, and rabbit polyclonal anti-NeuN antibody (Millipore cat#ABN78) for neurons.
4. Secondary antibodies: donkey anti mouse Alexa fluor® 488 (Thermo Scientific) and donkey anti rabbit Alexa 555 (Thermo Scientific).

3 Methods: Primary Cell Culture from Embryonic Mouse Brain

3.1 Plates and Medium Preparation

1. Prepare the 12-well coated plates: to each well add 0.7 mL of poly-L-ornithine diluted 1/10 in water. Leave for 15 min, remove by aspiration, and let the plates dry for 15 min.
2. Prepare the plating solution (see Materials: Enzymes and chemicals) and add 600 μL to each well. Leave the plates in the CO_2 incubator during dissection.
3. For immunofluorescence, 24-well plates with a round cover glass covering the bottom of each well are similarly treated, using 0.5 mL of poly-L-ornithine, and 0.3 mL plating solution.

3.2 Cerebral Cortex Dissection

1. Prepare the dissecting solution.
2. Euthanize pregnant dams with CO_2 at gestational day 17.5.
3. Extract the uterus discarding the peritoneum and ovaries, place in a beaker with PBS at room temperature (RT). Cut open the uterus on a Petri dish with dissecting solution at RT and separate the fetuses from placenta, allantois, and chorion.
4. Euthanize the fetuses by decapitation, extract the brains, and place them on a Petri dish with dissecting solution under a stereomicroscope. We use a Leica MZ 7.5 high performance stereomicroscope with magnifying power between 6.3× and 50×.
5. Make a cut at the median line of the brain separating the two hemispheres. Lift up the cortex separating it from the hippocampus with the help of tweeters. Discard the meninges. Place the two hemicortices from each fetus in a 2 mL Eppendorf tube with 1 mL dissecting solution on ice.

3.3 Digestion

1. Aspirate carefully the dissecting solution from the tubes containing the cortices and replace with 1 mL of the freshly prepared papain solution, preheated at 37 °C. Incubate at 37 °C for 10 min.
2. Replace the papain solution with 1 mL of DNAse I solution preheated at 37 °C. Disaggregate the tissue by passing up and down with a Gilson© P1000, and then with a Gilson© P200.
3. Centrifuge at 1200 rpm (about 150 × *g*) for 5 min at RT. Discard the DNAse I solution carefully. Resuspend the cells in 1 mL of NB27 medium preheated at 37 °C. Transfer the cell suspension to a sterile 10–15 mL tube, and add 3 mL of NB27 medium.
4. Place 10 μL of cell suspension in an Eppendorf tube and mix with 10 μL trypan blue. Count the cells in duplicate in a Neubauer Chamber.
5. Adjust the cell concentration to 7.5×10^5 cells/mL with NB27.

3.4 Plating the Cells

1. Add 0.8 mL of cell suspension to each well containing 600 μL plating solution to have around 6×10^5 cells per well (*see* **Note 2**).
2. Place the plates in a 5% CO_2 incubator for 2 h. The cells should then be attached to the bottom of the wells.
3. Remove the medium and wash the cells twice with NB27 medium preheated at 37 °C.
4. Add 1.5 mL of NB27 medium at 37 °C/well. Incubate the cells in the 5% CO_2 incubator with medium changes every 2–3 days for 9–12 days.

3.5 Analyzing the Cell Composition of the Cultures

1. Use the cells plated on glass coverslips (*see* **Note 2**).
2. Fixing the cells: Add absolute acetone and let it stand for 7 min at RT.

3. Permeabilizing the cells: Remove the acetone and add the cell permeabilizing buffer. Let it stand for 5 min at RT.
4. Blocking nonspecific antibody binding: Remove the permeabilizing buffer, add the blocking buffer. Let it stand for 1 h at RT.
5. Staining: prepare a 1/500 dilution of primary antibodies in PBS containing 0.2% triton X-100 (*see* **Note 3**).
6. Add 0.25 mL of the antibody solution to the cells and incubate overnight at 4 °C.
7. Wash the cells 3 times, 5 min each with PBS containing 0.2% triton X-100 at RT.
8. Add 0.25 mL of secondary antibody diluted 1/500 in PBS containing 0.2% triton X-100. Incubate for 1 h at RT.
9. Wash the cells with PBS 3 times 5 min each.
10. Dilute DAPI at a concentration of 0.1 mg/mL in PBS. Add to the cells and incubate for 10 min to label the nuclei.
11. Wash as in **step 9**.
12. Keep the cells in PBS at 4 °C in the dark, or mount at convenience.

3.6 Incubation with T3

1. After 9–12 days of starting the culture, remove the medium, wash the cells with NB medium, and incubate for 24 h in 1.5 mL NB (*see* **Note 4**).
2. Replace the medium with 1.5 mL NB containing 0.1% FCS deprived of thyroid hormones and T3 at the desired concentration. Incubate for the time required by the experiment (*see* **Note 5**).
3. If an experiment with CHX is planned: Prepare NB medium containing 0.1% FCS deprived of thyroid hormones and 8 μg/mL CHX. Add to the cells and preincubate for 30 min. Then add the T3 and incubate for a maximum of 6 h.
4. At the end of incubation place the cells on ice, remove the medium, wash with PBS, and proceed to RNA isolation.

3.7 RNA Isolation and Sequencing

1. Isolate total RNA. For RNA-Seq we use the RNeasy Plus Kit (Quiagen) following the manufacturer's procedure.
2. Quantify and check RNA purity. To quantify RNA we use a NanoDrop ND-1000 (Thermo Scientific, Waltham, MA, USA). To check for RNA integrity we use Agilent 2100 Bioanalyzer (Agilent Technologies, Santa Clara, CA, USA).
3. RNA-Seq is accomplished by a core facility. The technique involves the construction of the index-tagged cDNA libraries and the pair or single sequence-by-synthesis. Reads are further

processed to split Reads according to adapter indexes, and produce fastq files. Further processing of the fastq files by bioinformatic analysis may be carried out by the RNA-Seq core facility. An example of the procedure is described below.

3.8 Sequence Bioinformatics (See Note 6)

1. Quality control: Determine the Reads quality with the FastQC application: (http://www.bioinformatics.babraham.ac.uk/projects/fastqc/). This resource provides an overview of the data and identifies problems related to the sequences. If required, the sequences can be manipulated to produce better mapping results using the following tools: FASTX-Toolkit (http://hannonlab.cshl.edu/fastx_toolkit/) and Cutadapt To remove adapter sequences (https://cutadapt.readthedocs.io/en/stable/).
2. Preparation of the reference genome: Download the fasta file containing sequences of the referred genome, in this case mouse, from Ensembl (http://www.ensembl.org/Mus_musculus/Info/Index).
3. Indexing the genome from Bowtie (http://bowtie-bio.sourceforge.net/bowtie2/index.shtml): Indexing the genome through this application facilitates efficient alignment processes of the sequencing Reads to long reference sequences [5].
4. Mapping of the sequence Reads to the indexed genome: Align the sequence Reads to the indexed genome using TopHat (http://ccb.jhu.edu/software/tophat/index.shtml). It generates BAM output files [6]. Evaluate mapping data quality from Qualimap (http://qualimap.bioinfo.cipf.es/). It facilitates an overall view that helps to detect biases in the alignment sequencing data [7].
5. Counting the Reads: Convert the BAM files to SAM files using Samtools (http://www.htslib.org/). Quantify the number of Reads mapping to specific genes using the Python module HT-SEQ (http://www-huber.embl.de/users/anders/HTSeq/doc/count.html). This resource generates txt output files. The number of Reads that align to a particular gene is known as Counts [8].

   The number of Reads that align to the different transcript of the genes can also be reported.

3.9 Differential Expression Analysis (See Note 7)

1. Data exploration: Explore the *Count* data by Principal Component Analysis and Clustering methods. There are some other useful plots to evaluate *Count* distribution and type of detected features as for example the Bioconductor package NOISeq [9] (http://bioconductor.org/packages/release/bioc/html/NOISeq.html).
2. Estimating gene expression levels: The absolute expression level of a gene is expressed as RPKM or FPKM (*see* **Note 8**).

3.10 Analysis of Differential Expression Across Samples

1. Discard the genes with very low expression and normalize the data (*see* **Note 9**).

   To normalize the data the most commonly used method is the *Trimmed Means of M-values* (TMM). It assumes that only a few genes are going to be differentially expressed [10].

2. Difference of expression is expressed by the Fold change (FC). It is defined as the ratio of expression between the two conditions. The $\log_2$ ratios are generally used for analysis and visualization of fold changes. There are many specific packages from Bioconductor (such as DESeq, EdgeR, Cuffdiff) to detect genes that are differentially expressed and assign a significance of the measurement (*see* **Notes 10** and **11**).

3.11 Interpretation of Results

Selection of Thyroid hormone targets genes: Select the genes differentially expressed (DE) (false discovery rate < 0.05) after incubation with T3 in comparison with cells in the absence of T3. Sensitive genes can be up regulated (positive genes) and down regulated (negative genes) by T3. Among the DE genes the best candidates are those with the highest change (FC in absolute value), the most significant change (low FDR), and higher level of expression (high RPKM).

3.12 Validation of Gene Expression Changes

Validation must be performed in biological replicates using RNA from independent cultures. Transcriptomic analysis of a set of genes (up to 100) can be used to analyze differences of expression by qPCR.

3.13 Functional Profiling by Enrichment Analysis (See Note 12)

1. The question is whether as a result of gene expression changes there is an enrichment of certain biochemical pathways. Enrichment analysis is carried out for Gene Ontology terms using the Bioconductor package GOSeq (http://bioconductor.org/packages/release/bioc/html/goseq.html) or using web tools as Panther (http://pantherdb.org/). Multiple testing by the Benjamini-Hochberg procedure is required. These tools provide information about pathways, biological processes, molecular functions, and cellular components that are over or under represented in the DE gene list. Annotations for the pathways analysis are collected mainly in the KEGG Pathways Database (http://www.genome.jp/kegg/pathway.html).

2. Gene Set Enrichment analysis.

   The question is whether there is a significant enrichment in an a priori defined set of genes between the two experimental groups? The focus is on gen sets previously defined instead of looking on individual genes of the DE list. Applications for this task include GSEA (http://software.broadinstitute.org/gsea/index.jsp) and Babelomics (https://babelomics.bioinfo.cipf.es/).

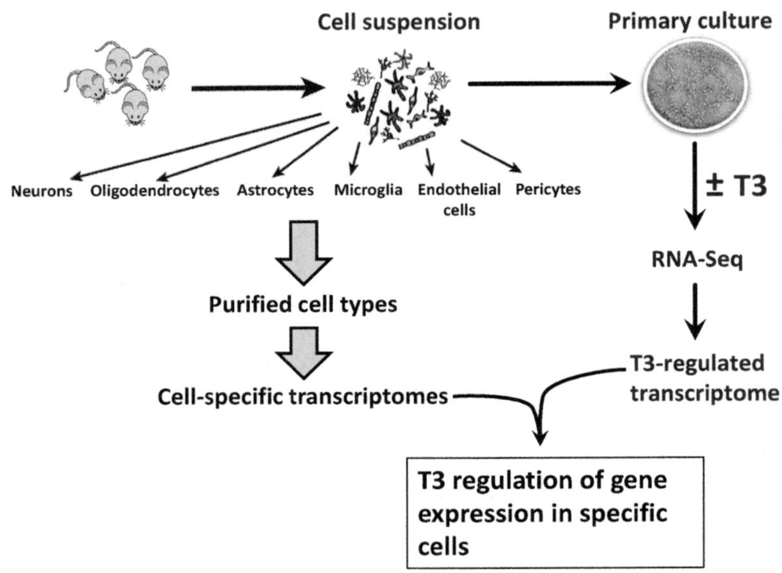

Fig. 2 Strategy to discover genes regulated by T3 in specific neural cells. Neural cells in primary culture, even if it is enriched for a particular cell type consist of a heterogeneous cell population. The T3 regulated transcriptome will contain genes from any individual cell type responsive to thyroid hormone present in the culture. Some of these genes will be enriched in specific cell types, and therefore will be markers of T3 responsiveness in these cells. To identify the subset of cell-enriched genes present in the T3-regulated transcriptome, the T3-regulated genes database can be compared with databases of cell-specific transcriptomes obtained from isolated cell types. In this way genes regulated by T3 in specific cells such as Cajal-Retzius cells, subplate cells, or cells from specific cortex layers can be identified. Adapted from [12]: Zhang Y, Chen K, Sloan SA, Bennett ML, Scholze AR, O'Keeffe S, Phatnani HP, Guarnieri P, Caneda C, Ruderisch N, Deng S, Liddelow SA, Zhang C, Daneman R, Maniatis T, Barres BA, Wu JQ (2014) An RNA-sequencing transcriptome and splicing database of glia, neurons, and vascular cells of the cerebral cortex. J Neurosci 34 (36):11929–11947

3.14 Analysis of Different Patterns of Response to T3

Cluster analysis and the Heatmap representation are very useful to detect sets of genes with similar expression patterns after T3 treatments.

3.15 Comparisons with Other Databases (See Note 13)

1. Differentiate between direct transcriptional responses or proximal T3-receptor complex responses, and indirect responses, i.e., distal T3-receptor complex responses. This can be approached by comparing the list of DE genes in the presence and absence of cycloheximide, and by comparing the DE gene list with genomic TR binding sites databases.
2. Enrichment of DE genes in different cell types (Fig. 2). This is especially relevant for the nervous system with many different cellular types. Even in more homogeneous preparations, such

as primary cells in culture, there is a representation of many cellular types [3]. The goal is to determine to what extent a gene is considered to be enriched in a given cell type, setting a lower limit of fold expression in a given cell type compared with the rest of cell populations. There are some user-friendly databases that can be used for this task: http://web.stanford.edu/group/barres_lab/brainseq2/brainseq2.html [11], http://web.stanford.edu/group/barres_lab/brain_rnaseq.html [12], http://linnarssonlab.org/cortex/ [13].

4 Notes

1. A practical way to add the DNAse I is as follows: take a P200 tip, cut with scissors or scalpel at about 7 mm from the tip. Insert 5 mm of the pipet into the DNAse powder. Mix the content of the tip with 5 mL PBS-BSA-glucose solution (dissecting solution).
2. Cerebral cortices from individual fetuses should originate two replicas for the cell culture, one with T3 and another without T3, each distributed in 2–3 wells. For immunofluorescence, plate 0.4 mL of the cell suspension (3×10^5 cells) in 24-well plates with a round cover glass of 12 mm diameter deposited in the bottom of the well.
3. To estimate the proportion of astrocytes and neurons present in the cultures, we use the following primary antibodies for immunofluorescence: monoclonal against mouse glial fibrillary acidic protein for astrocytes, and rabbit polyclonal anti-NeuN antibody for neurons. The secondary antibodies consist of donkey anti mouse IgG conjugated to Alexa fluor® 488 (green) to reveal the astrocytes and donkey anti rabbit IgG conjugated to Alexa 555 (red) to reveal the neurons.
4. Note that NB27 is replaced by NB in this step. This is because the B27 supplement contains T3, so the cells should be incubated for 24 h in the absence of B27 before T3 treatment.
5. T3 is active on gene expression in the range of 0.1 nM to 10 nM. Maximal effects are obtained between 24 and 48 h and 1–10 nM on most genes in the presence of 0.1% serum. Serum is present at this low concentration to stabilize the T3 and minimize binding to plasticware.
6. All the cited software tools for RNA-Seq analysis are running in command-line. For the differential expression analysis we use packages from Bioconductor (https://www.bioconductor.org/) that work in the free software environment R (https://www.r-project.org/). There are also open-source interfaces, web-based platform for these RNA-seq analysis

tools, e.g., the Galaxy Project (https://usegalaxy.org/), Babelomics (https://babelomics.bioinfo.cipf.es/) [14, 15].

7. Differential expression analysis refers to the comparison of gene expression between different conditions, for example, T3-treated versus untreated cultures or animals. To have meaningful results when comparing massive data as when using microarray or RNA-Seq it is important to start from an adequate number of biological replicates. At least six different biological samples should be compared especially when comparing tissue samples from experimental animals. When using primary cultures it is also advisable to compare individual cultures derived from at least six different embryos.

8. The expression of a gene can be quantified by the number of sequenced fragments from this gene. However, the number of sequenced fragments depends not only on the total number of Reads mapped on each gene, but also on the length of the gene. Obviously, large genes will have more Reads per RNA molecule than small genes. Therefore, a normalization method that takes this factor into account is needed. The results are expressed as RPKM (Reads Per Kilobase of the transcript per Million mapped Reads) in the case of single Reads, or FPKM (Fragments Per Kilobase of the transcript per Million mapped Reads) for paired-end sequencing.

$$\text{RPKM} = \frac{NumReads}{\left(\frac{GeneLength}{10^3}\right)\left(\frac{TotalNumReads}{10^6}\right)}, \quad \text{where}$$

NumReads: number of Reads mapped to a gene sequence; *GeneLength*: length of the gene sequence; *TotalNumReads*: total number of mapped Reads of a sample.

9. It is highly recommended to filter the genes, using the text file containing the Counts for each gene. The procedure consists of discarding the genes with very low expression, i.e., those having 0 or a very low number of *Counts*. To this end the *Counts* should be scaled by the number of sequenced fragments (CPM = Counts per million):$\text{CPM} = \frac{ReadsMappedToGene}{TotalNumReads * 10^6}$ (*ReadsMapsToGene* = number of Reads mapped to a selected gene. *TotalNumReads*—total number of mapped Reads of a sample).

10. High-throughput technologies enable us to test simultaneously thousands of genes for differential expression between biological conditions. The large number of comparisons makes it necessary to correct P values (P_{adjust}) for the total number of comparisons. Among the several methods used is the Benjamini & Hochberg correction, which gives the false discovery rate, or FDR [16].

11. The RNA-Seq data must be deposited in Data repositories, e.g., the National Center for Biotechnology Information (NCBI) Gene Expression Omnibus (GEO) (http://www.ncbi.nlm.nih.gov/geo).
12. Nowadays, free web tools, e.g., Babelomic, and commercial softwares, such as Genomatix and Ingenuity, provide further information about interactions between the DE genes, transcription factors involving in the regulation of the DE genes, metabolic consequences of these changes, related diseases, bibliographic relationship among the genes, etc.
13. A useful way to analyze the results is to compare the DE gene list with other databases. In such a way for example, information on cellular expression of the genes of interest can be obtained. Venn diagrams are commonly used to display this kind of comparison. To compare different gene sets there is an interactive tool "Venny," developed by JC Oliveros (Venny: an interactive tool comparing lists with Venn's diagrams (http://bioinfogp.cnb.csic.es/tools/venny/index.html)).

Acknowledgments

Supported by grants SAF2014-54919-R from the Plan Estatal de Investigación Científica y Técnica y de Innovación, Spain and by the Center for Research on Rare Dieseases (Ciberer) under the frame of E-Rare-2, the ERA-Net for Research on rare Diseases. The contribution of Drs Pilar Gil-Ibañez and Mónica M. Belinchón is gratefully acknowledged.

References

1. Chatonnet F, Flamant F, Morte B (2015) A temporary compendium of thyroid hormone target genes in brain. Biochim Biophys Acta 1849(2):122–129. https://doi.org/10.1016/j.bbagrm.2014.05.023
2. Gil-Ibanez P, Bernal J, Morte B (2014) Thyroid hormone regulation of gene expression in primary cerebrocortical cells: role of thyroid hormone receptor subtypes and interactions with retinoic acid and glucocorticoids. PLoS One 9(3):e91692. https://doi.org/10.1371/journal.pone.0091692
3. Gil-Ibanez P, Garcia-Garcia F, Dopazo J, Bernal J, Morte B (2015) Global Transcriptome analysis of primary Cerebrocortical cells: identification of genes regulated by Triiodothyronine in specific cell types. Cereb Cortex 27(1):706–717. https://doi.org/10.1093/cercor/bhv273
4. Samuels HH, Stanley F, Casanova J (1979) Depletion of L-3,5,3′-triiodothyronine and L-thyroxine in euthyroid calf serum for use in cell culture studies of the action of thyroid hormone. Endocrinology 105(1):80–85. https://doi.org/10.1210/endo-105-1-80
5. Langmead B, Trapnell C, Pop M, Salzberg SL (2009) Ultrafast and memory-efficient alignment of short DNA sequences to the human genome. Genome Biol 10(3):R25. https://doi.org/10.1186/gb-2009-10-3-r25
6. Trapnell C, Pachter L, Salzberg SL (2009) TopHat: discovering splice junctions with RNA-Seq. Bioinformatics 25(9):1105–1111. https://doi.org/10.1093/bioinformatics/btp120
7. Garcia-Alcalde F, Okonechnikov K, Carbonell J, Cruz LM, Gotz S, Tarazona S, Dopazo J, Meyer TF, Conesa A (2012) Qualimap: evalu-

ating next-generation sequencing alignment data. Bioinformatics 28(20):2678–2679. https://doi.org/10.1093/bioinformatics/bts503

8. Anders S, Pyl PT, Huber W (2015) HTSeq--a Python framework to work with high-throughput sequencing data. Bioinformatics 31(2):166–169. https://doi.org/10.1093/bioinformatics/btu638
9. Tarazona S, Garcia-Alcalde F, Dopazo J, Ferrer A, Conesa A (2011) Differential expression in RNA-seq: a matter of depth. Genome Res 21(12):2213–2223. https://doi.org/10.1101/gr.124321.111
10. Robinson MD, Oshlack A (2010) A scaling normalization method for differential expression analysis of RNA-seq data. Genome Biol 11(3):R25. https://doi.org/10.1186/gb-2010-11-3-r25
11. Zhang Y, Sloan SA, Clarke LE, Caneda C, Plaza CA, Blumenthal PD, Vogel H, Steinberg GK, Edwards MS, Li G, Duncan JA 3rd, Cheshier SH, Shuer LM, Chang EF, Grant GA, Gephart MG, Barres BA (2016) Purification and characterization of progenitor and mature human astrocytes reveals transcriptional and functional differences with mouse. Neuron 89(1):37–53. https://doi.org/10.1016/j.neuron.2015.11.013
12. Zhang Y, Chen K, Sloan SA, Bennett ML, Scholze AR, O'Keeffe S, Phatnani HP, Guarnieri P, Caneda C, Ruderisch N, Deng S, Liddelow SA, Zhang C, Daneman R, Maniatis T, Barres BA, Wu JQ (2014) An RNA-sequencing transcriptome and splicing database of glia, neurons, and vascular cells of the cerebral cortex. J Neurosci 34(36):11929–11947. https://doi.org/10.1523/JNEUROSCI.1860-14.2014
13. Zeisel A, Munoz-Manchado AB, Codeluppi S, Lonnerberg P, La Manno G, Jureus A, Marques S, Munguba H, He L, Betsholtz C, Rolny C, Castelo-Branco G, Hjerling-Leffler J, Linnarsson S (2015) Brain structure. Cell types in the mouse cortex and hippocampus revealed by single-cell RNA-seq. Science 347(6226):1138–1142. https://doi.org/10.1126/science.aaa1934
14. Garber M, Grabherr MG, Guttman M, Trapnell C (2011) Computational methods for transcriptome annotation and quantification using RNA-seq. Nat Methods 8(6):469–477. https://doi.org/10.1038/nmeth.1613
15. Goecks J, Nekrutenko A, Taylor J, Galaxy T (2010) Galaxy: a comprehensive approach for supporting accessible, reproducible, and transparent computational research in the life sciences. Genome Biol 11(8):R86. https://doi.org/10.1186/gb-2010-11-8-r86
16. Benjamini Y, Hochberg Y (1995) Controlling the false discovery rate: a practical and powerful approach to multiple testing. J R Stat Soc Ser B 57(1):289–300

Chapter 4

In Vitro Approaches to Identify Thyroid Hormone Receptor-Dependent Transcriptional Response

Maria Sirakov and Michelina Plateroti

Abstract

The thyroid hormones act through their nuclear receptor TRs that are T3-modulated transcription factors. To elucidate the molecular mechanisms at the basis of specific physiological responses to TH-TR, in vitro approaches are commonly used to demonstrate the activation or repression of target genes. These approaches allow identifying direct transcriptional targets of TRs which can eventually be confirmed by using in vivo chromatin binding assays.

Here, we describe two classical approaches in vitro and in cell lines such as Electro-Mobility Shift Assay and thyroid-hormone-responsive Luciferase-reporter assay (EMSA and Luc, respectively) we have been largely using to investigate TRα1-driven TH-TRs intestinal epithelium cell autonomous action.

Key words DNA binding, Electro Mobility Shift Assay (EMSA), Luciferase assay (Luc), Target gene promoters, Thyroid hormone Response Element (TRE), Transcriptional activity

1 Introduction

Thyroid hormone (THs) action results in different outcomes depending on the cellular context [1, 2]. It is largely documented that they act via the Thyroid hormone nuclear Receptors TRs, which are transcription factors whose activity can be modulated by the local availability of the hormone T3. In fact, the T3-induced physiological response depends on the expression of specific TR isoform(s), on the TH-transporters and iodothyronine deiodinase selenoenzymes that collectively control the local level of T3, thus TR activity [3, 4]. From a molecular point of view, TRs can act as both transcriptional activators and repressors [5]. They heterodimerize with retinoid X receptors (RXRs) and bind to Thyroid hormone Response Elements (TREs) located within the regulatory regions of target genes. In the absence of T3, TRs bind the TREs and interact with co-repressor proteins to inhibit positively regulated target genes. In the classical model of T3-dependent activation

Michelina Plateroti and Jacques Samarut (eds.), *Thyroid Hormone Nuclear Receptor: Methods and Protocols*, Methods in Molecular Biology, vol. 1801, https://doi.org/10.1007/978-1-4939-7902-8_4,
© Springer Science+Business Media, LLC, part of Springer Nature 2018

of target genes, in the absence of T3, TR/RXR bound to TREs, interacts with co-repressor proteins to inhibit transcription; in the presence of T3 the co-repressors are displaced whereas co-activator proteins are recruited to the ligand-bound TR/RXR complex [6]. Besides this classical model, other mechanisms of TR action have been described. For example, both TRβ1 and TRα1 can interact with β-catenin [7, 8], the intracellular mediator of the Wnt pathway [9], to form a complex in thyrocytes and in the intestinal epithelium, respectively. Interestingly, the action of THs via TRs on the Wnt pathway results in opposite outcomes: TRα1 activates whereas TRβ1 blocks the Wnt pathway [7, 8]. It remains unclear whether this opposite effect is due to the intrinsically different properties of TRα and TRβ or is due to the different cellular contexts in which their interaction with β-catenin occurs.

Several parameters modulate THs-TRs action in the different cellular contexts: the amount of bioavailable hormone [4, 10], the levels of different TR isoforms expressed [11], their post-transcriptional modifications [12, 13], the hetero-dimerization partner retinoid X receptors (RXRs) [14, 15], the interaction with co-repressors and co-activators [6, 16–20], and the structure of TREs [5].

TH-TRs have been shown to play pleiotropic actions in several organs and tissues, where they regulate the balance between cell proliferation and cell differentiation both during development and adulthood [21]. This function is in accordance with the classical pleiotropic action described during amphibian metamorphosis, triggered by an increase of circulating THs levels and of TRβ expression [22]. Moreover, in Mammals, TRs play a major role in controlling the development of the nervous system and the epidermis [23–26] as well as the intestinal postnatal maturation [27, 28]. Regardless of our enriched knowledge, THs functions still appear puzzling moving from one cellular context to another, possibly do to several factors including cell patho-physiological state and, finally, the physical/functional interaction with other signaling pathways in a cell-specific manner [2, 29, 30].

Because of this pleiotropic action occurring in vivo, analyzing in detail the molecular mechanisms of THs-TRs action needs simple approaches such as in vitro and in cell lines. We describe here thyroid-hormone-responsive Luciferase-reporter assay (Luc) and Electro-Mobility Shift Assay (EMSA) to monitor cellular response and DNA binding respectively.

2 Luciferase-Reporter Assay

The Luciferase-reporter assay (Luc) is an extremely rapid, simple, not expensive, and sensitive assay that allows us to monitor the activity of a transcription factor on its specific target sequences.

Usually, reporter vectors are composed of promoter fragments retaining the DNA binding element which drive the expression of heterologous reporter genes that encode enzymes that can be quantified using highly sensitive assays [31–34]. In such reporter experiments, the protein's activity within a transfected cell population is proportional to its mRNA level and thus to transcriptional regulation. A commonly used reporter gene is the *Luciferase* from the firefly *Photinus pyralis*, which encodes a 61 kDa enzyme that oxidizes D-luciferin in the presence of ATP, oxygen, and Mg(++). This process yields a luminescent product that can be quantified by measuring the released light. In our studies we have used the Dual-Luciferase® Reporter Assay System provided by Promega, in which the TRE-Luc (a DR4-type of TRE) activity is measured and normalized to the activity of another co-transfected unresponsive reporter vector expressing the *Luciferase* gene from *Renilla reniformis* or sea pansy. Briefly, the firefly TRE-dependent Luc reporter activity is measured first by adding its specific substrate, the reaction is then quenched and the substrate for the *Renilla* Luc reaction is added permitting the quantification of the control-Luc activity. Performing both the assays sequentially from a single sample provides rapid quantitation of both reporters either in transfected cells or in cell-free transcription/translation reactions. In our hands this system allowed us to study TRα1/T3-dependent activity on a TRE (i.e., DR4) and the effect on this activity by co-transfection with WNT effectors such as β-catenin and TCF4 [8].

2.1 Materials

- Cell line of interest.
- Appropriated culture medium (e.g., DMEM).
- Serum (e.g., heat-inactivated fetal calf serum) normal and thyroid hormone-depleted serum, following the protocol from Samuels et colleagues [35].
- The hormone T3, stock solution 10^{-3} M.
- Appropriated transfection reagent (e.g. Exgene).
- Expression vectors: pGS5-TRα1.
- T3-TR responsive luciferase reporter vectors: pGl2-DR4-Luc and its negative control pGl2-mutDR4-Luc [8, 37].
- T3-TR unresponsive luciferase control reporter vector pRL-CMV, to normalize firefly luciferase in each well.
- Luciferase cell culture lysis buffer, usually present in commercial kits (e.g., Passive Lysis Buffer 5×).
- Opaque multiwell plates or luminometer-adapted tubes.
- Dual-Luciferase® Reporter Assay System Components (Luciferase Assay Substrate, Luciferase Assay Buffer II, Stop & Glo® Substrate, Stop & Glo® Buffer, all from Promega).
- Luminometer.

2.2 Methods

- DAY 0: Replicate the cells and plate them at approximately 70% confluence in 24 multiwell dishes.
- DAY 1: Transfect the cells with expression and reporter vectors (*see* **Note 1**).
- DAY 2: Treatment with T3 10^{-6} M (final concentration) or vehicle, according to the experimental setup.
- DAY 3: Cell lysis, luciferase assay and luciferase activity quantification.

In our typical studies [8] we used the human Caco2 adenocarcinoma cell line (from the American Type Culture Collection), plated 5×10^4 cells/well in 24 multiwell plates, and cultured in DMEM supplemented with 10% heat-inactivated fetal calf serum. Expression vectors were transfected at 100 ng/well, reporter vectors were transfected 200 ng/well, and Renilla reporter was transfected at 1 ng/well. The total amount of transfected vectors was adjusted by adding the pSG5 empty vector. The relative quantity of expression versus reporter vectors was optimized as exemplified in Fig. 1. T3 (10^{-6} M) or vehicle alone was added to the culture medium 24 h before the end of the culture. Luciferase activity was measured 48 h after transfection.

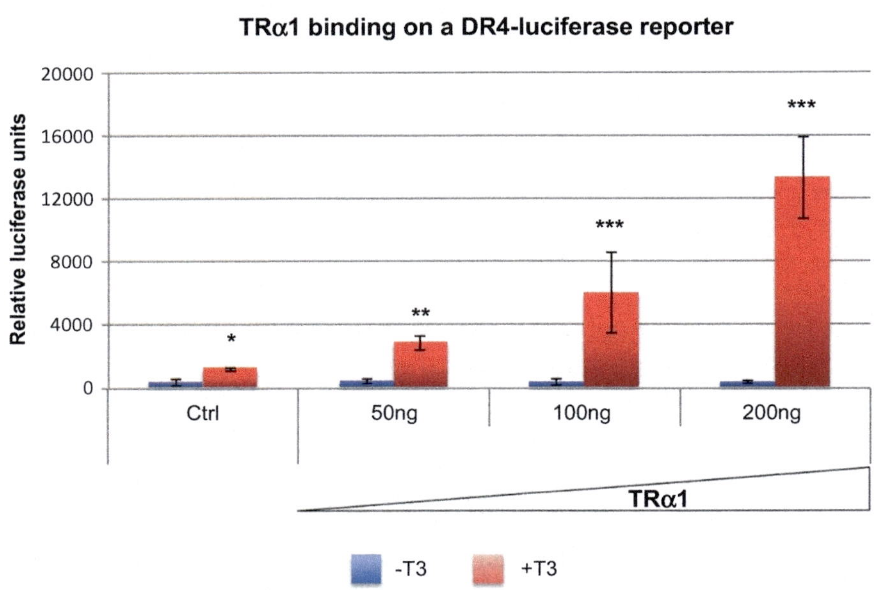

Fig. 1 Luciferase activity in response to increasing amount of TRα1 expression vector. The DR4-luc *Firefly* luciferase reporter vector was co-transfected with *Renilla* control reporter vector into Caco2 cells. Different amounts of TRα1-epressing vector were transfected as indicated. Cells were maintained in T3-depleted serum with or without supplementation with T3 as indicated. Histograms represent mean ± SD, $N = 3$. $^*P < 0.05$, $^{**}P < 0.01$ and $P < 0.001$ compared with the untreated condition (−T3)

3 Electrophoretic Mobility Shift Assay (EMSA)

The Electrophoretic Mobility Shift Assay (EMSA) (also referred to as mobility shift electrophoresis, gel shift assay, gel mobility shift assay, band shift assay, or gel retardation assay) is an affinity electrophoresis technique commonly used to study protein–DNA or protein–RNA interactions [36]. The assay can be performed using nuclear extract, purified protein, or in vitro transcribed/translated proteins. The nucleic acid sequence of interest is radioactively labeled and protein/DNA binding is obtained in vitro using the previously optimized binding condition. The complexes are separated and visualized on a native acrylamide gel, given that they migrate slowly than the free probe. Several controls can be performed to verify the specificity of the protein/DNA binding, such as addition of unlabeled-cold probe that competes with the labeled probe for protein binding. A well-characterized DNA-target sequence is also necessary to be used as positive control. We successfully used this approach in several studies [29, 37], in addition to luciferase-based or in vivo ChIP analyses to demonstrate that putative TREs identified in silico by Nubiscan (http://www.nubiscan.unibas.ch/) or Vista (http://genome.lbl.gov/vista/rvista/submit.shtml) were indeed bona fide TREs. In each experiment a canonical DR4 sequence was used as positive control.

3.1 Materials

- Equipment for vertical gel electrophoresis (*see* **Note 2**).
- Gel apparatus and power supply for vertical running.
- Radioactive labeled probes, prepared as follows: the single-strand oligonucleotides containing the sequence of interest are labeled by [γ-^{32}P]ATP with T4 polynucleotide kinase. After annealing with the complementary cold oligonucleotides, the labeled probes are purified on a 10% acrylamide gel and the specific bands eluted overnight at 4 °C in Tris-EDTA (final concentration: 10 mM Tris–HCl pH 8, 1 mM EDTA). Measure the radioactivity with a Scintillator counter and eventually dilute the labeled DNA up to 20,000 cpm/μL. We strongly recommend working in a radioactive safely equipped lab (*see* **Note 3**).
- Full-length cDNA cloned in expression vectors containing SP6 or T7 promoters for the in vitro transcription/translation.
- TNT® Coupled Reticulocyte Lysate Systems (Promega).
- Thermostat bath.
- X-ray film and cassette for exposure or phosphorimaging instrumentation.

- Nucleic Acid Binding Buffer, final concentrations: 10% glycerol, 10 mM HEPES, 30 mM KCl, 4 mM spermidine, 0.1 mM EDTA, 0.25 mM dithiothreitol, 1 mM Na_2PO_4, and poly(dI-dC, 0.5 mg/mL).
- Eventually specific antibody recognizing the transcription factor bound to DNA, resulting in a super-shifted band.

3.2 Methods

3.2.1 Synthesis of Proteins

- Heat the thermostat bath at 30 °C; thaw TNT® reagents on ice.
- Mix 25 μL of reticulocyte lysate, 2 μL of TNT buffer, 1 μL of amino acids mix (0.5 μL of mix depleted in Met +0.5 μL depleted in Cyst or Leu), 1 μL of polymerase, 1 μg of expression plasmid and H_2O up to 50 μL final volume.
- Incubate 1 h 30 at 30 °C.

3.2.2 Acrylamide Gel

- 5 mL of acrylamide/bis-acrylamide 29: 1, acrylamide 40%.
- 2 mL TBE 5× (1 mL if 10×).
- 33 mL of H_2O.
- 40 μL of TEMED.
- 400 μL 10% APS added just before casting.

Perform a 30 min prerun before loading samples (*see* **Note 4**).

3.2.3 Labeled DNA/Protein Binding Reaction

The amount of different components reported below refers to a single reaction. Prepare multiple of this mix, according to the number of DNA/protein binding reactions you are analyzing.

- Prepare solution A (for a single reaction): 2 μL of each protein synthesized, 1 μL of 1 μg/μL dldC, 6 μL of binding buffer (4 μL if you use a second protein). Gently mix all the components without vortexing and incubate for 10 min at room temperature.
- Prepare solution B (for a single reaction): 1 μL of radioactive probe (corresponding to 20,000 cpm/μL), 1 μL of 0.1 μg/μL Salmon Sperm DNA, H_2O up to 7 μL. Eventually 1 μL of cold competitor (not radioactive probe) can be added to the mix in order to confirm specificity of the DNA/protein binding (*see* **Note 5**) and incubate for 10 min at room temperature.
- Mix 9 μL of solution A (containing the protein(s)) to 7 μL of mixture B (containing the probe) and incubate for 15 min at room temperature.
- An additional control includes the incubation of the probe/protein complex with a specific antibody. For this perform a supplementary incubation for 30 min on ice. The amount of

antibody can be evaluated experimentally if datasheet does not contains this info.

- Load the entire reaction mixture, 16 μL.
- Migrate for approximately 1 h 30-2 h at 180 V in TBE buffer 0.5× (*see* **Note 6**).

3.2.4 Fixation, Drying, and Exposure of the Gel

- Remove gel from apparatus and remove the smaller, back plate. Be careful not to stretch the gel.
- Fix the gel in a 10% acetic acid/20% methanol solution for 10 min at room temperature and then wash it with dH_2O. Usually, the gel dethatches from the second plate during this procedure, you can decide to put it back on the plate or to manipulate the gel with the hands directly.
- Place the gel on a sheet of Whatman® 3MM filter paper. Cover with plastic wrap and dry on a gel dryer at 80 °C for 1 h.
- Remove Whatman/Gel/SaranWrap from the gel dryer. Be careful not to pull off the Saran Wrap from the gel when lifting the plastic sheet of the gel dryer off.
- Expose the gel to X-ray film from 1 h to overnight at −80 °C with an intensifying screen. Alternatively, analyze the gel using phosphorimaging instrumentation.

This approach allowed us to confirm TRα1-specific binding on TREs present in the promoter or intronic regions of TRα1 targets identified in the intestinal epithelium such as *Ctnnb1*, *sFRP2,* and *Jag1* [29, 37, 38]. Full-length cDNA cloned in pSG5 vectors coding for mouse TRα1 and mouse RXRα were used. Both cDNAs were under the control of the T7 promoter for the in vitro transcription/translation experiments. To demonstrate the specificity of TRα1 binding on labeled TRE sequences, we used an anti-TRα1 custom made antibody (raised against a C-terminal peptide and affinity purified with the same peptide) and observed a presence of a supershifted band of higher molecular size. Figure 2 illustrates an example of the results obtained with a DR4 containing probe.

4 Notes

1. For transfection, eliminate culture medium containing serum and add 500 μL of medium without serum. 4 h after starting the transfection add 500 μL of medium. Block the transfection after a total of 6 h by removing the transfection medium and adding 1 mL of medium containing T3-depleted serum.

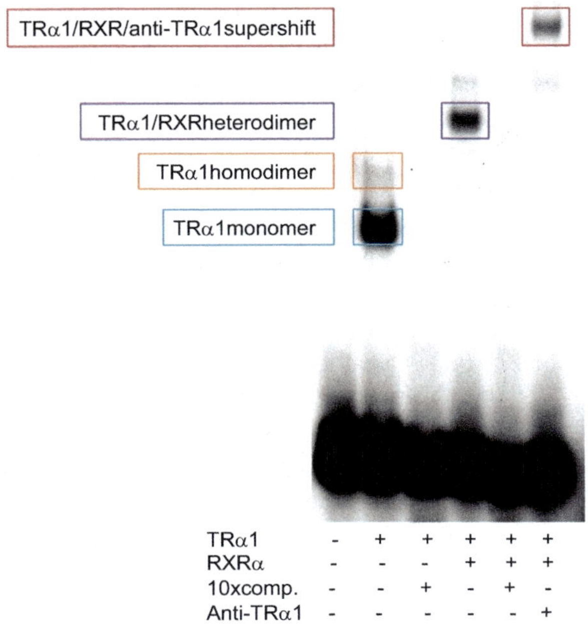

Fig. 2 Electrophoretic mobility shift assay analysis using a canonical DR4 labeled probe and in vitro transcribed/translated proteins. Addition of specific cold sequences at 10× molar excess (comp) was used to assess the specificity of the binding. +, present; −, absent. Rectangles indicate the TRα1 monomer and homodimer or the TRα1/RXR heterodimer. The addition of an anti-TRα1 antibody is able to induce a supershift of the TRα1-bound probe

2. Radioactive safely equipped lab must include Geiger counter and safety screens as well as all devices and supplies necessary for the entire experiment.
3. Ensure that gel plates, spacers, and comb are clean. Use preferably distillated water without soap. Remove any fingerprints or other residue from glass plates with a swab saturated with methanol or 70% ethanol. Dry all the components before casting the gel.
4. It is sometimes useful to wash the wells with a syringe before loading, for example when there are pieces of acrylamide.
5. Make sure that the quantity of the radioactive probe within the mix is as homogeneous as possible.
6. The electrophoresis conditions described here have been optimized for TRα1 and/or RXR and for specific consensus probes. Other TRs and DNA probes may require substantially different conditions. We recommend optimizing electrophoresis conditions and exposure times whenever a new system is examined.

References

1. Holzer G, Roux N, Laudet V (2017) Evolution of ligands, receptors and metabolizing enzymes of thyroid signaling. Mol Cell Endocrinol 459:5–13
2. Sirakov M, Skah S, Nadjar J, Plateroti M (2013) Thyroid hormone's action on progenitor/stem cell biology: new challenge for a classic hormone? Biochim Biophys Acta 1830:3917–3927
3. Kinne A, Schulein R, Krause G (2011) Primary and secondary thyroid hormone transporters. Thyroid Res 4(Suppl 1):S7
4. Bianco AC, Kim BW (2006) Deiodinases: implications of the local control of thyroid hormone action. J Clin Invest 116:2571–2579
5. Oetting A, Yen PM (2007) New insights into thyroid hormone action. Best Pract Res Clin Endocrinol Metab 21:193–208
6. Cheng SY (2000) Multiple mechanisms for regulation of the transcriptional activity of thyroid hormone receptors. Rev Endocr Metab Disord 1:9–18
7. Guigon CJ, Zhao L, Lu C, Willingham MC, Cheng SY (2008) Regulation of beta-catenin by a novel nongenomic action of thyroid hormone beta receptor. Mol Cell Biol 28:4598–4608
8. Sirakov M, Skah S, Lone IN, Nadjar J, Angelov D et al (2012) Multi-level interactions between the nuclear receptor TRalpha1 and the WNT effectors beta-catenin/Tcf4 in the intestinal epithelium. PLoS One 7:e34162
9. Clevers H (2006) Wnt/beta-catenin signaling in development and disease. Cell 127:469–480
10. Bianco AC, Larsen PR (2005) Cellular and structural biology of the deiodinases. Thyroid 15:777–786
11. Cheng SY (2005) Isoform-dependent actions of thyroid hormone nuclear receptors: lessons from knockin mutant mice. Steroids 70:450–454
12. Chen SL, Chang YJ, Wu YH, Lin KH (2003) Mitogen-activated protein kinases potentiate thyroid hormone receptor transcriptional activity by stabilizing its protein. Endocrinology 144:1407–1419
13. Lin HY, Hopkins R, Cao HJ, Tang HY, Alexander C et al (2005) Acetylation of nuclear hormone receptor superfamily members: thyroid hormone causes acetylation of its own receptor by a mitogen-activated protein kinase-dependent mechanism. Steroids 70:444–449
14. Rastinejad F (2001) Retinoid X receptor and its partners in the nuclear receptor family. Curr Opin Struct Biol 11:33–38
15. Szanto A, Narkar V, Shen Q, Uray IP, Davies PJ et al (2004) Retinoid X receptors: X-ploring their (patho)physiological functions. Cell Death Differ 11(Suppl 2):S126–S143
16. Manteuffel-Cymborowska M (1999) Nuclear receptors, their coactivators and modulation of transcription. Acta Biochim Pol 46:77–89
17. McKenna NJ, Lanz RB, O'Malley BW (1999) Nuclear receptor coregulators: cellular and molecular biology. Endocr Rev 20:321–344
18. Sadow PM, Chassande O, Koo EK, Gauthier K, Samarut J et al (2003) Regulation of expression of thyroid hormone receptor isoforms and coactivators in liver and heart by thyroid hormone. Mol Cell Endocrinol 203:65–75
19. Chen JD, Evans RM (1995) A transcriptional co-repressor that interacts with nuclear hormone receptors. Nature 377:454–457
20. Onate SA, Tsai SY, Tsai MJ, O'Malley BW (1995) Sequence and characterization of a coactivator for the steroid hormone receptor superfamily. Science 270:1354–1357
21. Puzianowska-Kuznicka M, Pietrzak M, Turowska O, Nauman A (2006) Thyroid hormones and their receptors in the regulation of cell proliferation. Acta Biochim Pol 53:641–650
22. Su Y, Damjanovski S, Shi Y, Shi YB (1999) Molecular and cellular basis of tissue remodeling during amphibian metamorphosis. Histol Histopathol 14:175–183
23. Pascual A, Aranda A (2013) Thyroid hormone receptors, cell growth and differentiation. Biochim Biophys Acta 1830:3908–3916
24. Mohan V, Sinha RA, Pathak A, Rastogi L, Kumar P et al (2012) Maternal thyroid hormone deficiency affects the fetal neocorticogenesis by reducing the proliferating pool, rate of neurogenesis and indirect neurogenesis. Exp Neurol 237:477–488
25. Bernal J (2007) Thyroid hormone receptors in brain development and function. Nat Clin Pract Endocrinol Metab 3:249–259
26. Santisteban P, Bernal J (2005) Thyroid development and effect on the nervous system. Rev Endocr Metab Disord 6:217–228
27. Sirakov M, Plateroti M (2011) The thyroid hormones and their nuclear receptors in the gut: from developmental biology to cancer. Biochim Biophys Acta 1812:938–946
28. Sun G, Roediger J, Shi YB (2016) Thyroid hormone regulation of adult intestinal stem cells: implications on intestinal development

and homeostasis. Rev Endocr Metab Disord 17:559–569
29. Sirakov M, Boussouar A, Kress E, Frau C, Lone IN et al (2015) The thyroid hormone nuclear receptor TRalpha1 controls the notch signaling pathway and cell fate in murine intestine. Development 142:2764–2774
30. Skah S, Uchuya-Castillo J, Sirakov M, Plateroti M (2017) The thyroid hormone nuclear receptors and the Wnt/beta-catenin pathway: an intriguing liaison. Dev Biol 422:71–82
31. O'Kane DJ, Woodward B, Lee J, Prasher DC (1991) Borrowed proteins in bacterial bioluminescence. Proc Natl Acad Sci U S A 88:1100–1104
32. Wood KV, de Wet JR, Dewji N, DeLuca M (1984) Synthesis of active firefly luciferase by in vitro translation of RNA obtained from adult lanterns. Biochem Biophys Res Commun 124:592–596
33. Matthews JC, Hori K, Cormier MJ (1977) Purification and properties of Renilla reniformis luciferase. Biochemistry 16:85–91
34. Farr A, Roman A (1992) A pitfall of using a second plasmid to determine transfection efficiency. Nucleic Acids Res 20:920
35. Samuels HH, Stanley F, Casanova J (1979) Depletion of L-3,5,3′-triiodothyronine and L-thyroxine in euthyroid calf serum for use in cell culture studies of the action of thyroid hormone. Endocrinology 105:80–85
36. Hellman LM, Fried MG (2007) Electrophoretic mobility shift assay (EMSA) for detecting protein-nucleic acid interactions. Nat Protoc 2:1849–1861
37. Plateroti M, Kress E, Mori JI, Samarut J (2006) Thyroid hormone receptor alpha1 directly controls transcription of the beta-catenin gene in intestinal epithelial cells. Mol Cell Biol 26:3204–3214
38. Kress E, Rezza A, Nadjar J, Samarut J, Plateroti M (2009) The frizzled-related sFRP2 gene is a target of thyroid hormone receptor alpha1 and activates beta-catenin signaling in mouse intestine. J Biol Chem 284:1234–1241

Chapter 5

Posttranslational Modification of Thyroid Hormone Nuclear Receptor by Phosphorylation

Yan-Yun Liu and Gregory A. Brent

Abstract

TR phosphorylation promotes TR binding to DNA and heterodimerization with RXR. TRβ phosphorylation is induced by thyroid hormone on the cell membrane and phosphorylation by extracellular signal-regulated kinases (ERK), presumably at serine 142. TRα1 N-termini harbors two phosphorylation sites at serine 12 and serine 28/29. Serine 12 is phosphorylated by casein 2 and serine 28/29 by protein kinase A. Mutation analysis of TRα2 identified 2 serine sites, S472 and S473, as the substrates for casein kinase II. Phosphorylated TRα2 does not bind to DNA and dephosphorylated TRα binds to DNA and antagonizes TRα1 binding. Phosphorylation of TR is critical for TR function and T3 signaling and approaches to detection and analysis of phosphorylated TR are described.

Key words Thyroid hormone receptor (TR), In vivo (cells) phosphorylation, In vivo (tissue) phosphorylation

1 Introduction

TR phosphorylation influences DNA binding [1, 2], is stimulated by ERK signaling [3], and occurs on specific TRα1 residues [4, 5], and TRα2 [6].

Protein phosphorylation can be detected using various methods, including, in vitro and in vivo ^{32}P incorporation, electrophoretic mobility shift, immunoprecipitation, or direct detection in Western blot with a phosphor-motif antibody, affinity chromatographic isolation of phosphorylated protein followed by western blot, and phosphate binding molecule incorporation (Phso-Tag) [7]. Each method has strengths and weaknesses. ^{32}P incorporation definitively shows that a protein is phosphorylated. However, the method uses large amounts of ^{32}P, at a time when most laboratories want to reduce the use of radiation. Electrophoretic mobility shifts not only detect phosphorylated proteins, but can also show the stoichiometry of phosphorylated and unphosphorylated protein.

Michelina Plateroti and Jacques Samarut (eds.), *Thyroid Hormone Nuclear Receptor: Methods and Protocols*, Methods in Molecular Biology, vol. 1801, https://doi.org/10.1007/978-1-4939-7902-8_5,
© Springer Science+Business Media, LLC, part of Springer Nature 2018

However, phosphorylated protein may not have distinct detectable mobility shift, since the shift resulting from phosphorylation is not the property of changing molecular weight [8]. A newer development in the isolation of phosphorylated proteins such as PhosphoProtein™, (Qiagen Inc) and phos-tag Western blotting [9] allows for the easy separation of phosphorylated proteins. For mapping and analysis of the phosphorylation sites, the mass spectrometry-based method is frequently used [8, 10–12]. If the goal is to identify the phosphorylation site in a target of interest, there are available databases (http://www.phosphonet.ca/; http://www.phosphosite.org/homeAction.do; and www.phodphoSitePlus.org) that list both experimentally confirmed and predicted phosphorylation sites.

Here, we describe basic approach using an anti-phosphor motif antibody-based method in the detection of a protein known to be phosphorylated, which is widely used in monitoring phosphorylation change in response to cellular signals, specific diseases, and mutations.

2 Materials

1. 1× RIPA lysis buffer for phosphor protein: 50 mM Tris (pH 7.4), 150 mM NaCl, 1 mM EDTA, 1 mM EGTA, 1% NP-40, 0.5% sodium deoxycholate, 0.1% SDS, 2 mM sodium orthovandate (Na_3VO_4) and 5 mM NaF. 1 mM β-glycerphosphate, and 2 mM sodium pyrophosphate. The buffer must have sufficient chelating agents (EDTA and EGTA) to inhibit kinase and phosphatase inhibitors to inhibit dephosphorylation.
2. Phosphatase inhibitors: Halt™ protease and phosphatase inhibitor cocktail (Thermofisher Scientific) and PhosStop™ cocktail (Sigma–Aldrich). If making cocktail by yourself, *see* **Note 1** for recipe.
3. 100× protease inhibitors cocktail: dissolve four tablet (cOmplete™ protease inhibitor, Roche Inc.) in 2 mL water or 100 mM phosphate buffer, pH 7.0. Aliquot and store at −80 °C. The stock solution is stable for at least 4 months in −80 °C.
4. Tissue homogenizer (Dounce homogenizer or motorized tissue grinder).
5. Orbital and end-to-end rotating shaker.
6. Precasted 10% mini SDS gel.
7. SDS gel running buffer: 25 mM Tris 190 mM glycine 0.1% SDS in distilled water, pH 8.3 if needed.

8. Western blot transfer buffer: 20% methanol in Tris-glycine buffer. Add SDS to final concentration 0.1% for proteins larger than 100 kDa.
9. Anti-specific phosphor-motif antibody: e.g., anti-TRβ serine 124 or the factors influenced by TR action, such as anti-phsophor AKT.
10. TBST buffer: 20 mM Tris pH 7.6, 150 mM NaCl, 0.1% Tween 20.
11. Blocking the buffer: 5% BSA (Not Milk) in TBST (*see* **Note 2**).
12. Stripping buffer: 15% glycine, 0.1% SDS, 1% tween 20 (V/V), pH 2.2, in distilled water.
13. HRP-linked 2ndary antibody.
14. Imaging instrument (for chemiluminescence imaging).

3 Methods

Phosphorylation may be induced before analysis. For example, cells may be treated with or without ligand in order to detect T3-induced TR phosphorylation. For animals (wild type and TR mutation carrying animals) may be stimulated with thyroid hormone agonist/antagonist or other stimulus in order to evaluate TR phosphorylation and the phosphorylation of other factors affected by TR and TR mutations. The entire process should be performed at 0 °C –4 °C unless otherwise indicated.

3.1 Preparation of Lysate

1. For tissue, weigh 10–50 mg fresh or liquid N_2 snap frozen tissue, add 1 mL ice-cold 1× RIPA buffer containing protease inhibitors and phosphatase inhibitors, homogenize using tissue grinder or Dounce homogenizer.

   If using more tissue in a small volume of lysis buffer, make sure to add an adequate amount of protease inhibitors and phosphatase inhibitors. The ideal protein concentration for IP is 1–5 mg/mL.
2. For cell culture, trypsinize the cells (adherent cells), centrifuge at 100 × *g* for 4 min, wash cells pellet once with cold PBS (without Calcium and Magnesium), remove the PBS, and add RIPA buffer (10^6 cells/100 μL) containing protease inhibitors and phosphatase inhibitors to the cells. Pipette lysate up and down a few times to disburse the cell clumps, and then keep on ice for 30–60 min with constant agitation using an end-end rotator. Alternatively, simply agitate by hand every 5 min.

3.2 Sonication (Necessary for Tissue, Optional for Cells)

1. Partially submerge the lysate tube in an ice-water bath, sonicate at medium setting for 4–6 times with 15–30 s intervals to cool the lysate on ice-water. Power settings of sonicators vary

depending on the brand, so follow the manufacturer's recommendations.

2. Centrifuge the lysate at 12,000 × *g* for 10 min at 4 °C. Transfer the supernatant to a prechilled tube.
3. Determine the protein concentration using BAC or Bradford method and adjust the concentration uniformly in all samples with lysis buffer containing phosphatase inhibitors and protease inhibitors.

3.3 Western Blot without Immunoprecipitation

1. Aliquot 30 μL of whole cell extract sample containing 30–50 μg protein, add 30 μL of 2× SDS sample buffer (sufficient for 2 gels). Heat to 95 °C for 5 min to denature the protein. If protein concentrate is low in the lysate, use 50 μL of lysate sample and mix with 4×–6× SDS sample buffer.
2. Load samples (30 μL) onto a mini SDS-gel and separate the protein under standard condition (110–150 volts for a mini-gel).
3. Pre-wet the PVDF membrane with methanol. Proceed the proteins transfer from gel to PVDF membrane. Block the membrane with 5% (w/v) BSA (factor V) in TBST at room temperature for 1 h.
4. Add primary antibody (specific anti-phosphor antibody) to recommended dilution with 5–10 mL 5% BSA in TBST. Incubate at room temperature for 2 h with gentle shaking on an orbital shaker or at 4 °C overnight.
5. Remove the primary antibody and gently wash the membrane three times with TBST, for 5 min each, at room temperature.
6. Incubate with HRP-linked secondary antibody at dilution 1:10,000 or according to the manufacturer's recommendation in blocking buffer for 1 h at room temperature. The target species of HRP-linked secondary antibody must match the primary antibody. For example, primary anti-TR is a rabbit antibody, and then secondary is HRP-linked anti-rabbit IgG.
7. Rinse the membrane in TBST three times for 5 min each time.
8. Perform enhanced Chemiluminescence (ECL) to detect the phosphor protein (*see* the example in Fig. 1).

3.4 Immunoprecipitation Followed by Western Blotting

The purpose of immunoprecipitation is to enrich the phosphorylated protein of interest prior to the detection by WB. Immunoprecipitation should use the antibody specific to the protein of interest, but not the anti-phosphor-antibody. The non-anti-phosphor antibody precipitates both phosphorylated and non-phosphorylated protein species (total protein). The total protein band indicates possible changes in total protein level and can be used as a baseline to compare the levels of phosphorylation under experimental condition. It is not recommended to determine

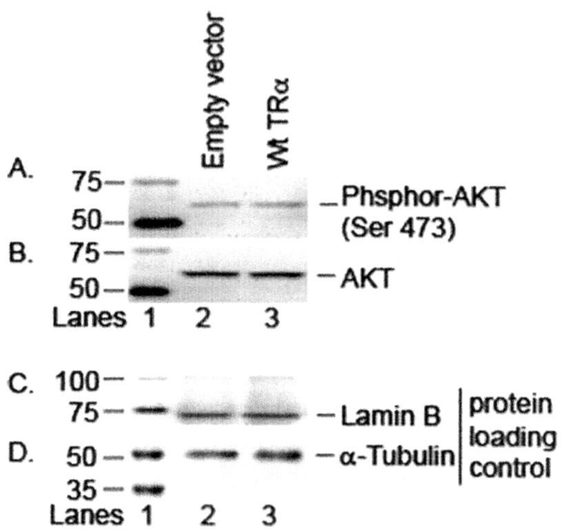

Fig. 1 Western Blot detection of phosphorylated AKT S473 proteins in nuclear fraction of cell lysate. The cells were transfected with expression vectors expressing empty vector (lane 2), and TRα (lane 3). Cells were serum-starved and then re-fed with the medium supplemented with 10% serum for 8 h. Total cell lysate was obtained by lysing cells in RIPA buffer with phosphatase inhibitors and protease inhibitors. Cell nuclear fraction was isolated using nuclear extraction kit (Abcam Inc) in the presence of phosphatase inhibitors and protease inhibitors. Two 10% mini-SDS gel gels were used. Thirty micrograms of the nuclear protein and total cell lysate were loaded onto each gel. After the proteins are transferred to PVDF membranes, the membranes were blocked with 5% BAS in TBST for 1 h. Then the membranes (**a**) were incubated with anti-AKT Ser473 (1: 250, rabbit antibody, Cell signaling Tech) and membrane (**b**) with anti-AKT (1: 250 rabbit antibody, cell signaling) for 2 h at room temperature with constant agitation. After wash three times with TBST, membrane was incubated with HRP-linked goat anti-rabbit IgG at1:10,000 dilution in blocking buffer for 1 h following three washes and performed ECL detection using Luminata™ Western chemiluminescent HRP substrates (Millipore Life Science). Western blots were imaged using BioSpectrum 515 (UVP Inc). For loading control, membranes were stripped for 15 min using stripping buffer (Bio-rad), then incubated in blocking buffer containing anti-laminin B1 (1: 500, mouse IgG, Abcam Inc) as nuclear protein loading control (**c**) and anti-α tubulin (1:500, mouse IgG, Abcam Inc) as total lysate loading control (**d**). The conditions for incubation, wash, and ECL are the same as described above. The secondary antibody used is HRP-linked rabbit-anti mouse. Lanes: 1, protein molecular mass marker; 2, Cell transfected with empty vector and 3, Cells transfected with TRα

stoichiometry of phosphorylation by this method, unless a 100% phosphorylated control can be produced [13].

1. Pre-clear (optional)

   Add 5–10 μL of normal IgG matching the species and the isotype (if possible) of the antibody to be used in IP, Incubate at 4 °C for 30 min with constant agitation.

2. Add 100 μL of pre-washed beads slurry (protein A/G magnetic beads or regular Protein A/G beads) to 1 mL of lysate. Incubate for 10–30 min at 4 °C with gentle agitation. It is necessary to use a large amount of bead slurry to completely remove the normal IgG and prevent the competition with specific antibody for the antigen.

3. Use a magnetic stand to remove the beads and transfer the pre-cleared lysate to a pre-chilled new tube.
4. Add an appropriate amount of antibody to the lysate according to the manufacturer's recommended dilution. Incubate samples for a minimum of 4 h or overnight at 4 °C with constant mixing.
5. Add 40 μL of Protein A/G magnetic beads and continue incubating for 1 h.
6. Collect beads using the magnetic stand, wash beads three times with TBS buffer.
7. Add 30 μL of SDS sample buffer to the beads and heat to 95 °C for 5 min to elute and denature the protein.
8. Run 10% Mini-SDS gel (*see* Subheading 3.3). It is optimal to run two identical gels side-by-side, to separately blot for total and phosphorylated protein.
9. Western blotting should follow the steps described in Subheading 3.3.
10. If two blot membranes are produced, use the antibody against specific phosphor-motif in one blot and the antibody against the total protein of interest (phosphorylated and nonphosphorylated) in another blot. If only one blot, use the antibody against the phosphorylated protein first and image it. After lightly stripping, apply the second antibody.

3.5 Isolation of Phosphor Proteins Followed by Western Blot

In many cases, a specific phosphor-motif antibody is not available. In order to detect particular phosphorylated protein, the reliable method is to isolate entire phosphorylated proteins from the cells or tissues of interests and then detect the phosphorylated protein of interest using regular antibody against the protein of interest. However, this method cannot identify the site of phosphorylation.

1. Lyse the cell and tissue samples as described in Subheadings 3.1 and 3.2.
2. Use commercial available kit (phsophorProtein, Qiagen) following the instructions to isolate the protein. This method involves chromatographic filtration. It is not recommended to use a self-made cartridge for protein purification.
3. After obtaining the phosphor protein, run an SDS gel and proceed to Western blot detection, as described in Subheading 3.3.

4 Notes

1. Commonly used phosphatase inhibitors and stock solutions are listed below.

Sodium Fluoride (NaF) inhibits acidic phosphatase. Stock solution 500 mM (100×): Dissolve NaF in H_2O and store at room temperature.

Sodium orthovandate sodium (Na_3VO_4) is a general inhibitor of alkaline phosphatase and tyrosine phosphatase. It is used with combination of NaF in all buffers for phosphor-protein and kinase studies. Stock solution 100 mM (100×): dissolve in H_2O, adjust PH to 10.0 with NaOH, solution is yellow/orange-ish color. Heat the solution at 100 °C until colorless, cool to room temperature, check the PH, and adjust it to pH 10.0. If color changed after cool, repeat the process until solution remains colorless and pH 10.0. Store the solution at −80 °C.

β-glycerophosphate is a classical serine-threonine phosphatase inhibitor. Stock solution 100 mM (100×): dissolve in H_2O and freeze in aliquot at −20 °C.

Sodium pyrophosphate ($Na_2P_2O_4$) inhibits serine-threonine phosphatase. Stock solution 100 mM (100×): dissolve in H_2O and freeze in aliquot at −20 °C.

2. For the detection of phosphorylated protein, it is not recommended to use milk in the blocking buffer, since milk contains a high amount of casein, a phosphoprotein which can be recognized by phosphor-specific antibody, and increases the background in a western blot.

Acknowledgments

This work was supported by NIH grant RO1DK98576 and VA Merit Review funds.

References

1. Tzagarakis-Foster C, Privalsky ML (1998) Phosphorylation of thyroid hormone receptors by protein kinase a regulates DNA recognition by specific inhibition of receptor monomer binding. J Biol Chem 273:10926–10932
2. Bhat MK, Ashizawa K, Cheng SY (1994) Phosphorylation enhances the target gene sequence-dependent dimerization of thyroid hormone receptor with retinoid X receptor. Proc Natl Acad Sci U S A 91:7927–7931
3. Davis PJ, Shih A, Lin HY, Martino LJ, Davis FB (2000) Thyroxine promotes association of mitogen-activated protein kinase and nuclear thyroid hormone receptor (TR) and causes serine phosphorylation of TR. J Biol Chem 275:38032–38039
4. Glineur C, Bailly M, Ghysdael J (1989) The c-erbA alpha-encoded thyroid hormone receptor is phosphorylated in its amino terminal domain by casein kinase II. Oncogene 4:1247–1254
5. Goldberg Y, Glineur C, Gesquiere JC, Ricouart A, Sap J, Vennstrom B, Ghysdael J (1988) Activation of protein kinase C or cAMP-dependent protein kinase increases phosphorylation of the c-erbA-encoded thyroid hormone receptor and of the v-erbA-encoded protein. EMBO J 7:2425–2433
6. Katz D, Reginato MJ, Lazar MA (1995) Functional regulation of thyroid hormone receptor variant TR alpha 2 by phosphorylation. Mol Cell Biol 15:2341–2348
7. Kinoshita E, Kinoshita-Kikuta E, Takiyama K, Koike T (2006) Phosphate-binding tag, a new

tool to visualize phosphorylated proteins. Mol Cell Proteomics 5:749–757

8. Dephoure N, Gould KL, Gygi SP, Kellogg DR (2013) Mapping and analysis of phosphorylation sites: a quick guide for cell biologists. Mol Biol Cell 24:535–542
9. Horinouchi T, Terada K, Higashi T, Miwa S (2016) Using Phos-tag in western blotting analysis to evaluate protein phosphorylation. Methods Mol Biol 1397:267–277
10. Jensen ON (2006) Automated phosphorylation site mapping. Nat Biotechnol 24:1226–1227
11. Zappacosta F, Huddleston MJ, Annan RS (2004) Comparative phosphorylation site mapping from gel-derived proteins using a multidimensional ES/MS-based approach. Methods Mol Biol 284:91–110
12. Elias JE, Haas W, Faherty BK, Gygi SP (2005) Comparative evaluation of mass spectrometry platforms used in large-scale proteomics investigations. Nat Methods 2:667–675
13. Johnson H, Eyers CE, Eyers PA, Beynon RJ, Gaskell SJ (2009) Rigorous determination of the stoichiometry of protein phosphorylation using mass spectrometry. J Am Soc Mass Spectrom 20:2211–2220

Chapter 6

Posttranslational Modification of Thyroid Hormone Nuclear Receptor by Sumoylation

Yan-Yun Liu and Gregory A. Brent

Abstract

Thyroid hormone has a broad range of biological effects, both during development and in the adult. Nuclear thyroid hormone action is mediated by thyroid hormone receptor (TR) α and β. Thyroid hormone also has nongenomic actions at the membrane, which are less well characterized. Both TRα and TRβ undergo posttranslational modification, including phosphorylation, acetylation, and recently identified sumoylation. These posttranslational modifications have been shown to influence thyroid hormone signaling by altering TR DNA binding, TR interaction with cofactors, and TR-mediated transcription. The best characterized modification, with respect to gene regulation, is sumoylation, which plays an important role in nutrient regulators-mediated gene expression. We present an approach to study posttranslational modification of TR by small ubiquitin-like modifier (SUMO), a process referred to as sumoylation.

Key words Thyroid hormone receptor (TR), In vitro sumoylation, In vivo (cells) sumoylation, In vivo (tissue) sumoylation, Phosphorylation

1 Introduction

1.1 Sumoylation of TR

Thyroid hormone has broad actions mediated by TR [1–3], and TR undergoes postranlational modification by phosphorylation and acetylation [4–9], as well as by sumoylation [8]. Sumoylation involves the covalent attachment of a SUMO to a specific lysine in the substrate (protein) and requires several enzymes that act in a stepwise process (Fig. 1). SUMO attaches to a specific lysine, located in a consensus motif, Ψ-K-x-D/E, where Ψ is a hydrophobic residue and x any amino acid [9]. A consensus motif is necessary, but not sufficient, since there are sumoylation consensus sites in proteins that do not conjugate to SUMO. This suggests that other factors, in addition to a consensus motif, are necessary for sumoylation. SUMO conjugation to a specific site on a protein must be evaluated directly by mutational analysis of the putative site.

Michelina Plateroti and Jacques Samarut (eds.), *Thyroid Hormone Nuclear Receptor: Methods and Protocols*,
Methods in Molecular Biology, vol. 1801, https://doi.org/10.1007/978-1-4939-7902-8_6,

© Springer Science+Business Media, LLC, part of Springer Nature 2018

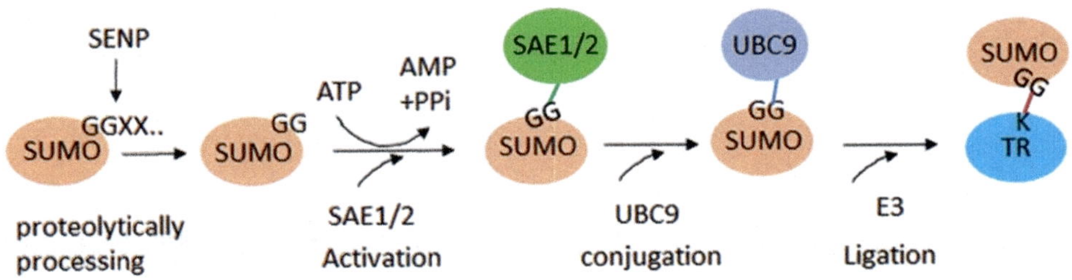

Fig. 1 The Process of Sumoylation. SUMO is first processed by SUMO-specific isopeptidases (sentrin-specific proteases; SENPs) that remove several C-terminal amino acids and expose the C-terminal Gly-Gly conjugation motif. Four amino acids are removed from SUMO1, 11 amino acids from SUMO2, and 2 amino acids from SUMO3. Mature SUMO is activated by SUMO-activating enzymes SAE1 and SAE2 (SAE1/2) in an ATP-dependent manner [10]. SUMO is then conjugated to the enzyme UBC9. Finally, SUMO is transferred to the substrate by UBC9. UBC9 delivers SUMO to the substrates and forms an isopeptide bond between a lysine residue in the substrate and a glycine in the C-terminus of SUMO. The final step may require an E3 ligase, depending on the substrate. Sumoylated TR is a substrate for SENPs that then ensures the reversible and dynamic nature of sumoylation

The molecular consequence of sumoylation is known, in general, to altered protein-protein interaction, protein stability and protein conformation [10]. TRα1 has two sumoylation sites, lysines 283 and 389. Lysine 288 can also be sumoylated, but only if lysine 283 is blocked. TRβ has three sumoylation sites, lysines 50, 146, and 443 [2]. These sumoylation sites are confirmed by SUMO conjugation using specific site mutations. Sumoylation of TR regulates TR/T3-mediated gene transcription, influences TR interaction with coactivation/corepressor, and promotes TR interaction with signaling pathways regulated by other transcription factors, such as the PPAR signaling pathway [2, 11].

In this chapter, we describe three approaches for detecting sumoylated TR and how it can be applied in different experimental models: using in vitro synthesized and radiolabeled TR, cells with exogenous expression of the protein of interest and using tissue samples.

2 Materials

2.1 Detection of TR Sumoylation Using In Vitro Synthesized and Radiolabeled TRα or TRβ

1. Plasmid expression vector: For in vitro synthesis of TRα or TRβ, the plasmid vector must have T7 or SP6 promoter to allow for in vitro transcription.
2. TNT® Quick Coupled Transcription/Translation System (Rabbit reticulocyte system) (Promega Inc).
3. Purified enzymes SAE1/SAEII, UBC9, SUMO1.
4. 10× sumoylation reaction buffer: 200 mM Hepes pH 7.5, 50 mM $MgCl_2$, 20 mM ATP.

5. Sumoylation kit, including purified proteins: is commercially available.
6. [S^{35}]-methionine (PerkinElmer Inc).
7. Geiger counter.
8. Radiation shield and Radioisotope waste container.
9. Pre-casted 10% SDS minigel.
10. Electrophoresis mini-gel system (Bio-rad).
11. Gel dryer (Bio-rad).
12. Phosphor-imager screen, cassette and imaging system (GE Healthcare, or other similar types).

2.2 Detection of Sumoylated Proteins in Cell Culture with Exogenously Expressed Protein of Interest

1. Cell line (example: HeG2 cells, ATCC, USA).
2. Cell culture medium for HepG2 cells: EMEM containing 10% FBS and penicillin-streptomycin antibiotics.
3. 1× Dulbecco's Phosphate-Buffered Saline (DPBS) without calcium and magnesium: 8 g NaCl, 0.2 g KCl, Na_2HPO_4 0.2 g KH_2PO_4 in 800 mL of distilled water, check pH 7.4 ± 0.1. Or use premade 1× DPBS without calcium and magnesium.
4. PBS (1×): 8 g NaCl, 0.2 g KCl, 1.44g Na_2HPO_4 0.2 g KH_2PO_4 in 800 mL distilled water, adjust pH to 7.4 with HCl, then add distilled water up to 1 L.
5. Mammalian expression vectors expressing SUMO or tagged-SUMO, TR or tagged-TR and UBC9.
6. Nucleofection device 4DX (Lonza Inc): program EH-100 for HepG2 cells (*see* **Note 1**).
7. Nucleofactor for transfection (Lonza Inc): HepG2 cells use SF kit. Each kit contains cell-specific P1 solution and the Supplemental solution (*see* **Note 2**).
8. Non-denaturing IP lysis buffer: 25 mM Tris–HCl pH 7.4, 150 mM NaCl, 1 mM EDTA, 1% NP-40, and 5% glycerol.
9. RIPA buffer: 25 mM Tris–HCl pH 7.6, 150 mM NaCl, 1% NP-40, 1% sodium deoxycholate, 0.1% SDS.
10. Protease inhibitor cocktail (*see* **Note 3**).
11. N-Ethylmaleimide (NEM) stock 2 M solution (200×): 0.25 g NEM (Mr = 125.13) in 1 mL ethanol. Make it fresh for each experiment (*see* **Note 4**).
12. 0.5 mg/mL (stock solution in water) SUMO-2 aldehyde, a SUMO-specific peptidase (SENP) inhibitor. Working solution ranges from 0.1 to 2.0 μM (*see* **Note 5**).
13. SDS gel running buffer: 25 mM Tris 190 mM glycine 0.1% SDS in distilled water, pH pH 8.3 if needed.

14. Western blot transfer buffer: 20% methanol in Tris-glycine buffer. Add SDS to final concentration 0.1% for proteins larger than 100 kDa.
15. Antibody for IP and Western blot: anti-SUMO1, anti-TRα, anti-TRβ, anti-tag if cells transfected with vector expressing tagged-SUMO1 or tagged-TR.
16. Antibody for IP negative control: use isotype-matched IgG from the same species as antibody used for IP.
17. HRP conjugated secondary antibody for Western blot: the target species must be same as primary antibody. For example, primary antibody is anti-SUMO1, a rabbit antibody. The secondary antibody should be HRP-linked anti-rabbit antibody.
18. Protein A/G magnetic beads and magnetic stand (Bio-rad).
19. 10% pre-casted mini SDS gel.
20. Pre-stained Protein ladders and Western blotting protein standards (Bio-rad).
21. Transfer buffer: 0.1% SDS, 50 mM Tis base, 380 mM glycine and 20% methanol.
22. TBS buffer (1×): 20 mM Tris, pH 7.5, 150 mM NaCl, pH to 7.6 with HCl.
23. TBST buffer: add 0.1% Tween 20 in TBS buffer.
24. Ponceau red (1×): 0.1% Ponceau S in 5% Glacial acetic acid. Pre-made stock solution is commercially available.
25. Clean Blot IP detection Kit (Thermofisher Scientific Inc).
26. Luminata™ Western chemiluminescent HRP substrates (Millipore Life Science) or other type of ECL kit.
27. Imaging system (BioSpectrum, UVP) or other types.

2.3 Detecting Sumoylated TR in Tissue Samples

1. Mouse tissue of interest, 100–200 mg tissue.
2. Mortar and pestle or other cryo-grinding devices.
3. Tissue homogenizer with Teflon pestle.
4. Sonicator.
5. IP and Western blot: same as that described in Subheading 2.2.

3 Methods

3.1 Detection of Sumoylated TR Using In Vitro Synthesized and Radio-Labeled TRα or TRβ

Sumoylated transcription factors represent a small proportion of the total transcription factors in the nucleus, which makes detection of sumoylated protein very challenging. A confirmative and efficient method to explore whether a protein is subjected to SUMO modification is to use in vitro sumoylation with a radiolabeled protein.

3.1.1 In Vitro Synthesis TRα and TRβ Protein

All procedure should be performed according to institutional radiation safety regulations.

1. Place the radiation vial containing [S^{35}]-methionine behind a radiation shield. Thaw [S^{35}]-methionine at room temperature. After thawing, leave the radiation vial on ice.
2. Thaw one vial of rabbit reticulocyte on ice, transfer 40 μL of reticulocyte to a PCR tube, add 100 ng plasmid (50 ng/μL), 1 μL of [S^{35}]-methionine (500 μci (18.5 MBq) and PCR grade water up to 50 μL.
3. The reaction mixture is incubated for 60 min at 37 °C.
4. Once the reaction is complete, maintain on ice or store at −80 °C, if not used immediately.

3.1.2 Sumoylation Reaction

1. Label tubes for sumoylation reactions (20 μL/each reaction).
2. Add reagents in the following sequence (*see* **Note 6**):
   First, add 2 μL in vitro synthesized radio-labeled TR protein to the tube, then SUMO1 protein (final 50 μg/mL), 2 μL 10× reaction buffer, enzymes SAEI (final 7.5 μg/mL) and SAEII (final 50 μg/mL), and protein UBC9 (final 50 μg/mL). At final step, add extra 2 μL of fresh 2 mM ATP and water up to 20 μL. For control, add all reagents except SUMO1.
3. Incubate at 37 °C for 1 h.
4. Terminate the reaction by the addition of 5 μL 4× SDS buffer, heat at 90 °C for 5 min. The samples should be stored at −80 °C if not used immediately.

3.1.3 SDS-PAGE and Imaging

1. Load 25 μL of reaction onto a 10% mini-SDS gel. Run gel at constant 100 V for 30 min and then increase to 120 volts, until blue dye reaches the bottom of the gel.
2. If samples are frozen, thaw at room temperature and then warm briefly at 90 °C before loading to the gel.
3. Remove one plate from the gel and gently rinse gel adhered to the remaining plate with H_2O.
4. Place a Whatman filter paper on the top of the gel and lift the paper and gel together from the plate.
5. Place the gel with the Whatman filter paper on bottom onto the gel dryer, cover the gel with plastic wrap, and dry the gel for 1 h at 80 °C.
6. Check radioactivity in the dried gel using a Geiger counter to confirm successful protein synthesis.
7. Mark the gel lanes on the top of the filter paper and transfer the dried gel to a phosphor-imaging cassette. Exposure time is variable depending on the radioactivity. Generally, 24 h exposure to the phosphor screen is required for imaging.
8. Scan radiolabeled protein image using a phosphor-imager (*see* the example in Fig. 2).

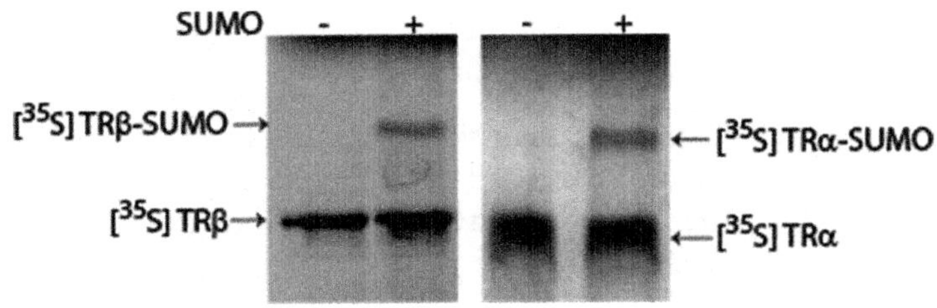

Fig. 2 Detection of TR sumoylation using in vitro synthesized and radiolabeled TR. The top protein bands are TR-SUMO conjugated protein. The bottom bands shown are in vitro synthesized radio-labeled TR proteins. [^{35}S]-TRα and -TRβ were incubated with E1, UBC9, and SUMO1 in sumoylation buffer at 37 °C for 1 h and then separated by 10% SDS-PAGE. The gel is dried (Bio-Rad Inc), and exposed to a phosphor imager screen overnight. Then, the imager screen is scanned using phosphor-imager (GE Healthcare Life Science). The sumoylation shown in the figure was performed using a sumoylation kit (LAE Biotech International Inc)

3.2 Detection of Sumoylated TR in In Vitro by Immunoprecipitation (IP)

In vitro sumoylation, as detected in cell culture, is a very useful method to determine whether a protein is a SUMO substrate (Fig. 3). However, in vitro sumoylation usually occurs without the involvement of an E3 ligase. E3 ligase is thought to be protein-specific, so must be identified in the tissue environment that the individual protein of interest is expressed. In vivo, E3 ligase plays a role in protein sumoylation, such that more sumoylated protein species are often observed in an in vivo sumoylation assay, than are seen in an in vitro cell culture sumoylation system for the same protein. The fraction of sumoylated proteins in a cell culture system is very small, therefore, it is a common approach to enhance the amount of sumoylated protein by transfecting a vector that expresses SUMO and the conjugation enzyme UBC9. Expression of a tagged protein of interest is preferred for easy detection [12, 13].

3.2.1 Cell Culture and Transfection

1. HepG2 cells are grown in 10 cm dish to 60–70% confluence.
2. Remove medium, wash cells with 10 mL DPBS, and then remove the DPBS.
3. Add 1–2 mL trypsin to the cells and incubate the cells for 3–5 min to allow complete detachment of cells from the dish. Examine the plate under a microscope to determine if cells are detached. If residual cells are still attached to the dish, lightly tap the bottom of culture dish and gently shake the dish horizontally to aid cell detachment.
4. Add 10 mL medium to neutralize the trypsin and pipette up and down to dissociate any cell clusters.
5. Transfer cell suspension to a 15 mL tube and count cells using a hemocytometer.
6. Aliquot ~2 × 10^{6} million cells to a microfuge tube and centrifuge in a desktop centrifuge at 500 rpm for 5 min.

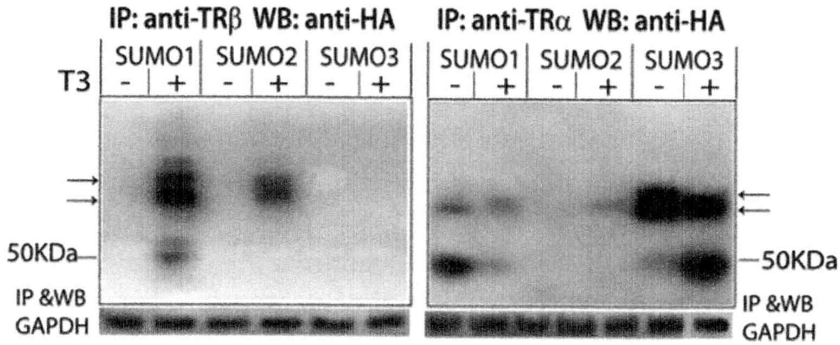

Fig. 3 Detection of in vivo (cells) sumoylation of TRα and TRβ. HepG2 cells were transfected with expression vectors expressing TRβ, TRα and SUMO-HA. Cells were lysed and immunoprecipitated with anti-TRβ or anti-TRα (Sigma-Aldrich Inc. and Thermofisher Scientific). For the Western blot, a primary antibody anti-HA tag (host species, mouse, EMD Millipore Inc) was used at 1: 500 and a secondary antibody, HRP-linked anti-mouse IgG (Cell Signal Biotech) at dilution 1:15,000. The Blot is imaged using BioSpectrum 515 chemiluminescence imager (UVP Inc) [2]. Arrows indicate the SUMO-TRα (right panel) and SUMO-TRβ (left panel). Protein mass is close to 50 KDa for the unsumoylated TRα (right panel) and TRβ (left panel). Membranes were stripped and re-probed with anti-GAPDH to show protein loading

7. Prepare Nucleofection mixture for transfection in a tube by mixing 82 μL P1 solution and 18 μL Supplement. Add 2 μg of each plasmid (1–2 μg/μL) to be transfected to the tube and mix well.
8. Continue from **step 6**, after centrifuge, remove as much supernatant as possible, and gently resuspend the cell pellet in the Nucleofection mixture (*see* **Note 7**).
9. Transfer cell suspension to an electroporation cuvette. Turn on the 4D-X link device and follow the screen prompter to complete the electroporation.
10. Upon completion of transfection, add 0.5 mL pre-warmed (37 °C) cell medium to the cuvette and transfer the cells to a 10 cm culture dish containing 10 mL pre-warmed culture medium.
11. Grow cells in the incubator for 48–72 h to reach 70–80% confluence. The number of cells required for each IP is a minimum of 1×10^7 cells.

3.2.2 Collect Cells

1. Trypsinize the cells as described in Subheading 3.2.1. Transfer the cells into a 15 mL conical tube. Centrifuge the cells at $250 \times g$ or 800 rpm for 4 min (*see* **Note 8**).
2. Wash the cells once with 10 mL cold PBS, pellet the cells by centrifugation as described above.
3. Remove the PBS, gently tap the tubes using fingertip to loosen pellet, and then leave the pellet on ice.

3.2.3 Lyse Cells

1. Add IP or RIPA lysis buffer to cell pellet (100 μL/10^6 cells) containing protease inhibitors, 1 mM PMSF, 20 mM NEM, and 2.0 μM SUMO-2 aldehyde.
2. Incubate samples on ice for 60 min with agitation.
3. Centrifuge the lysate at 12,000 × *g* at 4C for 10 min on a benchtop centrifuge.
4. Transfer the supernatant to a pre-chilled tube and leave on ice for the next step.
5. Measure the protein concentration in the lysate (*see* **Note 9**). Then, adjust protein concentration uniformly among different samples using IP lysis buffer or RIPA buffer containing protease inhibitors, NEM and SUMO-2 aldehyde and 1 mM PMSF. Protein concentration should be 5–10 mg/mL for IP detection of sumoylated protein.
6. Remove 30 μL lysate from each sample and mix with 30 μL of 2× SDS samples buffer. Heat at 90 °C for 5 min. This sample is the input.

3.2.4 Pre-Clear

The purpose of pre-clear is to remove nonspecific proteins that bind to the Protein A or G beads, which reduces the background noise of the IP, especially when high protein concentration is used. To detect sumoylated proteins, a high protein concentration is usually required, since the sumoylated protein represents a very small fraction of the total protein present.

1. Wash magnetic protein A/G beads three times with cold PBS. Incubate Beads with 1% BSA for 1 h with constant agitation and then followed by three washes with cold PBS.
2. Resuspend beads in PBS to make a 50% bead slurry.
3. Add 100 μL of protein A/G beads (50% slurry) to 1 mL of lysate and incubate at 4 °C for 30 min with agitation.
4. Remove beads using a magnetic stand. Transfer the lysate to pre-chilled new tubes, kept on ice and ready for immunoprecipitation.

3.2.5 Immuno-precipitation (IP)

1. For a negative control, add 2 μL (1 mg/mL) normal IgG to 1 mL pre-cleared lysate. The normal IgG should be species and isotype-matched to the antibody used for IP.
2. For IP, antibody to the pre-cleared lysate at 1:100 dilution or according to the pilot study and/or recommendation by the manufacturer (*see* **Note 10**).
3. Incubate the IP reaction at 4 °C with constant rotating for at least 4 h, or overnight.
4. Add Protein A/G magnetic beads (40–60 μL, 50% beads slurry) to the IP samples.

5. Incubate at 4 °C for 1–2 h with constant agitation.
6. Use a magnetic stand to collect the beads and discard the lysate. Wash Protein A/G magnetic beads three times with 1 mL of cell lysis buffer (IP or RIPA buffer) containing protease inhibitors and then two times with 1 mL PBS. Alternatively, wash beads three times with 0.5% NP40 in PBS containing protease inhibitors.
7. After final wash, collect Protein A/G magnetic beads. Add 30–40 μL of 2× SDS sample buffer to the beads and heat the beads to 90 °C for 5 min to elute the immunoprecipitated protein. Transfer the eluted protein to a new tube for SDSPAGE and Western blot. Samples can be stored at −80 °C for later use.

3.2.6 SDS-PAGE and Transfer Protein to PVDF Membrane

Various types of devices and buffers have been developed for Immunoblotting. Follow the manufacturer's instructions when using the device for immunoblotting. The quality of the primary antibody is the key factor for successful blotting. Here, we describe the basic techniques without using a specific blotting device.

1. Separate immunoprecipitated proteins in 10% mini-SDS gel. Run gel at 100–150 constant voltage until the 35 KDa protein marker reaches the gel bottom, or to the desired position.
2. Pour the 100% methanol to a shallow container. Immerse the PVDF membrane in methanol briefly to activate the membrane. Then equilibrate the membrane in transfer buffer for a few minutes.
3. Transfer proteins to PVDF membrane using a Mini Trans-blot^R cell (Bio-rad) for Western blot. The transfer cassette is color-coded: Black (negative) and Red (positive). Assemble the transfer "sandwich" in the following order: black (negative) at bottom, then Scotch brite pad> Whatman paper > gel > PVDF > Whatman paper > Scotch brits pad, close the transfer assembly with red (positive).
4. Inert the transfer assembly into the Mini Trans-blot^R cell with the orientation of the red side of transfer cassette facing the positive electrode and black side facing the negative.
5. Pour ice-cold transfer buffer into the transfer cell and proceed with the protein transfer at a constant 180 mApm for 45 min to 1 h, or at constant voltage 100 V for 2 h. The transfer is carried out at 0–4 °C.

3.2.7 Visualization of the Transferred Proteins Using Ponceau Red

1. Rinse the PVDF membrane with TBST.
2. Cover the membrane with Ponceau Red for 5 min with gentle shanking at room temperature.

3. Wash the membrane in distilled water until the background becomes clear. Photograph the proteins and then proceed to the next step.

3.2.8 Immunoblotting

1. Block the membrane with 5% Nonfat milk in TBST for 1 h.
2. Dilute the primary antibody in blocking buffer at 1:250 (*see* **Note 11**) and incubate for 2 h at room temperature or for a longer period at 4 °C. Preferably, the antibody should be different from that used in the IP. If using anti-TR in IP, then use anti-SUMO or anti-tag for a tagged-SUMO expressed in transfected cells as the primary antibody for Western blotting.
3. Wash membrane with TBST three times, 5 min for each wash.
4. Incubate membrane in blocking buffer with HRP conjugated secondary antibody (usually 1: 5000 to 1: 10,000) for 1 h. Choose the species-matched secondary antibody as the primary antibody. For example, Primary antibody is a rabbit antibody, then use anti-rabbit IgG, HRP-linked as secondary antibody.
5. Wash the membrane three times with TBS, perform ECL, and image the data (*see* the example in Fig. 2).

3.3 Detection of Sumoylated TR in Tissues

Protein sumoylation and desumoylation occurs in response to cellular processes, including growth, differentiation, senescence, and apoptosis. The study of native sumoylated protein is important to understand the mechanism of how sumoylation regulates transcription factor-mediated gene transcription. It is very difficult to detect sumoylated protein in tissues. This is because the sumoylated protein is at very low concentration in the tissue, in addition, desumoylation can occur quickly during tissue isolation and processing. Care must be taken in tissue homogenization. After the tissue lysate is obtained, the IP step is the same as previously described.

3.3.1 Tissue Homogenization

1. Remove the tissue and rinse with PBS.
2. Cut tissue into small pieces (10–20 mg/piece) and snap freeze in liquid nitrogen (alternately, tissues can be frozen in dry ice).
3. Pre-cool mortar, pestle, spatula and a brush on dry ice.
4. Grind 100–400 mg of tissue in the mortar with the pestle to fine powder and brush all the tissue powder together (*see* **Note 12**). Alternatively, use cryogrinding device, if available.
5. Transfer tissue powder, using a spatula, to a 15 mL tube containing 5–10 mL ice-cold RIPA buffer with added protease inhibitors, 1 mM PMSF, 20 mM NEM and 2 μM SUMO-2 aldehyde. When transferring tissue powder to RIPA buffer, transfer a small amount of tissue powder each time and let it dissolve in the RIPA buffer, then repeat until all the tissue powder is transferred to the tube.

3.3.2 Sonication

1. Transfer the lysate to smaller tubes (1.5–2 mL). Keep lysate volume at maximum of 2/3 of tube volume to avoid spills during sonication.
2. Sonicate the lysate at the appropriate settings (*see* **Note 13**) for a period 10 s, 4–6 times with 30 s interval between sonications in order to cool samples in ice water.
3. Centrifuge at 14,000 rpm (12,000 × *g*) for 10 min at 4 °C in a microfuge. Transfer the supernatant to a pre-chilled tube and discard the pellet.
4. Determine protein concentration and adjust protein concentration uniformly among the samples using the lysis buffer containing protease inhibitors and NEM.

3.3.3 IP and Western Blot

The IP and Western blotting are the same as described in Subheading 3.2.

4 Notes

1. The electroporation program for Nucleofection depends on the cell line used and the model of instrument. The transfection database for different cell lines can be found in www.Lonza.com website.
2. Nucleofection is electroporation-based method and gives high transfection efficiency. Alternatively, a regular transfection method can also be used. If using a regular transfection method, follow the manufacturer's instructions.
3. Make 100× stock solution of cOmplete™ protease inhibitor solution: dissolve four tablet in 2 mL water or 100 mM phosphate buffer, pH 7.0. Aliquot and store at −80 °C. The stock solution is stable for at least 4 months in −80 °C.
4. Make NEM stock solution (400 mM): dissolve 50 mg NEM in 1 mL ethanol. NEM is less soluble in water (1 mg/mL), but highly soluble in ethanol (50 mg/mL). NEM working concentration varies from 5 to 50 mM. Reducing agents (DTT, β-mercaptoethanol) inactivate NEM. Make sure no DTT or β-mercaptoethanol is contained in the lysis buffer.
5. SUMO-2 aldehyde is the SENP-specific inhibitor available from Enzo Lifesciences, although there is no specific information about the selectivity of this inhibitor.
6. Sumoylation reaction ingredients must be added in sequence, as indicated. The sequence is according to the sumoylation reaction shown in Fig. 1.

7. When transfecting using the nucleofection method, resuspend cells in nucleofection mixture with minimum pipetting up and down to prevent cell membrane damage. Do not leave cells in nucleofection mixture for more than 5 min. Perform electroporation as soon as possible.
8. Cells can be lysed directly on the cell culture dish without cell collection. However, considering the small amount of IP or RIPA lysis buffer used (10^6 cells/100 μL lysis buffer), collecting cells is a better choice for lysing cells completely in a tube.
9. To detect sumoylated protein using IP, more protein (total 10–20 mg) is needed, but the protein concentration should not be more than 5 mg/mL. High concentrations of protein may increase nonspecific protein binding to the Protein A/G beads.
10. Using a high amount of antibody in IP may cause the IgG antibody to be co-eluted with the target protein. Crosslinking the antibody with beads will reduce the background. Using Clean-blot™ in post IP Western blot reduces the background as well. Clean blot™ binds to the whole IgG of the antibody used in IP, including heavy chain (~50 kDa) and light chain (~25 KDa) IgG.
11. Antibody dilution should be pre-determined, prior to use.
12. Grinding tissue on dry ice protects protein from degradation and is much more efficient than using a tissue homogenizer or Dounce homogenizer. Liver tissue can be easily grinded to very fine powder. However, adipose tissue appeared to be sticky and sonication is sometimes necessary.
13. The power of the sonicator depends on the model. It is recommended that pilot experiments be carried out to determine the output power and time used for different tissue sonications. The extent of tissue breakdown can be assessed by centrifugation. The presence of a substantial number of small tissue pieces indicates that tissue is not sufficiently broken down.

Acknowledgments

This work was supported by NIH grant RO1DK98576 and VA Merit Review funds.

References

1. Mullur R, Liu YY, Brent GA (2014) Thyroid hormone regulation of metabolism. Physiol Rev 94:355–382
2. Liu YY, Kogai T, Schultz JJ, Mody K, Brent GA (2012) Thyroid hormone receptor isoform-specific modification by small ubiquitin-like modifier (SUMO) modulates thyroid hormone-dependent gene regulation. J Biol Chem 287:36499–36508
3. Bernal J (2007) Thyroid hormone receptors in brain development and function. Nat Clin Pract Endocrinol Metab 3:249–259
4. Tzagarakis-Foster C, Privalsky ML (1998) Phosphorylation of thyroid hormone receptors by protein kinase a regulates DNA recognition by specific inhibition of receptor monomer binding. J Biol Chem 273:10926–10932
5. Bhat MK, Ashizawa K, Cheng SY (1994) Phosphorylation enhances the target gene sequence-dependent dimerization of thyroid hormone receptor with retinoid X receptor. Proc Natl Acad Sci U S A 91:7927–7931
6. Goldberg Y, Glineur C, Gesquiere JC, Ricouart A, Sap J, Vennstrom B, Ghysdael J (1988) Activation of protein kinase C or cAMP-dependent protein kinase increases phosphorylation of the c-erbA-encoded thyroid hormone receptor and of the v-erbA-encoded protein. EMBO J 7:2425–2433
7. Lin HY, Hopkins R, Cao HJ, Tang HY, Alexander C, Davis FB, Davis PJ (2005) Acetylation of nuclear hormone receptor superfamily members: thyroid hormone causes acetylation of its own receptor by a mitogen-activated protein kinase-dependent mechanism. Steroids 70:444–449
8. Becares N, Gage MC, Pineda-Torra I (2017) Post-translational modifications of lipid-activated nuclear receptors: focus on metabolism. Endocrinology 158(2):213–225
9. Flotho A, Melchior F (2013) Sumoylation: a regulatory protein modification in health and disease. Annu Rev Biochem 82:357–385
10. Geiss-Friedlander R, Melchior F (2007) Concepts in sumoylation: a decade on. Nat Rev Mol Cell Biol 8:947–956
11. Liu YY, Ayers S, Milanesi A, Teng X, Rabi S, Akiba Y, Brent GA (2015) Thyroid hormone receptor sumoylation is required for preadipocyte differentiation and proliferation. J Biol Chem 290:7402–7415
12. Tatham MH, Rodriguez MS, Xirodimas DP, Hay RT (2009) Detection of protein SUMOylation in vivo. Nat Protoc 4:1363–1371
13. Hwang J, Kalejta RF (2011) In vivo analysis of protein sumoylation induced by a viral protein: detection of HCMV pp71-induced Daxx sumoylation. Methods 55:160–165

Chapter 7

Demonstration of the Receptor Site for Thyroid Hormone on Integrin αvβ3

Hung-Yun Lin, Shaker A. Mousa, and Paul J. Davis

Abstract

Integrin αvβ3 is one of the 24 heterodimeric structural proteins of the plasma membrane of animal cells. The extracellular domain of αvβ3 comprises about 80% of its mass and is devoted largely to interactions with extracellular matrix (ECM) proteins, such as vitronectin and fibronectin, and cell-cell interaction. The binding of specific ECM proteins is conveyed to the cell interior by signal transduction. In the past decade, integrin αvβ3 has been shown to bind small molecules, such as nonprotein hormones—thyroid hormone analogues and dihydrotestosterone—and the stilbene, resveratrol. The αvβ3-thyroid hormone interaction signal generated at the integrin is transduced into intracellular protein trafficking, nucleoprotein phosphorylation, and expression of specific genes. The hormone receptor on the integrin bears no structural homologies with nuclear thyroid hormone receptors (TRs). Because integrin αvβ3 is generously expressed on the surface of cancer cells and rapidly dividing endothelial cells, thyroid hormone actions initiated at the receptor on integrin not unexpectedly relate to cell proliferation, cancer cell survival pathways and to angiogenesis. In this chapter, we present methods for the definition of the receptor, for monitoring certain of its functions and for the downregulation of the gene for the integrin.

Key words Integrin αvβ3, L-thyroxine (T_4), 3,5,3′-triiodo-L-thyronine (T_3), Tetraiodothyroacetic acid (tetrac), Immunoblotting

1 Introduction

The molecular basis for the intracellular actions of thyroid hormone is the intranuclear binding of 3,5,3′-triiodo-L-thyronine (T_3) with nuclear thyroid hormone receptor (TR) proteins, consequent DNA binding of the TR-hormone complex with coactivator or co-repressor proteins and modulation of specific gene expression [1, 2]. T_3 is the principal intracellular thyroid hormone and is derived by deiodination from tetraiodothyronine, (L-thyroxine, T_4).

Michelina Plateroti and Jacques Samarut (eds.), *Thyroid Hormone Nuclear Receptor: Methods and Protocols*, Methods in Molecular Biology, vol. 1801, https://doi.org/10.1007/978-1-4939-7902-8_7,
© Springer Science+Business Media, LLC, part of Springer Nature 2018

In the last decade, it has become clear that a specific receptor for thyroid hormone exists on the surface of certain cells, particularly cancer cells and dividing endothelial cells [1, 3]. At this receptor, T_4 is the principal ligand at physiological concentrations. The receptor is in the extracellular domain of a structural heterodimeric protein of the plasma membrane, integrin αvβ3. There are two dozen integrins and they are involved, through discrete binding domains, in cell-extracellular matrix (ECM) protein interactions and cell-cell interactions. Such binding results in the generation of intracellular signals that are important to cell orientation and mobility and to tissue stability. The small molecule receptor on αvβ3 for T_4 and T_4 analogues (such as tetraiodothyroacetic acid (tetrac)), however, is capable of generating downstream signals that regulate specific intracellular protein trafficking, phosphorylation of specific nucleoproteins and modulation of expression of a large number of genes involved in control of the cell cycle, of cancer cell survival pathways and in angiogenesis [3].

The trafficking within the cell of proteins such as TRs, of nuclear estrogen receptors (ERs) and mitogen-activated protein kinase (MAPK; ERK1/2) from cytoplasm to the nuclear compartment is regulated from αvβ3. Cancer cell survival pathways involving p53 and other apoptosis-relevant proteins and mitochondrial permeability can be modulated from the thyroid hormone receptor on αvβ3. Finally, a number of control mechanisms for new blood vessel formation are regulated by T_4 and tetrac via the integrin. Examples of these mechanisms are (a) control of expression of genes for vascular endothelial growth factor (VEGF) and basic fibroblast growth factor (bFGF), for matrix metalloproteinases and for epidermal growth factor receptor (EGFR), (b) regulation of the activities of platelet-derived growth factor (PDGF) and EGF and (c) control of endothelial cell migration in response to protein cues.

In this chapter, we present techniques for (1) demonstrating the presence of the receptor for thyroid hormone and hormone analogues on integrin αvβ3, (2) quantitating αvβ3 in cells by immunoblotting and (3) suppressing transcription of αvβ3 in cancer cells.

2 Materials

A Protein Purification Facility (http://wolfson.huji.ac.il/purification/) is required for certain of the procedures described below, e.g., Subheading 3.2.

2.1 Definition of Receptor for Thyroid Hormone on Integrin αvβ3, Quantitation of the Receptor and Inhibition by shRNA of Expression of the Receptor Binding Protein (αvβ3)

1. PRISM software bundle (GraphPad, San Diego, CA, USA).
2. BioSpectrum® Imaging System (UVP, LLC, Upland, CA, USA).
3. illustra™ RNAspin Mini RNA Isolation Kit (GE Healthcare Life Sciences, Buckinghamshire, United Kingdom).
4. RevertAid H Minus First Strand cDNA Synthesis Kit (Life Technologies Corporation, Carlsbad, CA, USA).
5. QuantiNovaTM SYBR® Green PCR Kit (Qiagen, Valencia, CA, USA).
6. Connect™ Real-Time PCR Detection System (Bio-Rad Laboratories, Inc., Hercules, CA, USA).
7. RNeasy kit (Qiagen) forward 5′-TCCGATTCCA AACTGGGAGC-3′ and reverse 5′-AAGGCCACTGAAGAT GGAGC-3′ (Accession No.: NC_000002.12); homo sapiens integrin β3 (ITG β3), forward 5′-CTGGTGTTTA CCACTGATGCCAAG-3′ and reverse 5′-TGTTGA GGCAG GTGGCATTGAAGG-3′ (Accession No.: NM_000212.2); homo sapiens 18S ribosomal RNA (18S), forward 5'-GTAACCCGTTGAACCCCATT-3′ and reverse 5'-CCATCCAATCGGTAGTAGCG-3′ (Accession No. NR_ 003286).
8. Millipore Immobilon™-PSQTransfer PVDF membranes (Millipore, Billerica, MA, USA).
9. Mini Trans-Blot® Cell (Bio-Rad Laboratories, Inc.).
10. Primary antibodies to integrin αv (Santa Cruz Inc., Santa Cruz, CA, USA), integrin β3 (Santa Cruz), and glyceraldehyde 3-phosphate dehydrogenase (GAPDH) (GeneTex International Corporation, Hsinchu City, Taiwan).
11. Short hairpin (sh)-RNA for scrambled RNA (2 μg/well) (National RNAi Core Facility, Academia Sinica, Taipei, Taiwan).
12. Opti-MEM™ I medium (Invitrogen, Carlsbad, CA, USA).
13. HRP-conjugated secondary antibodies and Immobilon™ WesternHRP Substrate Luminol Reagent (Millipore).

3 Methods

3.1 Demonstration of the Presence of the Receptor on the Integrin

1. Incubate 2 μg of commercially available purified integrin αvβ3 with ^{125}I-T_4 or ^{125}I-tetrac. Run the mixture on a non-denaturing polyacrylamide gel. Binding by integrin αvβ3 of radiolabeled T_4 is reduced in the presence of unlabeled T_4 or tetrac, which is added to integrin αvβ3 before the incubation with ^{125}I-T_4.

2. The assay is performed following the basic-native gel protocol from the Protein Purification Facility (http://wolfson.huji.ac.il/purification/Protocols/PAGE_Basic.html).
3. Dilute all test compounds to their final concentration in 0.04 N KOH with 0.4% polyethylene glycol to ensure that the effect is independent of the solvent used.
4. Mix 2 μg of purified integrin αvβ3 (stock concentration, 0.3–0.5 μg/μL) with the indicated concentrations of test compounds and allow to incubate for 30 min at room temperature. $^{125}I\text{-}T_4$ (2 μCi) is then added, and incubate the mixture for an additional 30 min at room temperature. Mix the samples with sample buffer [50% glycerol, 0.1 M Tris–HCl (pH 6.8), and bromophenol blue] and run on a 5% basic-native gel for 24 h at 45 mA in a cold room at 4 °C. Disassemble the apparatus and place the gels on filter paper, wrap in plastic wrap, and expose to film. Determine band intensity with the VersaDoc 5000 Imaging system.
5. Determine the dissociation constant (Kd) and EC50 using PRISM software. Determine Kd with nonlinear regression, using the programmed homologous competitive binding curve with one class of binding sites equation. The nonspecific binding is held constant at 15. Set the constant Hot nM at 0.13, as determined by the equation Hot nM = hot cpm/(specific activity × incubation volume × 1000). Determine EC50 using nonlinear regression with the programmed equation for sigmoidal dose response.

3.2 Quantitation of Integrin in Cell Samples

3.2.1 RNA Isolation and qPCR

1. Total RNA is extracted and genomic DNA is also eliminated with illustra RNAspin Mini RNA Isolation Kit. One μg of DNase I-treated total RNA is reverse-transcribed with RevertAid H Minus First Strand cDNA Synthesis Kit into cDNA and used as the template for real-time PCR reactions and analysis.
2. Real-time PCR reactions are performed using QuantiNovaTM SYBR® Green PCR Kit on CFX Connect™ Real-Time PCR Detection System. This involves an initial denaturation at 95 °C for 5 min, followed by 45 cycles of denaturing at 95 °C for 5 s and combined annealing/extension at 60 °C for 10 s, as described in the manufacturer's instructions.
3. Total RNA is extracted from cell cultures 72 h after transfection using the RNeasy kit according to the manufacturer's instructions. Primers are based on published species-specific sequences: homo sapiens integrin αv (ITG αv), forward 5′-TCCGATTCCAAACTGGGAGC-3′ and reverse 5′-AAGGCCACTGAAGAT GGAGC-3′ (Accession No.: NC_000002.12); homo sapiens integrin β3 (ITG β3), forward 5′-CTGGTGTTTACCACTGATGCCAAG-3′ and reverse 5′-TGTTGA GGCAGGTGGCATTGAAGG-3′ (Accession

No.: NM_000212.2); homo sapiens 18S ribosomal RNA (18S), forward 5'-GTAACCCGTTGAACCCCATT-3′ and reverse 5'-CCATCCAATCGGTAGTAGCG-3′ (Accession No. NR_003286); Calculations of relative gene expression (normalized to 18S reference gene) are performed according to the ΔΔCT method. Fidelity of the PCR reaction is determined with melting temperature analysis.

3.2.2 Western Blotting

1. Harvest the whole cell lysate.
2. Load 20 μg of protein into each well with 5× sample buffer, and resolve the protein samples with electrophoresis at 100 V for 2 h.
3. Transfer the resolved proteins from the polyacrylamide gel to Millipore Immobilon-PSQTransfer PVDF membranes with the Mini Trans-Blot® Cell.
4. Block the membranes with a solution of 2% bovine serum albumin in Tris-buffered saline. Incubate membranes with primary antibodies to integrin monomer αv, integrin monomer β3, and GAPDH at 4 °C overnight and wash. The proteins are detected with HRP-conjugated secondary antibodies and Immobilon™ WesternHRP Substrate Luminol Reagent.
5. Visualize and record the western blots with the BioSpectrum® Imaging System.

3.3 Inhibition of the Expression of Integrin

1. Seed cells in 6-well tissue culture plates at 60%–80% confluence and maintain in the absence of antibiotic for 24 h before transfection.
2. Remove the culture medium before transfection, and wash the cells once with PBS, then transfect with shRNA for scrambled RNA (2 μg/well) from a commercial source, using lipofectamine™ 2000 (2 μg/well) in an Opti-MEM I medium according to the manufacturer's instructions. After transfection, incubate cultures at 37 °C for 4 h and then place in a fresh culture medium. After an additional 72 h, use the cells in experiments.

References

1. Cheng SY, Leonard JL, Davis PJ (2010) Molecular aspects of thyroid hormone actions. Endocr Rev 31:139–170
2. Brent GA (2012) Mechanisms of thyroid hormone action. J Clin Invest 122:3035–3043
3. Davis PJ, Goglia F, Leonard JL (2016) Nongenomic actions of thyroid hormone. Nat Rev Endocrinol 12:111–121

Chapter 8

The Deiodinase Trio and Thyroid Hormone Signaling

Antonio C. Bianco and Rodrigo R. da Conceição

Abstract

Thyroid hormone signaling is customized in a time and cell-specific manner by the deiodinases, homodimeric thioredoxin fold containing selenoproteins. This ensures adequate T3 action in developing tissues, healthy adults and many disease states. D2 activates thyroid hormone by converting the pro-hormone T4 to T3, the biologically active thyroid hormone. D2 expression is tightly regulated by transcriptional mechanisms triggered by endogenous as well as environmental cues. There is also an on/off switch mechanism that controls D2 activity that is triggered by catalysis and functions via D2 ubiquitination/deubiquitination. D3 terminates thyroid hormone action by inactivation of both T4 and T3 molecules. Deiodinases play a role in thyroid hormone homeostasis, development, growth and metabolic control by affecting the intracellular levels of T3 and thus gene expression on a cell-specific basis. In many cases, tight control of these pathways by T3 is achieved with coordinated reciprocal changes in D2-mediated thyroid hormone activation D3-mediated thyroid hormone inactivation.

Key words Thyroid hormone, Deiodinase, Selenoprotein

1 Introduction

The thyroid gland takes up and concentrates iodide that is used to synthesize biologically active molecules known as thyroid hormones. Their biological activity and other intrinsic properties depend on the number as well as spatial localization of the iodide atoms within the molecule of thyroid hormone. Thyroxine (T4) is the main hormone produced by the thyroid gland. It has a relatively long half-life (~8 days) and is only minimally active. T3 (3,5,3′-triiodothyronine) is a fully active thyroid hormone; it has a much shorter half-life (~12 h), but is produced mostly outside the thyroid gland through deiodination of T4. In contrast, removal of the inner T4 or T3 ring iodine produces reverse T3 (3,3′,5′-triiodothyronine) or T2 (3,3′-T2), respectively, which are biologically inactive molecules. Deiodinases are enzymes that catalyze these activating and inactivating reactions. They are expressed across multiple organs and tissues, constituting enzymatic pathways that

Michelina Plateroti and Jacques Samarut (eds.), *Thyroid Hormone Nuclear Receptor: Methods and Protocols*, Methods in Molecular Biology, vol. 1801, https://doi.org/10.1007/978-1-4939-7902-8_8,

© Springer Science+Business Media, LLC, part of Springer Nature 2018

initiate or terminate thyroid hormone action. The deiodinases play a role in various aspects of vertebrate physiology such as development [1], TSH and TRH feedback regulation [2–4], and diseases states [5]. Please refer to recent reviews for a comprehensive discussion of this topic [6–8].

2 Fundamental Aspects of the Deiodinases

Types 1 and 2 deiodinases (D1 and D2) activate thyroid hormone whereas the type 3 deiodinase (D3) inactivates both T4 and T3 [9]. Sequence analysis coupled with hydrophobic cluster analysis (HCA) revealed that the deiodinases share an overall 50% sequence similarity. Most similarity lies within the conserved thioredoxin-fold (TRX) domain composed of βαβ and ββα motifs [10]. Deiodinases are anchored in cell membranes via a single transmembrane domain with its catalytic globular domain facing the cytosol [11, 12]. This model was further refined by crystallization of mouse D3 globular domain [13]. The cDNA encoding the three deiodinases predict a molecular weight between 29 and 33 kDa [14]. However, in gel filtration studies deiodinases are found in higher molecular weight complexes, between 44 and 200 kDa [15], suggesting that they could be part of multi-protein complexes or be organized in multimeric forms. Indeed, deiodinases are homodimers and dimerization is required for catalytic activity [11, 16–18]. Dimerization involves the trans-membrane and globular domains of the deiodinases, and may be mediated by disulfide bonds [17].

The deiodinase active center is a pocket within the globular domain defined by a βαβ motif and a highly conserved intervening element that is also found in α-L-iduronidase, a lysosomal enzyme. In this pocket is the rare amino acid selenocysteine (Sec), critical for the nucleophilic attack that takes place during the deiodination reaction [10]. Given the topology of all three deiodinases, thyroid hormones need to enter cells in order to be metabolized by deiodinases. Cell entry is mediated by membrane transport proteins [19], with the monocarboxylate 8 (MCT8) likely being the most relevant one. Mutations in the MCT8 gene are associated with neurodevelopmental problems and endocrine dysfunctions [19–21]. T3 binds two nuclear thyroid hormone receptors (TRs), TRα and β, that modify transcription of multiple sets of thyroid hormone-responsive genes, initiating the biological effects of thyroid hormone [22]. Given that TR saturation depends on the concentration of intracellular T3, it is also recognized that D2 and D3 play antagonist roles in thyroid hormone action by inversely affecting intracellular T3 levels. D2 produces T3 and thus increases thyroid hormone signaling whereas D3 inactivates T4 and T3, silencing thyroid hormone signaling. Both pathways are relatively independent of circulating T4 or T3 levels [23].

A corollary of such studies is that thyroid hormone action can be customized on a time- and cell-dependent fashion via the coordinated expression of D2 and D3 [1, 24]. A well-known model that illustrates such interplay is vertebrate development. In most developing tissue D3 activity predominates, preventing rapidly developing cells from being exposed to thyroid hormone. As the embryo matures, D3 expression is minimized and D2 increases, enhancing thyroid hormone signaling. D1, on the other hand, does not contribute significantly to the local control of thyroid hormone signaling. This is explained by its localization in the plasma membrane, which facilitates rapid exit of D1-generated T3 back to the circulation. Indeed, in rodents D1 is responsible for approximately half of the daily extrathyroidal production of T3, but in humans D1 plays a lesser role in maintaining circulating T3 [25, 26]. In addition, studies with D1 knock out (KO) mice suggest that D1 also plays a scavenger role, preferentially deiodinating sulfated forms of iodothyronines in the process of being eliminated in the bile and urine [27].

2.1 Regulation of the D2 Pathway

Intrinsic biological properties make D2 a key enzyme regulating local thyroid hormone signaling. For example, D2 has a relatively short half-life (approximately 40 min) that is due to ubiquitination and destruction in the proteasomes, a process that is accelerated by interaction with T4, its natural substrate [28]. This provides rapid control of intracellular T3 production and TR occupancy.

D2 and the ubiquitin-proteasome system: Subcellular localization studies indicate that D2 is an ER resident protein by virtue of being retained in the ER [12, 29]. Residency in the ER is critical for D2's ability to contribute to thyroid hormone signaling. Physical proximity with the nucleus allows for D2-generated T3 to gain easy access to the TR-containing nuclear compartment. In tissues where D2 is expressed, at least 50% of the TR-bound nuclear T3 is made intracellularly via the D2 pathway [9]. Notably, ER residency places D2 physically close to an array of proteins that interact and modify the D2 molecule via ubiquitination and targeting to the proteasomal system, explaining its relatively short half-life. In addition, interaction with its natural substrate T4 accelerates disposal of D2 via the proteasomal system [30]. Substrate interaction with the enzyme's catalytic site triggers D2 ubiquitination [28]. Two Lys residues in D2 are involved in this process, K237 and K244. Although a mutation in either one of these residues does not affect D2-ubiquitination, mutation of both prevented D2 ubiquitination and prolonged D2 half-life [18].

Both ubiquitin conjugases UBC6 and or UBC7 interact with D2 and support D2 ubiquitination [31, 32]. In addition, two E3 ligases mediate D2 ubiquitination: the sonic-hedgehog (Shh) inducible WD-40 repeat and SOCS box-containing 1 (WSB-1 or SWiP) and the ERAD enzyme TEB4 (the human ortholog of yeast Doa10).

WSB-1 is part of the ECS^{WSB1} multi-protein catalytic complex that contains Elongin B and C, Cullin5, RBX1 and the E2 ligase UBC7 [33]. The D2-WSB-1 interaction requires the 18 amino acid "instability loop" in D2 and the WSB-1 WD-40 propeller-shaped domain. In the developing chicken bone growth plate, perichondrial/periosteal cells express both D2 and WSB-1 [33]. In this setting, Indian hedgehog (Ihh) secreted from chondrocytes leaving the proliferative pool induce expression of WSB-1. This in turn accelerates D2 ubiquitination, decreasing thyroid hormone signaling, which in turn induces PTHrP expression [33]. The other D2 ubiquitin ligase is TEB4, an ER resident protein that is a component of the ER associated degradation (ERAD) machinery. TEB4 was shown to be a D2-interacting protein in human cells, and when co-expressed with D2 it decreases D2 protein levels and enzymatic activity. In contrast, cellular knock down of TEB4 slows down D2 protein/activity turnover [34]. Like WSB-1, TEB4 requires the instability loop in D2 in order to interact and mediate D2 ubiquitination [34].

Ubiquitin moieties can be removed from targeted proteins by specific enzymes, avoiding proteasomal destruction. These enzymes are known as deubiquitinases or DUBs. Two DUBs, USP33 (or VDU-1) and USP20 (or VDU-2), interact with and regulate D2 [35]. USP33 and USP20 share a high degree (~59%) of sequence homology at the carboxyl and amino termini, being ubiquitously expressed in human and mouse tissues [36, 37]. Co-expression of USP33/20 decreases the levels of ubiquitinated D2 molecules and prolonges D2 protein half-life [35]. USP33 is up regulated in BAT by cold-exposure, which supports the several-fold increase in D2 activity. USP33-mediated deubiquitination of D2 can also take place in the brain, as USP33 and D2 are co-expressed in astrocytes and tanycytes [38].

An example of how D2 ubiquitination plays important physiological roles is the existence of tissue-specific differences in D2 ubiquitination [39]. These differences were implicated in the relatively higher T4/T3 plasma ratio observed in adult L-T4-treated thyroidectomized rats. While treatment with L-T4 decreased whole-body D2-dependent T3 production, D2-mediated conversion of T4 to T3 in the hypothalamus was only minimally affected by L-T4. In vivo studies in mice harboring an astrocyte-specific Wsb1 deletion as well as in vitro analysis of D2 ubiquitination driven by tissue extracts indicated that D2 ubiquitination in the hypothalamus is relatively less efficient than what it is observed in the rest of the body. Because T4-induced down-regulation of D2 is turned off in the hypothalamus, the TRH-producing neuron is wired to have increased sensitivity to T4. Therefore, tissue-specific differences in D2 ubiquitination are an inherent property of the TRH/TSH feedback mechanism [39].

D2 and ER stress: ER stress is recognized as an important condition that can be caused by accumulation of unfolded/misfolded proteins within the ER. A number of physiological and pathological conditions have been recognized as exhibiting different levels of ER stress [40]. To neutralize the effects of ER stress, cells trigger a conserved ER-to-nucleus signaling cascade that (1) stops global mRNA translation, (2) increases expression of chaperone proteins such as HSP40 and BiP, and (3) accelerates input of misfolded proteins into the ER stress-induced (ERAD) machinery [41–43]. D2 expression is greatly affected by ER stress [44]. Exposure of D2-expressing cells to different ER stressors reduced D2 activity in as little as 1 h, with significant increase in ER stress markers [44]. This down regulation was independent of transcriptional modulation of the D2 gene (Dio2), as ER stress did not change Dio2 mRNA levels. This loss of D2 activity leads to a significant decrease in D2-mediated T3 production, which shows that D2-expressing cells under ER stress are hypothyroid [44].

Molecules regulating the D2 pathway: Dio2 is a cAMP responsive gene [45–47]. Thus, any signaling pathway or molecule that increases cAMP concentration will stimulate D2 activity. For example, flavonols are a class of small polyphenolic compounds widely found in today's diet, induce D2 expression and intracellular D2-generated T3 via activation of cAMP production [48]. Kempferol treatment increased D2 activity up to 50-fold in various cell models via transcriptional mechanisms, while increasing oxygen consumption by 30% and the expression of key metabolic genes such as PGC1α, CPT-1, uncoupling protein 3 (UCP-3) and mitochondrial transcription factor 1 (mTFA) [48]. Another example is bile acids, which have been linked with activation of the mitogen-activated kinase pathways, FXRα nuclear receptors and the G-coupled protein receptor TGR5 [49, 50]. Diet supplementation with cholic acid protects against diet-induced obesity and reversed the body weight gain in animals kept on a high-fat diet for 120 days via TGR5 activation [50]. Dietary bile acid supplementation also increased oxygen consumption that coincided with an increase in BAT expression of PGC1α and β, CPT-1, UCP-3 and D2 [50].

Two chemical chaperones, tauroursodeoxycholic acid (TUDCA) and 4-phenyl butyric acid (4-PBA) were shown to increase D2 expression and intracellular T3 production [51]. D2 expression was induced via transcriptional activation of the *Dio2* gene [51]. Similar results were also observed in the setting of in vitro differentiated primary brown adipocytes, leading to acceleration in cellular oxygen consumption, lasting up to 72 h. Strikingly, the TUDCA- or 4-PBA-induced oxygen consumption was reduced or lost in primary brown adipocytes isolated from $Dio2^{-/-}$ mice, indicating a D2-dependent effect on oxygen consumption [51].

2.2 Physiological Roles of D2

D2 in developmental settings. D2 expression is low in most tissues during development, only increasing toward birth and the perinatal period [52]. The availability of D2 knockout mice (D2KO) allowed a much better understanding of D2's role in developing tissues [1, 53, 54]. Examples of developing tissues in which D2 plays well established roles include cochlea [55], bone [33, 56–59], brown adipose tissue (BAT) [60]. Although the subject of much attention, it is less clear that D2 plays a role in muscle development [61, 62]. The liver was recently found to be a site of important developmental roles played by D2 [63]. In mice, at around the first day of life, there is a transient surge in hepatocyte D2 that modifies the expression of ~165 genes involved in broad aspects of hepatocyte function, including lipid metabolism. Hepatocyte-specific D2 inactivation revealed that the surge in D2 expression causes hundreds of differentially methylated local DNA regions (DMR) that map to sites of active/suppressed chromatin, thus qualifying as epigenetic modifications. These DMRs underlie a dramatic metabolic phenotype that involves adult susceptibility to diet-induced steatosis, hypertriglyceridemia, and obesity [63].

Regulation of the hypothalamus-pituitary-thyroid axis: TSH (and TRH) stimulates thyroid activity and hormone production/secretion. Both circulating T4 and T3 play a role in the negative feedback loop that inhibits both TRH and TSH secretion [64]. However, T4 is only minimally active. Thus, D2 plays a central role in this regulatory loop, as it converts T4 to T3 inside thyrotrophs (cells that produce TSH) [2, 3]. Consequently, an increase in circulating T4 is expected to increase thyrotroph T3 concentration and to shut down TSHb gene expression, while a drop in circulating T4 results in an opposite effect [3]. Accordingly, inactivation of the *Dio2* gene in mice leads to central resistance to T4 [65], a phenotype that is also seen in mice treated with the D2 inhibitor amiodarone [66]. D2 also plays a role in TRH regulation [4]. D2-generated T3 in tanycytes, specialized ependimal cells located in the walls of the III ventricle, mediate the negative feedback on TRH neurons via paracrine mechanisms [67]. These two pathways function in a coordinated fashion, as elucidated in studies of two mouse strains with pituitary- and astrocyte-specific D2 knockdown (pit-D2 KO and astro-D2 KO mice, respectively). Such coordination of T4-to-T3 conversion between thyrotrophs and tanycytes is critical to maintain normal plasma T3 levels [68].

Adaptive thermogenesis and metabolic control: When exposed to cold, mammals minimize heat loss to the environment at the same time that accelerates energy expenditure (and consequently heat production), a process known as adaptative thermogenesis [69]. Release of norepinephrine (NE) in a number of tissues, including the brown adipose tissue (BAT), increases cAMP and up-regulates cAMP responsive genes, such as PGC-1α and Dio2. Although activation of all three β-adrenergic receptor subtypes increases cAMP production, they do play slightly different roles in metabolic control [70–72]. In the BAT, for example, the several-fold increase in D2 activity is critical for thermogenic function, as it potentiates cAMP production and directly induces uncoupling protein 1 (UCP-1) expression [47, 73, 74]. Concomitant induction of the ubiquitin-specific protease 33 (USP-33 or VDU-1) extends D2 half-life by deubiquitination and elevates further D2 activity [35]. D2KO mice are cold intolerant, developing hypothermia when exposed to cold [75].

D2KO animals develop obesity and severe hepatic steatosis when kept on a thermoneutral environment and fed on a high fat diet [76]. To understand the mechanisms underlying this phenotype, systemically euthyroid fat-specific (FAT), astrocyte-specific (ASTRO), or skeletal-muscle-specific (SKM) D2 knockout (D2KO) mice that were developed. The ASTRO-D2KO mice exhibit lower diurnal RQ and greater contribution of fatty acid oxidation to energy expenditure, but no differences in food intake. In contrast, the FAT-D2KO mouse exhibit greater contribution of carbohydrate oxidation to energy expenditure as illustrated by sustained (24 h) increase in RQ, food intake, tolerance to glucose, and sensitivity to insulin. Furthermore, FAT-D2KO animals that were kept on a high-fat diet gained more body weight and fat, indicating impaired BAT thermogenesis and/or inability to oxidize the fat excess. Acclimatization of FAT-D2KO mice at thermoneutrality dissipated both features of this phenotype. Notably, muscle D2 does not seem to play a significant metabolic role given that SKM-D2KO animals exhibited no metabolic phenotype [77]. These studies indicate that brain D2 plays a dominant albeit indirect role in fatty acid oxidation via its sympathetic control of BAT activity. In addition, D2-generated T3 in BAT accelerates fatty acid oxidation and protects against diet-induced obesity [77].

Food availability is another factor that plays a role in thyroid hormone signaling via D2-pathway. While the activity of the thyroid gland is stimulated by food intake (via leptin-induced TRH/TSH expression), food availability also stimulates D2-mediated T3 production in mouse skeletal muscle and in a cell model transitioning from 0.1 to 10% FBS [78]. Dio2 inhibition normally occurs via FOXO1 binding to the Dio2 promoter. Transcriptional de-repression occurs through insulin signaling, which takes place via the mTORC2 pathway. Therefore, FOXO1 represses DIO2 during fasting and that de-repression occurs via nutritional activation of the PI3K-mTORC2-Akt pathway [78].

D2 has a role in behavior and mood: Despite reduced T3 content in the neonatal D2KO brain, the D2KO mouse exhibits minimal neurological phenotype, suggesting the existence of potent compensatory mechanisms that minimize functional abnormalities caused by the absence of the D2 [79]. This includes diminished agility, and an altered global gait pattern (mice walked slower, with shorter strides and with a hindlimb wider base of support than wild-type mice). However, there was also impairment on coordination and prehensile reflex and a reduction of muscle strength [80]. However, D2 inactivation does seem to play a role in behavioral and mood processes in the adult mouse [81]. The ASTRO-D2KO mouse was found to exhibit anxiety-depression-like behavior despite normal serum T3 levels. This was found during a comprehensive battery of tests that included open field and elevated plus maze studies and when tested for depression using the tail-suspension and the forced-swimming tests. Despite normal neurogenesis, microarray gene expression profiling of the ASTRO-D2KO hippocampi identified an enrichment of three gene sets related to inflammation and impoverishment of three gene sets related to mitochondrial function and response to oxidative stress [81]. Increased anxiety and fear memory was also reported in the D2KO mouse [82]. These findings suggest that human defects in Dio2 expression in the brain could, potentially, result in mood and behavioral disorders.

D2 in the heart: The healthy human (but not rodent) myocardium expresses D2 and thus is potentially capable of generating T3 inside the muscle fiber. In fact, inhibition of T4 deiodination to T3 has been proposed as a contributory mechanism to the antiarrhythmic efficacy of amiodarone [83]. To better understand the role played by D2 in the myocardium, a transgenic mouse was created that expresses the human D2 gene in the myocardium under the α-myosin heavy chain (α-MHC) promoter [84]. This mouse has normal thyroid function tests but exhibits a discrete increase in myocardial T3 content and a gene expression profile compatible with increased thyroid hormone signaling, i.e., increased mRNA levels of HCN2 (an ionic channel that is key to the cardiac pacemaker) and decreased mRNA levels of β-MHC [84]. In perfused ex vivo studies, the α-MHC-D2 heart has about a 20% higher heart rate and decreased levels of phosphocreatine and ADP, indicating accelerated metabolic rates. This is supported by in vivo studies in which glucose uptake is increased by about 2.5-fold in the α-MHC-D2 heart [84]. These "thyrotoxic" effects are associated with an increased capacity of the α-MHC-D2 heart to generate cAMP in response to catecholamine stimulation [85]. Cardiac-specific increase in thyroid hormone signaling was confirmed in a second α-MHC-D2 mouse model conditionally expressing human D2 in the myocardium [86]. This model further demonstrated that myocardial D2 expression provides a functional advantage such as increased fractional shortening, velocity of circumferential fiber shortening, peak aortic outflow velocity and aortic velocity acceleration [86]. Thus, by virtue of accumulating in the myocardium and being a noncompetitive D2 inhibitor, amiodarone/DEA can potentially decrease thyroid hormone signaling in the heart.

Skeletal muscle: D2 expression is found in minimal amounts in both humans and murine muscle tissue [87]. However, D2 can be induced several-fold and play a role in muscle regeneration [88] and exercise [77, 89]. D2 expression in skeletal muscle is upregulated by acute treadmill exercise through a β-adrenergic receptor-dependent mechanism. Pharmacological block of D2 or disruption of the Dio2 gene in skeletal muscle fibers impaired acute exercise-induced PGC-1a expression and mitochondrial citrate synthase activity in chronically exercised mice [77].

2.3 Regulation of the D3 Pathway

D3 is an obligatory inner-ring deiodinase with a half-life of about 12 h [90, 91]. The D3 gene (*Dio3*) is unique amongst the deiodinases in being an imprinted gene [92] that is assembled in a cluster of genes that share a common regulatory element [93]. Typically, D3 is highly expressed in embryonic tissues, in the brain and in the placenta [23, 94]. However, D3 expression can be reactivated in multiple tissues during disease estates, contributing to the low circulating levels of T3 observed in these situations [95–97].

D3 traffics through different cell compartments: Like many other proteins anchored in the plasma membrane, D3 is internalized and becomes part of the endosomal vesicles [12]. These predominantly clathrin-coated vesicles are also capable of recycling internalized D3 back to the cell surface. Thus, under normal circumstances newly synthesized D3 transits to the plasma membrane and becomes part of the pool recycling between the plasma membrane and the early endosomes.

D3's intracellular trafficking can be modified in different settings. For example, hypoxia leads to nuclear import of D3 in neurons, without which thyroid hormone signaling and metabolism cannot be reduced. After unilateral hypoxia in the rat brain, D3 protein level is increased predominantly in the nucleus of the neurons in the pyramidal and granular ipsilateral layers, as well as in the hilus of the dentate gyrus of the hippocampal formation. In hippocampal neurons in culture as well as in a human neuroblastoma cell line, a 24 h hypoxia period redirects active D3 from the endoplasmic reticulum to the nucleus via the co-chaperone Hsp40. Preventing nuclear D3 import by Hsp40 knockdown resulted an almost doubling in the thyroid hormone-dependent glycolytic rate and quadrupling the transcription of thyroid hormone target gene ENPP2. In contrast, Hsp40 overexpression increased nuclear import of D3 and minimized thyroid hormone effects in cell metabolism, possibly functioning to reduce ischemia-induced hypoxic brain damage [98].

2.4 Physiological and Patho-physiological Roles of D3

D3 in developmental settings: Embryonic tissues express high levels of D3. This ensures that thyroid hormone signaling is kept at low levels during development. The timed and tissue-specific reduction of D3 during development coordinated with the slow increase in D2 activity provide tighter control of thyroid hormone action [9]. TGFβ and the Hedgehog family of proteins are key molecules that stimulate D3 expression [99, 100]. In fact, the Hedgehog family has been recently identified as a major player in determining thyroid hormone signaling through coordinated effects mediated via

D2 and D3. For example, in the chicken developing growth plate, Hedgehog signaling inhibits D2-mediated T3 production by inducing WSB-1, an ubiquitin ligase that inactivates D2 by transiently disrupting its dimeric conformation [18, 33]. At the same time, Hedgehog signaling stimulates *Dio3*, which will inactivate thyroid hormone and further decrease thyroid hormone action. The stimulation of *Dio3* by Gli proteins, which are downstream messengers of the Hedgehog cascade, has recently been characterized in keratinocytes from both normal skin and basal cell carcinomas, the most common human malignancy [100].

Pancreatic β-cells express D3: D3 plays a role in pancreatic islet function and glucose homeostasis [101, 102]. Dio3 expression and D3 protein is highly expressed in embryonic and adult pancreatic islets, predominantly in β-cells in both humans and mice. D3KO animals were found to be glucose intolerant due to in vitro and in vivo impaired glucose-stimulated insulin secretion, without changes in peripheral sensitivity to insulin. D3KO neonatal (postnatal day 0) and adult pancreas exhibited reduced total islet area due to reduced β-cell mass, insulin content, and impaired expression of key β-cells genes. D3 expression in perinatal pancreatic β-cells prevents untimely exposure to thyroid hormone, the absence of which leads to impaired β-cell function and subsequently insulin secretion and glucose homeostasis [102]. Studies in adult heterozygous mice with disruption of the Dio3 gene indicated that Dio3 is preferentially expressed from the maternal allele in pancreatic islets and that the inactivation of this allele is sufficient to disrupt glucose homeostasis by reducing the pancreatic islet area, insulin2 gene expression, and glucose-stimulated insulin secretion [101].

D3 in the hypertrophic heart: Severe illness that is associated with ischemia/hypoxia results in ectopic cardiac expression of D3, which inactivates thyroid hormone and causes localized hypothyroidism [103, 104]. D3 expression has also been observed in animal models of adverse remodeling such as myocardial infarction [105] and chronic pulmonary hypertension with right ventricular hypertrophy and ventricular failure (treatment with monocrotaline) [106, 107]. These studies served as the basis for a clinical trial that enrolled patients undergoing elective open heart surgery to assess thyroid hormone deiodination in the human heart [108]. Myocardial thyroid hormone metabolism was assessed by analyzing the difference in serum thyroid hormone levels between the aortic root (incoming blood) and the coronary sinus (outgoing blood) of patients undergoing cardiac surgery. Immediately before cardiopulmonary bypass, blood flowing through the myocardium of patients with aortic stenosis (with left ventricular hypertrophy) exhibited ~5% reduction in T3 and ~7% increase in rT3 levels, decreasing the serum T3/rT3 ratio by ~10%. In contrast, no myocardial thyroid hormone metabolism was observed in patients with coronary artery disease (no ventricular hypertrophy). These data indicate that there is accelerated thyroid hormone inactivation in the myocardium of patients with aortic stenosis, which is likely the result of D3 expression. Notably, no evidence to suggest thyroid hormone activation in the myocardium was obtained [108].

Drugs and tumors: Hemangiomas are common tumors of infancy that express variable levels of D3. Depending on the size of the tumor, D3 expression can be so intense that inactivates circulating thyroid hormone faster than the thyroid gland can secrete, resulting in what it is known as consumptive hypothyroidism [109]. A similar condition has been observed in patients with metastatic renal cell carcinoma or gastrointestinal stromal tumors (GISTs) receiving treatment with the tyrosine kinase inhibitor sunitinib [110]. Hepatic D3 activity increased markedly in rats undergoing similar treatment with this kinase inhibitor, indicating that D3 induction plays a role in sunitinib induced hypothyroidism [110]. In fact, similar to hemangiomas, GISTs themselves can produce consumptive hypothyroidism caused by marked overexpression of D3 within the tumor [111].

3 Methods to Study Deiodinases

Deiodinases can be detected and quantified studying enzyme kinetics [112]. Deiodinase kinetics are typically studied in broken cell/tissue preparations in the presence of co-factors such as dithiothreitol (DTT) and radioactive iodothyronines as substrates followed by quantitation of the radioactive products per unit time. In many cases deiodinase protein can also be assessed by immunohistochemistry [112]. Expression level can also be determined by measuring mRNA levels by standard reverse transcriptase quantitative polymerase chain reaction (RT-qPCR) techniques [112].

For deiodinase reactions enzyme kinetics are measured based on production of tracer iodide or of a specific reaction product. Background controls with no deiodinase activity must always be included. For D1 kinetics the preferred substrate is rT3 and the reaction buffer is based on PBS containing 1–2 mM EDTA, 0.25–0.3 M sucrose and 10–50 mM DTT, whereas the substrate concentration is 0.1–2 μM. The amount of tissue sonicate is adjusted to ensure <30% deiodination in about 1 h. The assay for D2 is carried out similarly, except that the preferred substrate is T4 (0.5–2 nM) and 1 mM PTU is included to inhibit potential background D1 activity. The preferred substrate for D3 is T3 (0.5–2 nM), which can be used in a similar setting as for D2 studies.

Acknowledgment

The authors thank NIDDK for support.

References

1. Galton VA (2005) The roles of the iodothyronine deiodinases in mammalian development. Thyroid 15(8):823–834
2. Larsen PR (1982) Thyroid-pituitary interaction: feedback regulation of thyrotropin secretion by thyroid hormones. N Engl J Med 306(1):23–32
3. Christoffolete MA, Ribeiro R, Singru P, Fekete C, da Silva WS, Gordon DF et al (2006) Atypical expression of type 2 iodothyronine deiodinase in thyrotrophs explains the thyroxine-mediated pituitary thyrotropin feedback mechanism. Endocrinology 147(4):1735–1743
4. Fekete C, Gereben B, Doleschall M, Harney JW, Dora JM, Bianco AC et al (2004) Lipopolysaccharide induces type 2 iodothyronine deiodinase in the mediobasal hypothalamus: implications for the nonthyroidal illness syndrome. Endocrinology 145(4):1649–1655
5. Huang SA, Bianco AC (2008) Reawakened interest in type III iodothyronine deiodinase in critical illness and injury. Nat Clin Pract Endocrinol Metab 4(3):148–155
6. Peeters RP, Visser TJ (2000) In: De Groot LJ, Chrousos G, Dungan K, Feingold KR, Grossman A, Hershman JM et al (eds) Endotext. South Dartmouth, MA
7. Galton VA (2017) The ups and downs of the thyroxine pro-hormone hypothesis. Mol Cell Endocrinol 458:105–111
8. Arrojo EDR, Fonseca TL, Werneck-de-Castro JP, Bianco AC (2013) Role of the type 2 iodothyronine deiodinase (D2) in the control of thyroid hormone signaling. Biochim Biophys Acta 1830(7):3956–3964
9. Bianco AC, Salvatore D, Gereben B, Berry MJ, Larsen PR (2002) Biochemistry, cellular and molecular biology, and physiological roles of the iodothyronine selenodeiodinases. Endocr Rev 23(1):38–89
10. Callebaut I, Curcio-Morelli C, Mornon JP, Gereben B, Buettner C, Huang S et al (2003) The iodothyronine selenodeiodinases are thioredoxin-fold family proteins containing a glycoside hydrolase clan GH-A-like structure. J Biol Chem 278(38):36887–36896
11. Sagar GD, Gereben B, Callebaut I, Mornon JP, Zeöld A, Curcio-Morelli C et al (2008) The thyroid hormone-inactivating deiodinase functions as a homodimer. Mol Endocrinol 22(6):1382–1393
12. Baqui M, Botero D, Gereben B, Curcio C, Harney JW, Salvatore D et al (2003) Human type 3 iodothyronine selenodeiodinase is located in the plasma membrane and undergoes rapid internalization to endosomes. J Biol Chem 278(2):1206–1211
13. Schweizer U, Schlicker C, Braun D, Kohrle J, Steegborn C (2014) Crystal structure of mammalian selenocysteine-dependent iodothyronine deiodinase suggests a peroxiredoxin-like catalytic mechanism. Proc Natl Acad Sci U S A 111(29):10526–10531
14. Darras VM, Van Herck SL (2012) Iodothyronine deiodinase structure and function: from ascidians to humans. J Endocrinol 215(2):189–206
15. Safran M, Leonard JL (1991) Comparison of the physicochemical properties of type I and type II iodothyronine 5′-deiodinase. J Biol Chem 266(5):3233–3238
16. Leonard JL, Visser TJ, Leonard DM (2001) Characterization of the subunit structure of the catalytically active type I iodothyronine deiodinase. J Biol Chem 276(4):2600–2607
17. Curcio-Morelli C, Gereben B, Zavacki AM, Kim BW, Huang S, Harney JW et al (2003) In vivo dimerization of types 1, 2, and 3 iodothyronine selenodeiodinases. Endocrinology 144(3):937–946
18. Sagar GD, Gereben B, Callebaut I, Mornon JP, Zeold A, da Silva WS et al (2007) Ubiquitination-induced conformational change within the deiodinase dimer is a switch regulating enzyme activity. Mol Cell Biol 27(13):4774–4783
19. Visser WE, Friesema EC, Visser TJ (2011) Minireview: thyroid hormone transporters: the knowns and the unknowns. Mol Endocrinol 25(1):1–14
20. Dumitrescu AM, Liao XH, Best TB, Brockmann K, Refetoff S (2004) A novel syndrome combining thyroid and neurological abnormalities is associated with mutations in a monocarboxylate transporter gene. Am J Hum Genet 74(1):168–175
21. Friesema EC, Kuiper GG, Jansen J, Visser TJ, Kester MH (2006) Thyroid hormone transport by the human monocarboxylate transporter 8 and its rate-limiting role in intracellular metabolism. Mol Endocrinol 20(11):2761–2772
22. Cheng SY, Leonard JL, Davis PJ (2010) Molecular aspects of thyroid hormone actions. Endocr Rev 31(2):139–170
23. Gereben B, Zavacki AM, Ribich S, Kim BW, Huang SA, Simonides WS et al (2008) Cellular and molecular basis of deiodinase-regulated thyroid hormone signaling. Endocr Rev 29(7):898–938

24. Galton VA (1992) The role of thyroid hormone in amphibian metamorphosis. Trends Endocrinol Metab 3(3):96–100
25. Nguyen TT, Chapa F, JJ DS 3rd (1998) Direct measurement of the contributions of type I and type II 5′-deiodinases to whole body steady state 3,5,3′-triiodothyronine production from thyroxine in the rat. Endocrinology 139(11):4626–4633
26. Abdalla SM, Bianco AC (2014) Defending plasma T3 is a biological priority. Clin Endocrinol 81(5):633–641
27. Schneider MJ, Fiering SN, Thai B, Wu SY, St Germain E, Parlow AF et al (2006) Targeted disruption of the type 1 selenodeiodinase gene (Dio1) results in marked changes in thyroid hormone economy in mice. Endocrinology 147(1):580–589
28. Gereben B, Goncalves C, Harney JW, Larsen PR, Bianco AC (2000) Selective proteolysis of human type 2 deiodinase: a novel ubiquitin-proteasomal mediated mechanism for regulation of hormone activation. Mol Endocrinol 14(11):1697–1708
29. Baqui MM, Gereben B, Harney JW, Larsen PR, Bianco AC (2000) Distinct subcellular localization of transiently expressed types 1 and 2 iodothyronine deiodinases as determined by immunofluorescence confocal microscopy. Endocrinology 141(11):4309–4312
30. Steinsapir J, Bianco AC, Buettner C, Harney J, Larsen PR (2000) Substrate-induced down-regulation of human type 2 deiodinase (hD2) is mediated through proteasomal degradation and requires interaction with the enzyme's active center. Endocrinology 141(3):1127–1135
31. Kim BW, Zavacki AM, Curcio-Morelli C, Dentice M, Harney JW, Larsen PR et al (2003) Endoplasmic reticulum-associated degradation of the human type 2 iodothyronine deiodinase (D2) is mediated via an association between mammalian UBC7 and the carboxyl region of D2. Mol Endocrinol 17(12):2603–2612
32. Botero D, Gereben B, Goncalves C, De Jesus LA, Harney JW, Bianco AC (2002) Ubc6p and ubc7p are required for normal and substrate-induced endoplasmic reticulum-associated degradation of the human selenoprotein type 2 iodothyronine monodeiodinase. Mol Endocrinol 16(9):1999–2007
33. Dentice M, Bandyopadhyay A, Gereben B, Callebaut I, Christoffolete MA, Kim BW et al (2005) The Hedgehog-inducible ubiquitin ligase subunit WSB-1 modulates thyroid hormone activation and PTHrP secretion in the developing growth plate. Nat Cell Biol 7(7):698–705
34. Zavacki AM, Arrojo E, Drigo R, Freitas BC, Chung M, Harney JW, Egri P et al (2009) The E3 ubiquitin ligase TEB4 mediates degradation of type 2 iodothyronine deiodinase. Mol Cell Biol 29(19):5339–5347
35. Curcio-Morelli C, Zavacki AM, Christofollete M, Gereben B, de Freitas BC, Harney JW et al (2003) Deubiquitination of type 2 iodothyronine deiodinase by von Hippel-Lindau protein-interacting deubiquitinating enzymes regulates thyroid hormone activation. J Clin Invest 112(2):189–196
36. Li Z, Na X, Wang D, Schoen SR, Messing EM, Wu G (2002) Ubiquitination of a novel deubiquitinating enzyme requires direct binding to von Hippel-Lindau tumor suppressor protein. J Biol Chem 277(7):4656–4662
37. Komander D, Clague MJ, Urbé S (2009) Breaking the chains: structure and function of the deubiquitinases. Nat Rev Mol Cell Biol 10(8):550–563
38. Fekete C, Freitas BC, Zeöld A, Wittmann G, Kádár A, Liposits Z et al (2007) Expression patterns of WSB-1 and USP-33 underlie cell-specific posttranslational control of type 2 deiodinase in the rat brain. Endocrinology 148(10):4865–4874
39. Werneck de Castro JP, Fonseca TL, Ueta CB, McAninch EA, Abdalla S, Wittmann G et al (2015) Differences in hypothalamic type 2 deiodinase ubiquitination explain localized sensitivity to thyroxine. J Clin Invest 125(2):769–781
40. Hetz C, Saxena S (2017) ER stress and the unfolded protein response in neurodegeneration. Nat Rev Neurol 13(8):477–491
41. Schröder M (2008) Endoplasmic reticulum stress responses. Cell Mol Life Sci 65(6):862–894
42. Meusser B, Hirsch C, Jarosch E, Sommer T (2005) ERAD: the long road to destruction. Nat Cell Biol 7(8):766–772
43. Vembar SS, Brodsky JL (2008) One step at a time: endoplasmic reticulum-associated degradation. Nat Rev Mol Cell Biol 9(12):944–957
44. Arrojo e Drigo R, Fonseca TL, Castillo M, Salathe M, Simovic G, Mohacsik P et al (2011) Endoplasmic reticulum stress decreases intracellular thyroid hormone activation via an eIF2a-mediated decrease in type 2 deiodinase synthesis. Mol Endocrinol 25(12):2065–2075
45. Bartha T, Kim SW, Salvatore D, Gereben B, Tu HM, Harney JW et al (2000) Characterization of the 5′-flanking and 5′-untranslated regions of the cyclic adenosine 3′,5′-monophosphate-responsive human

type 2 iodothyronine deiodinase gene. Endocrinology 141(1):229–237

46. Canettieri G, Celi FS, Baccheschi G, Salvatori L, Andreoli M, Centanni M (2000) Isolation of human type 2 deiodinase gene promoter and characterization of a functional cyclic adenosine monophosphate response element. Endocrinology 141(5):1804–1813
47. Bianco AC, Kieffer JD, Silva JE (1992) Adenosine 3′,5′-monophosphate and thyroid hormone control of uncoupling protein messenger ribonucleic acid in freshly dispersed brown adipocytes. Endocrinology 130(5):2625–2633
48. da-Silva WS, Harney JW, Kim BW, Li J, Bianco SD, Crescenzi A et al (2007) The small polyphenolic molecule kaempferol increases cellular energy expenditure and thyroid hormone activation. Diabetes 56(3):767–776
49. Houten SM, Watanabe M, Auwerx J (2006) Endocrine functions of bile acids. EMBO J 25(7):1419–1425
50. Watanabe M, Houten SM, Mataki C, Christoffolete MA, Kim BW, Sato H et al (2006) Bile acids induce energy expenditure by promoting intracellular thyroid hormone activation. Nature 439(7075):484–489
51. da-Silva WS, Ribich S, Arrojo e Drigo R, Castillo M, Patti ME, Bianco AC (2011) The chemical chaperones tauroursodeoxycholic and 4-phenylbutyric acid accelerate thyroid hormone activation and energy expenditure. FEBS Lett 585(3):539–544
52. Bates JM, St Germain DL, Galton VA (1999) Expression profiles of the three iodothyronine deiodinases, D1, D2, and D3, in the developing rat. Endocrinology 140(2):844–851
53. Galton VA, Schneider M, Clark AS, Germain DL (2009) Life without T4 to T3 conversion: studies in mice devoid of the 5′-deiodinases. Endocrinology 150(6):2957–2963
54. St Germain DL, Hernandez A, Schneider MJ, Galton VA (2005) Insights into the role of deiodinases from studies of genetically modified animals. Thyroid 15(8):905–916
55. Ng L, Goodyear RJ, Woods CA, Schneider MJ, Diamond E, Richardson GP et al (2004) Hearing loss and retarded cochlear development in mice lacking type 2 iodothyronine deiodinase. Proc Natl Acad Sci U S A 101(10):3474–3479
56. Gouveia CH, Christoffolete MA, Zaitune CR, Dora JM, Harney JW, Maia AL et al (2005) Type 2 iodothyronine selenodeiodinase is expressed throughout the mouse skeleton and in the MC3T3-E1 mouse osteoblastic cell line during differentiation. Endocrinology 146(1):195–200
57. Miura M, Tanaka K, Komatsu Y, Suda M, Yasoda A, Sakuma Y et al (2002) Thyroid hormones promote chondrocyte differentiation in mouse ATDC5 cells and stimulate endochondral ossification in fetal mouse tibias through iodothyronine deiodinases in the growth plate. J Bone Miner Res 17(3):443–454
58. Bassett JH, Boyde A, Howell PG, Bassett RH, Galliford TM, Archanco M et al (2010) Optimal bone strength and mineralization requires the type 2 iodothyronine deiodinase in osteoblasts. Proc Natl Acad Sci U S A 107(16):7604–7609
59. Capelo LP, Beber EH, Huang SA, Zorn TM, Bianco AC, Gouveia CH (2008) Deiodinase-mediated thyroid hormone inactivation minimizes thyroid hormone signaling in the early development of fetal skeleton. Bone 43(5):921–930
60. Hall JA, Ribich S, Christoffolete MA, Simovic G, Correa-Medina M, Patti ME et al (2010) Absence of thyroid hormone activation during development underlies a permanent defect in adaptive thermogenesis. Endocrinology 151(9):4573–4582
61. Dentice M, Marsili A, Ambrosio R, Guardiola O, Sibilio A, Paik JH et al (2011) The FoxO3/type 2 deiodinase pathway is required for normal mouse myogenesis and muscle regeneration. J Clin Invest 120(11):4021–4030
62. Ignacio DL, Silvestre DH, Anne-Palmer E, Bocco BM, Fonseca TL, Ribeiro MO et al (2017) Early developmental disruption of type 2 deiodinase pathway in mouse skeletal muscle does not impair muscle function. Thyroid 27(4):577–586
63. Fonseca TL, Fernandes GW, McAninch EA, Bocco BM, Abdalla SM, Ribeiro MO et al (2015) Perinatal deiodinase 2 expression in hepatocytes defines epigenetic susceptibility to liver steatosis and obesity. Proc Natl Acad Sci U S A 112(45):14018–14023
64. Chiamolera MI, Wondisford FE (2009) Minireview: thyrotropin-releasing hormone and the thyroid hormone feedback mechanism. Endocrinology 150(3):1091–1096
65. Schneider MJ, Fiering SN, Pallud SE, Parlow AF, St Germain DL, Galton VA (2001) Targeted disruption of the type 2 selenodeiodinase gene (DIO2) results in a phenotype of pituitary resistance to T4. Mol Endocrinol 15(12):2137–2148
66. Rosene ML, Wittmann G, Arrojo e Drigo R, Singru PS, Lechan RM, Bianco AC (2010) Inhibition of the type 2 iodothyronine deiodinase underlies the elevated plasma TSH

associated with amiodarone treatment. Endocrinology 151(12):5961–5970
67. Freitas BC, Gereben B, Castillo M, Kalló I, Zeöld A, Egri P et al (2010) Paracrine signaling by glial cell-derived triiodothyronine activates neuronal gene expression in the rodent brain and human cells. J Clin Invest 120(6):2206–2217
68. Fonseca TL, Correa-Medina M, Campos MP, Wittmann G, Werneck-de-Castro JP, Arrojo-e-Drigo R et al (2013) Coordination of hypothalamic and pituitary T3 production regulates TSH expression. J Clin Invest 123(4):1492–1500
69. McAninch EA, Bianco AC (2014) Thyroid hormone signaling in energy homeostasis and energy metabolism. Ann N Y Acad Sci 1311:77–87
70. Preite NZ, Nascimento BP, Muller CR, Americo AL, Higa TS, Evangelista FS et al (2016) Disruption of beta3 adrenergic receptor increases susceptibility to DIO in mouse. J Endocrinol 231(3):259–269
71. Fernandes GW, Ueta CB, Fonseca TL, Gouveia CH, Lancellotti CL, Brum PC et al (2014) Inactivation of the adrenergic receptor beta2 disrupts glucose homeostasis in mice. J Endocrinol 221(3):381–390
72. Ueta CB, Fernandes GW, Capelo L, Fonseca TL, Maculan FD, Gouveia C et al (2012) beta1 adrenergic receptor is key to cold- and diet-induced thermogenesis in mice. J Endocrinol 214(3):359–365
73. Carvalho SD, Kimura ET, Bianco AC, Silva JE (1991) Central role of brown adipose tissue thyroxine 5′-deiodinase on thyroid hormone-dependent thermogenic response to cold. Endocrinology 128(4):2149–2159
74. Branco M, Ribeiro M, Negrao N, Bianco AC (1999) 3,5,3′-Triiodothyronine actively stimulates UCP in brown fat under minimal sympathetic activity. Am J Phys 276(1 Pt 1):E179–E187
75. de Jesus LA, Carvalho SD, Ribeiro MO, Schneider M, Kim SW, Harney JW et al (2001) The type 2 iodothyronine deiodinase is essential for adaptive thermogenesis in brown adipose tissue. J Clin Invest 108(9):1379–1385
76. Castillo M, Hall JA, Correa-Medina M, Ueta C, Won Kang H, Cohen DE et al (2011) Disruption of thyroid hormone activation in type 2 deiodinase knockout mice causes obesity with glucose intolerance and liver steatosis only at thermoneutrality. Diabetes 60(4):1082–1089
77. Fonseca TL, Werneck-De-Castro JP, Castillo M, Bocco BM, Fernandes GW, McAninch EA et al (2014) Tissue-specific inactivation of type 2 deiodinase reveals multilevel control of fatty acid oxidation by thyroid hormone in the mouse. Diabetes 63(5):1594–1604
78. Lartey LJ, Werneck-de-Castro JP, O-Sullivan I, Unterman TG, Bianco AC (2015) Coupling between nutrient availability and thyroid hormone activation. J Biol Chem 290(51):30551–30561
79. Galton VA, Wood ET, St Germain EA, Withrow CA, Aldrich G, St Germain GM et al (2007) Thyroid hormone homeostasis and action in the type 2 deiodinase-deficient rodent brain during development. Endocrinology 148(7):3080–3088
80. Barez-Lopez S, Bosch-Garcia D, Gomez-Andres D, Pulido-Valdeolivas I, Montero-Pedrazuela A, Obregon MJ et al (2014) Abnormal motor phenotype at adult stages in mice lacking type 2 deiodinase. PLoS One 9(8):e103857
81. Bocco BM, Werneck-de-Castro JP, Oliveira KC, Fernandes GW, Fonseca TL, Nascimento BP et al (2016) Type 2 deiodinase disruption in astrocytes results in anxiety-depressive-like behavior in male mice. Endocrinology 157(9):3682–3695
82. Barez-Lopez S, Montero-Pedrazuela A, Bosch-Garcia D, Venero C, Guadano-Ferraz A (2017) Increased anxiety and fear memory in adult mice lacking type 2 deiodinase. Psychoneuroendocrinology 84:51–60
83. Vassallo P, Trohman RG (2007) Prescribing amiodarone: an evidence-based review of clinical indications. JAMA 298(11):1312–1322
84. Pachucki J, Hopkins J, Peeters R, Tu H, Carvalho SD, Kaulbach H et al (2001) Type 2 iodothyronine deiodinase transgene expression in the mouse heart causes cardiac-specific thyrotoxicosis. Endocrinology 142(1):13–20
85. Carvalho-Bianco SD, Kim BW, Zhang JX, Harney JW, Ribeiro RS, Gereben B et al (2004) Chronic cardiac-specific thyrotoxicosis increases myocardial beta-adrenergic responsiveness. Mol Endocrinol 18(7):1840–1849
86. Trivieri MG, Oudit GY, Sah R, Kerfant BG, Sun H, Gramolini AO et al (2006) Cardiac-specific elevations in thyroid hormone enhance contractility and prevent pressure overload-induced cardiac dysfunction. Proc Natl Acad Sci U S A 103(15):6043–6048
87. Grozovsky R, Ribich S, Rosene ML, Mulcahey MA, Huang SA, Patti ME et al (2009) Type 2 deiodinase expression is induced by peroxisomal proliferator-activated receptor-gamma agonists in skeletal myocytes. Endocrinology 150(4):1976–1983

88. Dentice M, Marsili A, Ambrosio R, Guardiola O, Sibilio A, Paik JH et al (2010) The FoxO3/type 2 deiodinase pathway is required for normal mouse myogenesis and muscle regeneration. J Clin Invest 120(11):4021–4030
89. Bocco BM, Louzada RA, Silvestre DH, Santos MC, Anne-Palmer E, Rangel IF et al (2016) Thyroid hormone activation by type 2 deiodinase mediates exercise-induced peroxisome proliferator-activated receptor-gamma coactivator-1alpha expression in skeletal muscle. J Physiol 594(18):5255–5269
90. Baqui M, Botero D, Gereben B, Curcio C, Harney JW, Salvatore D et al (2003) Human type 3 iodothyronine selenodeiodinase is located in the plasma membrane and undergoes rapid internalization to endosomes. J Biol Chem 278(2):1206–1211
91. Salvatore D, Low SC, Berry M, Maia AL, Harney JW, Croteau W et al (1995) Type 3 lodothyronine deiodinase: cloning, in vitro expression, and functional analysis of the placental selenoenzyme. J Clin Invest 96(5):2421–2430
92. Hernandez A, Fiering S, Martinez E, Galton VA, St Germain D (2002) The gene locus encoding iodothyronine deiodinase type 3 (Dio3) is imprinted in the fetus and expresses antisense transcripts. Endocrinology 143(11):4483–4486
93. Structure HA (2005) function of the type 3 deiodinase gene. Thyroid 15(8):865–874
94. Dentice M, Salvatore D (2011) Local impact of thyroid hormone inactivation. J Endocrinol 209(3):273–282
95. Huang SA, Bianco AC (2008) Reawakened interest in type III iodothyronine deiodinase in critical illness and injury. Nat Clin Pract Endocrinol Metab 4(3):148–155
96. Huang SA (2005) Physiology and pathophysiology of type 3 deiodinase in humans. Thyroid 15(8):875–881
97. Fliers E, Bianco AC, Langouche L, Boelen A (2015) Thyroid function in critically ill patients. Lancet Diabetes Endocrinol 3(10):816–825
98. Jo S, Kallo I, Bardoczi Z, Arrojo EDR, Zeold A, Liposits Z et al (2012) Neuronal hypoxia induces hsp40-mediated nuclear import of type 3 deiodinase as an adaptive mechanism to reduce cellular metabolism. J Neurosci 32(25):8491–8500
99. Huang SA, Mulcahey MA, Crescenzi A, Chung M, Kim B, Barnes CA et al (2005) TGF-B promotes inactivation of extracellular thyroid hormones via transcriptional stimulation of type 3 iodothyronine deiodinase. Mol Endocrinol 19:3126–3136
100. Dentice M, Luongo C, Huang S, Ambrosio R, Elefante A, Mirebeau-Prunier D et al (2007) Sonic hedgehog-induced type 3 deiodinase blocks thyroid hormone action enhancing proliferation of normal and malignant keratinocytes. Proc Natl Acad Sci U S A 104(36):14466–14471
101. Medina MC, Fonesca TL, Molina J, Fachado A, Castillo M, Dong L et al (2014) Maternal inheritance of an inactive type III deiodinase gene allele affects mouse pancreatic beta-cells and disrupts glucose homeostasis. Endocrinology 155(8):3160–3171
102. Medina MC, Molina J, Gadea Y, Fachado A, Murillo M, Simovic G et al (2011) The thyroid hormone-inactivating type III deiodinase is expressed in mouse and human beta-cells and its targeted inactivation impairs insulin secretion. Endocrinology 152(10):3717–3727
103. Peeters RP, Wouters PJ, Kaptein E, van Toor H, Visser TJ, Van den Berghe G (2003) Reduced activation and increased inactivation of thyroid hormone in tissues of critically ill patients. J Clin Endocrinol Metab 88(7):3202–3211
104. Ueta CB, Oskouei BN, Olivares EL, Pinto JR, Correa MM, Simovic G et al (2012) Absence of myocardial thyroid hormone inactivating deiodinase results in restrictive cardiomyopathy in mice. Mol Endocrinol 26(5):809–818
105. Olivares EL, Marassi MP, Fortunato RS, da Silva AC, Costa-e-Sousa RH, Araujo IG et al (2007) Thyroid function disturbance and type 3 iodothyronine deiodinase induction after myocardial infarction in rats a time course study. Endocrinology 148(10):4786–4792
106. Wassen FW, Schiel AE, Kuiper GG, Kaptein E, Bakker O, Visser TJ et al (2002) Induction of thyroid hormone-degrading deiodinase in cardiac hypertrophy and failure. Endocrinology 143(7):2812–2815
107. Simonides WS, Mulcahey MA, Redout EM, Muller A, Zuidwijk MJ, Visser TJ et al (2008) Hypoxia-inducible factor induces local thyroid hormone inactivation during hypoxic-ischemic disease in rats. J Clin Invest 118(3):975–983
108. Paolino BS, Pomerantzeff PM, Dallan LAO, Gaiotto FA, Preite NZ, Latronico AC et al (2017) Myocardial inactivation of thyroid hormones in patients with aortic stenosis. Thyroid 27(5):738–745
109. Huang SA, Tu HM, Harney JW, Venihaki M, Butte AJ, Kozakewich HP et al (2000) Severe hypothyroidism caused by type 3 iodothyronine deiodinase in infantile hemangiomas. N Engl J Med 343(3):185–189

110. Kappers MH, van Esch JH, Smedts FM, de Krijger RR, Eechoute K, Mathijssen RH et al (2011) Sunitinib-induced hypothyroidism is due to induction of type 3 deiodinase activity and thyroidal capillary regression. J Clin Endocrinol Metab 96(10):3087–3094
111. Maynard MA, Marino-Enriquez A, Fletcher JA, Dorfman DM, Raut CP, Yassa L et al (2014) Thyroid hormone inactivation in gastrointestinal stromal tumors. N Engl J Med 370(14):1327–1334
112. Bianco AC, Anderson G, Forrest D, Galton VA, Gereben B, Kim BW et al (2014) American thyroid association guide to investigating thyroid hormone economy and action in rodent and cell models. Thyroid 24(1): 88–168

Chapter 9

Thyroid Hormones and Derivatives: Endogenous Thyroid Hormones and Their Targets

Josef Köhrle

Abstract

More than a century after the discovery of L-Thyroxine, the main thyroid hormone secreted solely by the thyroid gland, several metabolites of this iodinated, tyrosine-derived ancestral hormone have been identified. These are utilized as hormones during development, differentiation, metamorphosis, and regulation of most biochemical reactions in vertebrates and their precursor species. Among those metabolites are the thyromimetically active 3,3′,5-Triiodo-L-thyronine (T3) and 3,5-Diiodo-L-thronine, reverse-T3 (3,3′,5′-Triiodo-L-thyronine) with still unclear function, the recently re-discovered thyronamines (e.g., 3-Iodo-thyronamine), which exert in part T3-antagonistic functions, the thyroacetic acids (e.g., Tetrac and Triac), as well as various sulfated or glucuronidated metabolites of this panel of iodinated signaling compounds. In the blood most of these hydrophobic metabolites are tightly bound to the serum distributor proteins thyroxine binding globulin (TBG), transthyretin (TTR), albumin or apolipoprotein B100. Cellular import and export of these charged, highly hydrophobic amino acid derivatives requires a number of cell-membrane transporters or facilitators such as MCT8 or MCT10 and members of the OATP and LAT families of transporters. Depending on their structure, the thyroid hormone metabolites exert their cellular action by binding and thus modulating the function of various receptors systems (e.g., $\alpha\nu\beta3$ integrin receptor and transient receptor potential channels (TRPM8) of the cell membrane), in part linked to intracellular downstream kinase signaling cascades, and several isoforms of membrane-associated, mitochondrial or nuclear thyroid hormone receptors (TR), which are members of the c-erbA family of ligand-modulated transcription factors. Intracellular deiodinase selenoenzymes, which obligatory are membrane integrated enzymes, ornithine decarboxylase and monoamine oxidases control local availability of biologically active thyroid hormone metabolites. Inactivation of thyroid hormone metabolites occurs mainly by deiodination, sulfation or glucuronidation, reactions which favor their renal or fecal elimination.

Key words Thyroid hormone, Iodothyronine, Thyronamine, Thyroacetic acid, Thyronome, Deiodinase, Ornithine decarboxylase, Monoamine oxidase, Conjugation, Immunoassay, Mass spectrometry, Analogue, Inhibitor

1 Introduction

Thyroid hormones (TH) are powerful, endogenous, iodinated, amino acid-derived biomolecules with signaling properties identified in animals (metazoa), including even the most primitive vertebrates

Michelina Plateroti and Jacques Samarut (eds.), *Thyroid Hormone Nuclear Receptor: Methods and Protocols*, Methods in Molecular Biology, vol. 1801, https://doi.org/10.1007/978-1-4939-7902-8_9,

© Springer Science+Business Media, LLC, part of Springer Nature 2018

such as lamprey. Thyroid hormone receptor-mediated mechanisms of action and rapid, membrane-initiated, or kinase cascade-associated signaling of TH derivatives have been traced back in evolution to bilateria and aquatic life forms such as cnidarians, molluscs, and echinoderms [1–3]. Currently, the term TH is mainly used for the prohormone L-thyroxine (T4), an exclusive product synthesized by the follicular structures of the thyroid gland and its main thyromimetically active metabolite T3 (3,3′,5-triiodo-L-thyronine) (Fig. 1). T3, in part, is generated and secreted by the thyroid but mainly formed outside the thyroid gland by cellular enzymatic deiodination. Several other TH metabolites (THM) are known to exert biological effects mediated in part by T3 receptors or other target molecules in various subcellular compartments [4]. These compounds, also named metabolites or derivatives, mainly formed by various enzymatic reactions outside of the thyroid gland, recently received more attention due to their non-classical specific effects on various metabolic pathways and developmental processes [5]. Identification of two genes encoding several isotypes and isoforms of T3 receptors (TR) generated by use of different start sites, alternative splicing, as well as development- and tissue-specific expression [6] opened the possibility to design specific ligands for T3 receptor isoforms acting as agonists or antagonists [7]. Such analogues may not contain iodine or halogen atoms any more, and some of them representing "TH mimics" might have quite different structural and functional properties. Beyond those novel TH analogues, also a large series of nutritional or environmental compounds, both natural or man-made, have been identified recently, which interfere with TH binding sites of receptors, metabolizing enzymes, or target molecules of TH action. These are so-called endocrine disrupting compounds (EDC), some of them produced in large scale amount, raising concern about adverse interference with TH metabolism, transport and action during development and in adult and ageing individuals, including aquatic and terrestrial wildlife species [8–11]. The delicate balance between adequate availability of the essential trace element iodine for biosynthesis of endogenous TH and their distribution, transport, metabolism, and action reveals a complex and intricate relationship between TH, developmental processes in many species, and regulation of metabolism as well as reproduction of mature life forms. Especially, the crucial key role of active TH during development, metamorphosis, and metamorphosis-like processes in the whole animal kingdom including human development highlights the importance of both TH and their constituents, the essential trace element iodine, the amino acid tyrosine, and various factors and compounds involved in biosynthesis and metabolism.

T4, identified and isolated a century ago [12–14], is one of the most prescribed medical drugs worldwide with long and excellent clinical expertise [15]. T4 is mainly administered to patients suffering

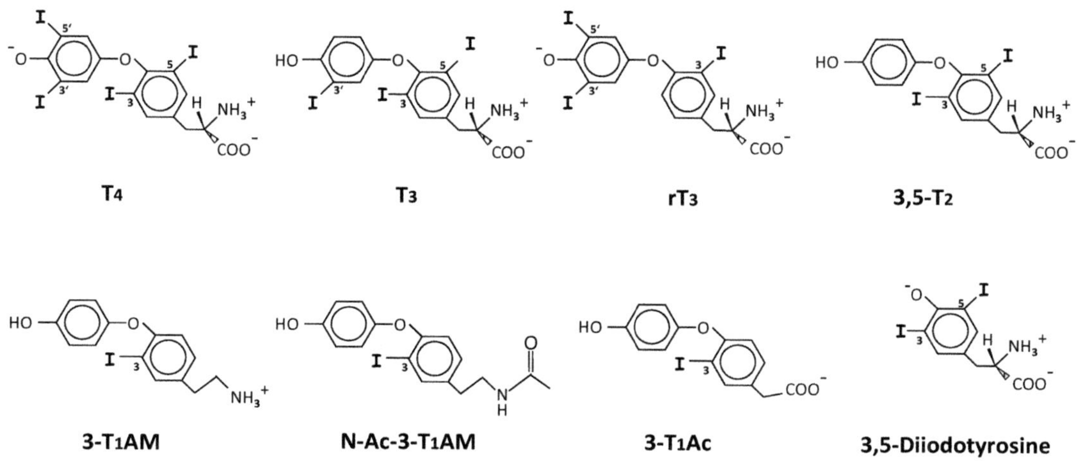

Fig. 1 Molecular structure of thyroid hormones and their derivatives

from hypothyroidism due to autoimmune thyroiditis, after surgical removal of the thyroid gland due to thyroid nodules, thyroid cancer, or Graves' disease, as well as for patients with congenital hypothyroidism, one of the most frequent genetic endocrine disorders with a worldwide incidence of 1:3.000 [16]. Limited success has been obtained with therapeutic application of the active thyroid hormone T3, either alone or in combination with T4 [17–19]. Unfortunately, ill-defined extracts from animal thyroids (sheep, bovine, pork) are still in use even 100 years after discovery of the natural endogenous parental hormone T4 [20]. However, synthetic T3 analogues have not yet made it into clinical practice due to various side effects observed during clinical trials [21, 22]. This, however, does not exclude widespread abuse of preparations of various thyroid hormone metabolites such as 3,5-T2 in the doping, body building, body weight control, and wellness scene, as such products or preparations containing badly defined and controlled active thyroid hormone metabolites and mixtures, are available via the internet market with no or poorly controlled restriction and without medical prescription [23–25]. Therefore, adverse effects occur and have been reported.

2 Thyroid Hormone Biosynthesis and Enzymatic Metabolism

The complex process of TH biosynthesis requiring the angiofollicular units of the thyroid gland [26] has been characterized in molecular and functional detail over the last decades. Most proteins, enzymes, molecular mechanisms, and subcellular structural requirements for this H2O2-driven synthesis pathway are now known [27]. For each individual step required for TH biosynthesis, storage, mobilization, metabolism, export, and secretion genetic defects have been identified which impair or block this process, resulting in thyroid hormone dysgenesis, congenital hypothyroidism, or susceptibility to thyroid-related diseases [28, 29]. Details of this highly specialized biosynthesis machinery and its regulation have recently been reviewed and will not be elaborated in more detail here [27]. However, apart from the TH biosynthesis occurring by posttranslational modification of the only known TH synthesis and storage protein thyroglobulin (Tg) at the apical surface inside the follicular lumen, the adequate provision of three essential trace elements is a limiting nutritional factor: Iodine, actively accumulated by the sodium-iodide symporter (NIS) across the basolateral plasma membrane, iron, an essential part of the heme prosthetic group of thyroperoxidase (TPO), and selenium, an essential trace element in the active site of selenoproteins involved in cellular redox control and anti-oxidative defense of H2O2-exposed thyrocytes are required in adequate amounts during the whole life [30]. The major product generated within the Tg polypeptide chain as precursor is T4, which is formed at three so-called "hormonogenic sites" of Tg. A minor but significant component is T3, generated in the T3 hormonogenic site of Tg [27, 31]. Typically, the human thyroid gland secretes T4 to T3 in a ratio of four to five molecules of T4 for one T3 molecule. Other THM with lower iodination grade such as rT3 and diiodo- (T2) and monoiodo-*th*yro*n*ines (T1) are no relevant secretion products of the normal thyroid gland. In addition, Tg contains significant amounts of mono- and diiodinated *t*yro*s*ine residues (MIT, DIT). These are also liberated during lysosomal degradation of Tg after its micropinocytosis and endocytotic processes at the apical thyrocyte membrane. However, MIT and DIT are typically not secreted by the thyroid gland into the blood stream but locally metabolized by a specific enzyme, iodotyrosine dehalogenase (DEHAL1), which liberates iodide from DIT and MIT to recover it for a sustainable re-use for subsequent Tg iodination and TH biosynthesis [32]. DIT and MIT are typically no major iodine-containing compounds in human blood except under condition of severe infection, activation of macrophages or phagocytic processes [33, 34], which lead to oxidative cleavage of T4 and T3 by these activated macrophages and monocytes expressing DEHAL1. For example, elevated DIT concentrations associated with T4 and T3 oxidative degradation have been reported in septic patients [35]. As TH are iodinated, hydrophobic derivatives of the amino acid tyrosine, they are charged molecules both at their amino acid side chain and the 4′ phenolic OH group (Fig. 1) and thus in contrast to previous biochemically unsubstantiated assumptions do *not* easily diffuse through membranes but require facilitating transporters for their export from thyrocytes into the blood stream and their import into target cells [36–38]. Here, during the last decade, several facilitating TH transmembrane transporters (THTT) of the MCT8, MCT10, OATP, and LAT families have been identified [39]. Apparently, MCT8 plays a key role as T3 and T4 transporter, as mutations of this molecule are causing the

Allan-Herndon-Dudley syndrome, a major developmental disorder affecting neuromuscular function and brain development. Currently, this disease cannot be prevented or treated, however, first clinical studies employing thyroid hormone analogues, which bypasss MCT8 barrier, are currently tested [21, 40].

3 Thyroid Hormone Metabolism: Deiodination

Soon after the discovery and isolation of T4, several metabolites and analogues have been synthesized and tested in various species for their biological actions on amphibian metamorphosis, stimulation of oxygen consumption or related developmental aspects of thyroid hormone action (for recent review *see* ref. 41). Only in 1952, the existence of 3,3′,5-triiodothyronine (T3) has been unequivocally demonstrated [42] after its existence had already been postulated by Frank Hird and colleagues in 1948 [43]. This discovery and its confirmation by several teams unraveled the true nature of the majority of TH action exerted mainly by transcriptional modulation after its binding to T3 receptors cloned in 1986 [44]. By 1970 it became clear that T3 not only is generated directly in the thyroid gland at the T3 hormonogenic site [30, 45] but that enzymatic, reductive, sequential deiodination of T4 specifically generates T3, and as subsequently demonstrated, several other, less iodinated T4 and T3 metabolites [46]. Meanwhile, the three enzymes involved in reductive deiodination leading to either activation of the prohormone T4 to the active T3, or inactivation of the prohormone T4 or its active product T3 have been functionally characterized and cloned. All three deiodinases belong to the family of selenocysteine-containing proteins with the redox-active, modified amino acids in their active center [47–49]. The highly reactive selenocysteine residue in the suitable biochemical environment of the active site is able to reductively cleave the carbon-iodide bond, a reaction that typically might be catalyzed by oxidative reaction conditions as in DIT and MIT biosynthesis [34, 37]. For a deiodinase-3 cysteine mutant, the structure has been solved by X-ray crystallography in the absence of ligand and revealed in detail the role of a thioredoxin fold and associated charged residues in the reaction mechanism [50]. Meanwhile, mechanisms of reaction of the three deiodinase isozymes have been clarified in more detail including their in vitro regeneration by reactive sulfhydryl regenerating co-factors. Whether in vivo regeneration is necessary or occurs for this Dio1 enzyme reaction dealing with a very limited low number of substrate molecules per day has to be clarified [51]. Single reaction or suicide-type of reaction kinetics cannot be excluded provided that the enzyme does not catalyze additional other reactions with much higher substrate concentrations apart from TH deiodination. Systematic analysis of structure-activity relationships for ligands acting either as substrate or inhibitors of deiodination has generated a relevant repertoire of agents selectively or globally interfering with deiodinase reactions, and some of these inhibitors might have clinically relevant potential by blocking especially deiodinase-3 activity [52–54] highly active in proliferating (tumor) cells or during development [55], while specific and selective inhibitors of either deiodinase-1 or -2 might be used as anti-thyroid drugs in preventing prohormone activation to the active T3. Recently also so-called deiodinase mimics based on selenium and tellurium analogues of TH with specific structural requirements in arrangement of functional groups have been generated [7, 54]. Some of these agents are able to mimic deiodinase reactions in vitro and to metabolize natural TH. The clinical and potential pharmacological relevance of these observations is not yet clear. Detailed comparative analysis of TH binding sites in TH distribution proteins in the blood, such as TBG, transthyretin and albumin, the cellular uptake and transport proteins, and the deiodinases as well as T3 receptor binding sites revealed some similarities in the active site [56] like a histidine clamp, especially relevant for T3 binding by MCT8 and T3 receptor while it also provided more specific insight into potential interference of nutritional and environmental agents interfering with TH binding to various target proteins, enzymes, and receptors. Such interference apparently provides the basis for strong adverse effects, which might impair thyroid hormone development in various species including humans [11]. Beyond that, these compounds might alter the equilibrium between bound and free thyroid hormone [57–59] leading to altered availability of both the active ligand T3 but also the other THM, which might act either as precursors of thyroid hormone action or exert specific, selective function on other target sites as postulated, for example, for 3,5-T2, TRIAC, thyronamines, or newly discovered acetylated forms of TH and thyronamines [41].

4 TH Side Chain Metabolism: Thyronamines and Thyroacetic Acids

4.1 Thyroacetic Acids (TAc)

TH are iodinated derivatives of the aromatic amino acid tyrosine. As such they undergo various metabolic pathways at their alanine side chain (*see* Fig. 1), such as decarboxylation, deamination and, recently discovered, also N-Acetylation. So far no enzymes specifically dedicated to metabolize TH at their side chain have been discovered and it is assumed that such reactions are catalyzed by enzymes also metabolizing aromatic amino acids such as tyrosine. The most abundant acetic acid derivative of TH is 3,3′,5,5′-tetraiodoacetic acid (Tetrac) which circulates in blood in low nanomolar concentrations [60] similar to those of the bioactive hormone T3. Tetrac shows high affinity for binding to TTR in serum [61].

During the last decade a specific hormone action has been attributed to Tetrac, i.e., binding to the cell membrane $\alpha\nu\beta3$ integrin receptor. This effect has been exploited to develop this ligand and nanoparticle bound Tetrac as a potential antagonistic agent to inhibit angiogenesis and growth of various tumors which may be stimulated by T4 and T3 via a plasma membrane initiated rapid signaling mechanism independent from classical nuclear T3 receptor action [62, 63]. The mechanism of synthesis and its site have not been clarified yet, neither it is known where 3,3′,5-Triiodothyronine (Triac), the deiodinated metabolite of Tetrac is formed, which also circulates in blood and is found in tissues [60]. Triac already is used in clinical practice based on its higher affinity for and activation of the classical TRβ transcription factor compared to T3. Triac (Tiratricol) has a short half-life compared to T3 but actively suppresses TSH in the negative feedback circuit of the HPT axis [64], while cardiac, hepatic, muscular and other TR target sites are almost unaffected. This qualifies Triac as important "natural" agent to ameliorate symptoms of TH resistance caused by mutations in TRβ [21]. Beyond this beneficial application Triac recently is tested in a multicentric clinical trial as one of the first drugs to improve catabolic clinical conditions in patients with the Allan-Herndon-Dudley Syndrome caused by mutations in the TH transporter MCT8 [65].

Whether 3,5,3′-triiodothyropropionic acid, another potential metabolite of the bioactive T3 occurs in vivo and plays a relevant role remains to be established, while its structure has been determined by X-ray crystallography [66].

Another acetic acid metabolite with remarkable biological function is 3-iodo-thyroacetic acid (3T1Ac), which is the oxidation product generated from 3-T1AM via monoamine oxidases (MAO, SSAO) [67, 70]. Whether 3T1Ac circulates in blood is unclear but this TH metabolite has been detected in various tissues and is rapidly formed after pharmacological administration of 3T1AM [68, 69]. Current evidence indicates that several of the central effects (e.g., learning, memory, fear, pain sensation in rodents), initially attributed to 3T1AM, might actually be exerted by its metabolite 3T1Ac, especially in the CNS. In contrast, several of the thermoregulatory, cardiac, metabolic, and thyroid effects of 3T1AM are not initiated after in vivo administration of 3T1Ac [70–72]. Remarkable in context of these minor and recently studied TH metabolites are their inhibitory effects on the expression of functional NIS in the thyroid [71], the suppression of insulin secretion by pancreatic beta cells via a mitochondrial mechanism already at low nanomomolar concentrations [72] and the hyperalgesia effect involving modulation of the histaminergic system after central application [70]. Whether tissue-specific action of 3T1AM or its oxidation product 3T1Ac is of physiological relevance and can be modulated by various available inhibitors of monoamino oxidases (MaoA, MaoB, SSA0, etc.) remains to be studied in more detail [41]. Similarly, the potential physiological role, biosynthesis, and metabolism of recently described N- and O-acetylated thyronamines remain to be established by further in vitro and in vivo studies ([68]; Höfig et al., unpublished data). Whether gastrointestinal metabolism, either by the host's mucosa [74] or the resident microbiota contributes to biosynthesis of thyronamines and their metabolites from L-thyroxine, orally administered to patients for therapeutic purposes, remains to be established and has recently been discussed [75].

4.2 Thyronamines (TAM)

Thyronamines (TAM), endogenous biogenic amines generated from iodinated TH, have attracted major attention after their biosynthesis and potential biological function has been proposed in a hypothesis proposed by Mary Dratman on the central mode of action of these TH metabolites [76]. She postulated that some of the (rapid) effects of TH, which she found accumulated in synaptosomal structures of nerve endings after administration to rodents, might arise from their metabolism to (false) neurotransmitters catalyzed by enzymes similar to those involved in generation of biogenic amines from aromatic amino acids, i.e., aromatic amino acid decarboxylase (AADC) (the AADC hypothesis has later on been rejected by Hoefig et al. [77], who identified ornithine decarboxylase as one enzyme generating thyronamines from TH [73, 74]. While existence of thyronamines and their biological effects have been described and studies soon after the discovery of T4 (*see* recent review [41]) detailed mechanistic studies of possible mode of action initially focused on 3,3′,5-Triiodothyronamine (Triam) [78, 79], which acted as powerful agonist at beta-adrenergic receptors but was a poor competitor for TR nuclear receptor binding or displacement for T3.

Two decades later a milestone publication by the Scanlan group revitalized thyronamine research after their demonstration of a rapid, powerful action of 3-iodothyronamine (3-T1AM) on thermoregulation, energy homeostasis and metabolism in rats [80]. These remarkable effects of 3T1AM, opposite to the major thermogenic and metabolic actions of thyromimetically active T3, sparked high interest and even funding of a priority research programme by the Deutsche Forschungsgemeinschaft in Germany ("ThyroidTransAct," http://www.thyroidtransact.de/) addressing issues of biosynthesis, metabolism, and action of these "rediscovered" "cool" TH metabolites. In the meanwhile, a variety of target molecules, binding sites, and receptors of high or medium affinity, with multiple downstream secondary messenger pathways have been characterized for 3T1AM, which proved to be a "dirty drug" with multiple modes of actions initiated either by 3T1AM itself or exerted by its metabolites (such as 3T1Ac, *see* above) (for recent review *see* ref. 41). Among those pharmacological actions tested in experimental animals, mainly rodents were reversible decrease in body temperature by 8–10 °C, shift from carbohydrate to lipid metabolism and oxidation, various cardiac, cardiovascular and metabolic effects, protection from adverse consequences of ischemia and stroke, and above mentioned effects on the CNS. So far it is not clear, whether endogenous 3T1AM or its metabolites may also provoke such action at physiological concentrations. Currently, concentration of endogenous 3T1AM and its metabolites in blood or tissues is a matter of debate. While a monoclonal-based chemiluminescence immunoassay detects low nanomolar 3T1AM concentrations in human (and mouse) serum [41, 81], analysis by liquid chromatography-tandem mass spectrometry (LC-MS/MS) found lower concentration of circulating 3T1AM [41, 82].

Among the receptors and downstream signaling cascades several targets and pathways have been partially characterized and novel synthetic, more stable 3T1AM analogues are exploited as potential drugs [83]. Interestingly, most of 3T1AM actions are exerted by cell membrane GPCR, TRP, TAAR or mitochondrial receptors and target molecules, while the classical T3 receptors are not affected [41, 84, 85]. The instability of the biogenic amine 3T1AM in vitro and in vivo, its hydrophobicity which requires adequate solvents for in vivo and in vitro experiments and controversies on its endogenous formation, metabolism and steady state concentrations remain significant obstacles in distinguishing effects of parental agent from those of its metabolites or even the solvents used which themselves might exert significant effects such as hypothermia observed in rodents.

5 3,5-T2: A New Thyromimetic Kid on the Block?

During the last century after discovery of T4 many naturally occurring and new synthetic TH metabolites, derivatives, and analogues have been identified and tested for potential TH (ant-)agonistic or direct pharmacological action, so far with limited success, considering that only the parental hormone T4, the thyromimetically active T3, and—for special pathological conditions (*see* above)—Triac have actually made it into clinical practice. Over the last years the hypothetical 3′-deiodination product of T3 (so far this enzymatic reaction could not yet be formally characterized in vitro, however!), i.e., 3,5-T2 has attracted major attention not only by researchers and clinicians but also by the internet, wellness, doping, and paramedical scene [94]. Rapid stimulatory metabolic effects on mitochondrial respiration and oxygen consumption have been demonstrated in mostly hypothyroid rodent models [95] and subsequently various mitochondrial and other subcellular targets and mechanisms of action have been proposed—mainly independent from classical (nuclear) T3 receptor action [93, 96]. One remarkable action attributed to 3,5-T2 was an efficient antisteatotic action in rodent models leading to the development of a synthetic mimic of 3,5-T2 which turned out to be not effective in a double-blind randomized controlled clinical trial on Insulin Sensitivity in male cardiometabolic patients [97]. In the meantime, several groups reported mainly for mouse models in part maintained on high fat diet but not hypothyroid conditions, that 3.5-T2 treatment effectively "cleared" steatotic livers from lipid deposits, however to the disadvantage of marked suppression of pituitary TSH secretion, interference with thyroid function and feedback regulation of the HPT axis, hepatic drug and steroid metabolism, and even—a major concern for its potential clinical application—adverse thyromimetic side effects on the heart [98–101]. Nevertheless, these and other authors come to the conclusion that part of 3,5-T2 action is distinct from that of classical T3 action [102–104], albeit some of the 3,5-T2 effects observed in rodent models overlap with those of T3 and might involve T3-like modulation of classical TR signaling in various tissues exposed to 3,5-T2. In this context it is of high interest that recently two forms of T3 receptors beta have been described in teleosts, which either prefer 3,5-T2 as ligand (TRβ-long) or T3 (TRβ-short) [105]. Apparently, different sets of genes are regulated by these two TR forms and their ligands and a joint coactivator, jun activation domain binding protein1 (Jab1), exerting dual coregulatory coactivator vs. corepressor properties at each receptor, might convey ligand specificity in targeting selective actions of the two thyromimetic ligands T3 and 3,5-T2 in this species [106].

Recent development of a monoclonal antibody-based chemiluminescence immunoassay allowed a more detailed analysis of 3,5-T2 concentrations in human serum [107]. Surprisingly, 3,5-T2 serum concentrations did not correlate to those of TSH, T4, or T3, the classical functional parameters used in clinical assessment of thyroid function. No major 3,5-T2 changes were observed in hyper- or hypothyroidism, while concentrations tended to be higher in patients on T4 supplementation compared to euthyroid controls with intact thyroid gland, a finding similar to that observed for 3T1AM (*see* above) [107–110]. However, under various pathophysiological conditions, such as non-survivor patients in critical illness on intensive care units, cardiac patients with postoperative atrial fibrillation, as well as patients with impaired renal function, elevated 3,5-T2 serum concentrations were observed [111, 112]. In sera collected during cross-sectional population-based studies of "healthy" adults in Northeast Germany ("SHIP"), serum concentrations of 3,5-T2 correlated to urinary metabolic parameters such trigonelline, hippurate, and 3-aminoisobutyrate [108, 109], a finding that so far is not explained but might also indicate a role for adequate renal function in metabolism and/or elimination of 3,5-T2.

6 Conjugated TH Metabolites

The aromatic phenolic iodinated tyrosine derivatives are also prone to be metabolized by conjugation of their rather acidic 4′-phenolic hydroxy group (*see* Fig. 1). Both sulfation and glucuronidation reactions have been described for TH, TAM, and probably TAc [33, 86, 88, 89]. While TH glucuronides are considered inactive metabolites, which however may undergo enterohepatic circulation after deconjugation by gastrointestinal microbiota [87], TH sulfate metabolites may still have inherent biological activity potentially binding to TR. Conjugation of TH decreases their biological activity because their increased solubility enhances elimination, biliary, urine, or fecal excretion [90]. However, the role of deconjugating host or microbial enzymes is not well studied and might lead to recirculation of liberated TH or their metabolites. Whether this leads to redistribution, storage, or different compartmentalization of THM as indicated by renal accumulation of T4-glucuronide [91] remains to be studied in more detail. Currently, no target molecules of receptors for conjugated TH metabolites have been characterized. Various sulfatases and glucuronidases may be involved in deconjugation of TH sulfates and glucuronides, thus contributing to recycling of active TH and their metabolites. Of interest, several of the enzymes involved in TH conjugation and deconjugation are also involved in steroid hormone metabolism. This might create interferences not only between the metabolic pathways of these two classes of hormones but also by drug metabolism of aromatic and phenolic compounds in general [92, 101] apart from competition of various drugs with serum distribution proteins for TH as well as carrier-mediated transport of TH across cellular membranes by various anion-, cation-, and multiple drug transporters. A sulfated metabolite of TH, the so called Compound W, which shows some crossreaction to 3,3′-T2-sulfate in immunoassays, has been postulated as a potential signaling compound in fetal-maternal communication [89, 113, 114]. The hypothesis has been proposed that concentrations of compound W in maternal blood my reflect (in-)adequate supply with TH and iodine to the developing fetus. However, the true chemical nature and regulation of fetal/placental/maternal sulfation of TH and transplacental transport of TH metabolites has not been clarified in detail.

7 Circadian Variations and Analytics of TH and THM by Immunoassays and Mass Spectrometry

Dynamic secretion of hormones is a key hallmark required for adequate feedback regulation, proper signaling and action of hormones and their active metabolites. While hypothalamic TRH and pituitary TSH show classical pulsatile and circadian profiles [115],

T4, the hormone with the longest half-life (7–8 days) due to its high affinity binding to four serum distributor proteins, only shows minimal circadian variations. Concentrations of serum T3, which has a shorter half-life of 1 day, exhibit circadian variation with small amplitude and some phase delay to TSH [116].

A retrospective analysis of a huge series of ca. half a million TSH and >100,000 free T4 measurements from healthy subjects aged 1–104 years revealed higher TSH and free T4 concentrations in females compared to males with ethnic differences in TSH circadian periodicity. For TSH no circannual rhythms were observed and reference ranges for free T4 concentrations were independent of factors such as age, sex, circadian, or circannual variation [117].

Classical analytics of serum thyroid hormones in clinical practice is based on immune assays for TSH, total and or free thyroid hormone determination, i.e., T4 and T3. Considering recent developments of detection of several thyroid hormone metabolites in human blood and tissues, which might exert specific biological actions or reflect TH status of target tissues, the new technology of liquid chromatography—tandem mass spectrometry (LC-MS/MS) conquers the major clinical diagnostic and research laboratories. While immunoassay-based analytics require multiple highly specific assays for each individual analyte based on (monoclonal) antibodies with minimal cross-reactivities to related TH metabolites, LC-MS/MS can exactly identify and quantify at the same time many different, but highly related metabolites from one single biological sample, a remarkable progress in up to date analytics! Of course both techniques have advantages and disadvantages. Immunoassays are robust, easy to perform, quick, and allow high sample throughput, provided assays have been adequately validated, use highly specific antibodies of minimal cross-reactivity for other related analytes of interest (if available at all). In case of TH and their metabolites multiplex immunoassays assays have not been developed, commercial assays are only available for T4, T3, and rT3 but not the other metabolites of interest. Beyond that the quality of some of these assays, especially for free T4 and even more so for free T3, is highly controversial at least in clinical settings for severely ill, children, adolescent, and aged individuals [57]. LC-MS/MS for TH and metabolites requires complex preanalytical sample workup, meticulous method validation, adequate use of stable isotope-labeled internal standards for each analyte of interest, and careful use of the isotope dilution principle, which together allow for precise validation of recovery, process efficiency, and possible matrix effects (i.e., interference by sample contaminants) [69, 118, 119]. This still expensive technique considering cost of high tech instruments, if adequately validated as any analytical method, meanwhile allows rapid analysis with high throughput and opens new avenues to simultaneously quantify panels of analytes, such as the "Thyronome" profile of patient sera, tissues, or cellular settings of in vitro experimental paradigms. During development, validation and first applications of LC-MS/MS technology in the thyroid hormone related clinical and research diagnostics a number of discrepancies between classical antibody-based analytics of TH and their metabolites with results obtained by the "more objective" LC-MS/MS' technique became obvious. Comparative analysis of TH metabolite profiles in human, experimental animal, wildlife, and in vitro cellular biological specimens [120–127] indicated that much more work is needed to improve and implement this powerful tool, which in part might replace some of classical immunoassay-based analytics, first in research and later on in clinical diagnostics. Such LC-MS/MS-based analytical technology will be essential to analyze and monitor the pharmacodynamic, pharmacokinetic, and metabolic fate of TH metabolites generated in vivo after administration of novel drugs based on synthetic peptide hormone-TH "hybrids," which allow tissue-specific targeting of classical or new TH analogues to target tissues by utilizing their peptide hormone cell membrane receptors and internalization machinery to selectively deliver the TH of interest without exposing other tissues to unwanted adverse side effects. Such a paradigm recently has been successfully established for a T3-glugagon conjugate administered to mice resulting in liver-specific expected T3 action without interfering with the feedback regulation of the HPT axis, the cardiac or CNS function [128].

Acknowledgment

This work has been supported by grants of the DFG funded Priority Programme 1629 "ThyroidTransAct" (Ko-922/16-2 and 17-2).

References

1. Taylor E, Heyland A (2017) Evolution of thyroid hormone signaling in animals: non-genomic and genomic modes of action. Mol Cell Endocrinol 459:14–20. https://doi.org/10.1016/j.mce.2017.05.019. [Epub ahead of print]. Review. PMID: 28549993
2. Holzer G, Roux N, Laudet V (2017) Evolution of ligands, receptors and metabolizing enzymes of thyroid signaling. Mol Cell Endocrinol 459:5–13. https://doi.org/10.1016/j.mce.2017.03.021. [Epub ahead of print]
3. Sower SA, Hausken KN (2017) A lamprey view on the origins of neuroendocrine regulation of the thyroid axis. Mol Cell Endocrinol 459:21–27. https://doi.org/10.1016/j.mce.2017.04.012. [Epub ahead of print] Review. PMID: 28412521
4. Moeller LC, Broecker-Preuss M (2011) Transcriptional regulation by nonclassical action of thyroid hormone. Thyroid Res 4 Suppl 1:S6. doi: https://doi.org/10.1186/1756-6614-4-S1-S6. PMID: 21835053
5. Brix K, Führer D, Biebermann H (2011) Molecules important for thyroid hormone synthesis and action—known facts and future perspectives. Thyroid Res 4 Suppl 1:S9. doi: https://doi.org/10.1186/1756-6614-4-S1-S9
6. Vella KR, Hollenberg AN (2017) The actions of thyroid hormone signaling in the nucleus. Mol Cell Endocrinol. doi: https://doi.org/10.1016/j.mce.2017.03.001. [Epub ahead of print] PMID: 28286327
7. Mondal S, Mugesh G (2017) Novel thyroid hormone analogues, enzyme inhibitors and

mimetics, and their action. Mol Cell Endocrinol. https://doi.org/10.1016/j.mce.2017.04.006. [Epub ahead of print] Review. PMID: 28408161

8. Boas M, Feldt-Rasmussen U, Main KM (2012) Thyroid effects of endocrine disrupting chemicals. Mol Cell Endocrinol 355(2):240–248. https://doi.org/10.1016/j.mce.2011.09.005. Epub 2011 Sep 10. Review.PMID: 21939731
9. Langer P (2010) The impacts of organochlorines and other persistent pollutants on thyroid and metabolic health. Front Neuroendocrinol 31(4):497–518. https://doi.org/10.1016/j.yfrne.2010.08.001. Epub 2010 Aug 24. Review. PMID: 20797403
10. Köhrle J (2008) Environment and endocrinology: the case of thyroidology. Ann Endocrinol (Paris) 69(2):116–122. https://doi.org/10.1016/j.ando.2008.02.008. Epub 2008 Apr 28. Review. PMID: 18440490
11. Murk AJ, Rijntjes E, Blaauboer BJ, Clewell R, Crofton KM, Dingemans MM, Furlow JD, Kavlock R, Köhrle J, Opitz R, Traas T, Visser TJ, Xia M, Gutleb AC (2013) Mechanism-based testing strategy using in vitro approaches for identification of thyroid hormone disrupting chemicals. Toxicol In Vitro 27(4):1320–1346. https://doi.org/10.1016/j.tiv.2013.02.012. Epub 2013 Feb 27. Review. PMID: 23453986
12. Kendall EC (1914) Studies in the chemistry and physiology of the thyroid. I. The determination of iodine in the thyroid. J Biol Chem 19:251–256
13. Kendall EC (1919) Isolation of the iodine compound which occurs in the thyroid. J Biol Chem 39:125–147
14. Harington CR (1926) Chemistry of thyroxine: isolation of thyroxine from the thyroid gland. Biochem J 20(2):293–299. No abstract available. PMID:16743658
15. Hennessey JV (2017) The emergence of levothyroxine as a treatment for hypothyroidism. Endocrine 55(1):6–18. https://doi.org/10.1007/s12020-016-1199-8. Epub 2016 Dec 16. Review.PMID: 27981511
16. Wassner AJ, Brown RS (2015) Congenital hypothyroidism: recent advances. Curr Opin Endocrinol Diabetes Obes 22(5):407–412. https://doi.org/10.1097/MED.0000000000000181
17. Okosieme O, Gilbert J, Abraham P, Boelaert K, Dayan C, Gurnell M, Leese G, McCabe C, Perros P, Smith V, Williams G, Vanderpump M (2016) Management of primary hypothyroidism: statement by the British Thyroid Association Executive Committee. Clin Endocrinol 84(6):799–808. https://doi.org/10.1111/cen.12824. Epub 2015 Jun 25
18. Wartofsky L (2013) Combination L-T3 and L-T4 therapy for hypothyroidism. Curr Opin Endocrinol Diabetes Obes 20(5):460–466. https://doi.org/10.1097/01.med.0000432611.03732.49. Review. PMID: 23974776
19. Jonklaas J (2016) Risks and safety of combination therapy for hypothyroidism. Expert Rev Clin Pharmacol 9(8):1057–1067. https://doi.org/10.1080/17512433.2016.1182019. Epub 2016 May 9
20. Hennessey JV (2015) Historical and current perspective in the use of thyroid extracts for the treatment of hypothyroidism. Endocr Pract 21(10):1161–1170. https://doi.org/10.4158/EP14477.RA. Epub 2015 Jun 29. Review. PMID: 26121440
21. Groeneweg S, Peeters RP, Visser TJ, Visser WE (2017) Therapeutic applications of thyroid hormone analogues in resistance to thyroid hormone (RTH) syndromes. Mol Cell Endocrinol 458:82–90. https://doi.org/10.1016/j.mce.2017.02.029. [Epub ahead of print] PMID: 28235578
22. Lammel Lindemann J, Webb P (2016) Sobetirome: the past, present and questions about the future. Expert Opin Ther Targets 20(2):145–149. https://doi.org/10.1517/14728222.2016.1090429. Epub 2015 Nov 13. PMID: 26565124
23. Regina A, Majlesi N (2016) Notes from the field. Thyrotoxicosis after consumption of dietary supplements purchased through the internet—Staten Island, New York, 2015. MMWR Morb Mortal Wkly Rep 65:353–354. https://doi.org/10.15585/mmwr.mm6513a4
24. Kang GY, Parks JR, Fileta B, Chang A, Abdel-Rahim MM, Burch HB, Bernet VJ (2013) Thyroxine and triiodothyronine content in commercially available thyroid health supplements. Thyroid 23(10):1233–1237. https://doi.org/10.1089/thy.2013.0101. Epub 2013 Sep 14
25. Germano S (2015) In Rebuff to Usada, WADA declines to ban thyroid medication. http://www.wsj.com/articles/in-rebuff-to-usada-wada-declines-to-ban-thyroid-medication-1443568862
26. Colin IM, Denef JF, Lengelé B, Many MC, Gérard AC (2013) Recent insights into the cell biology of thyroid angiofollicular units. Endocr Rev 34(2):209–238. https://doi.

org/10.1210/er.2012-1015. Epub 2013 Jan 24. Review. PMID: 23349248

27. Carvalho DP, Dupuy C (2017) Thyroid hormone biosynthesis and release. Mol Cell Endocrinol. https://doi.org/10.1016/j.mce.2017.01.038. [Epub ahead of print] PMID: 28153798
28. Szinnai G (2014) Clinical genetics of congenital hypothyroidism. Endocr Dev 26:60–78. https://doi.org/10.1159/000363156. Epub 2014 Aug 29. Review. PMID: 25231445
29. Dumitrescu AM, Refetoff S (2015) Impaired sensitivity to thyroid hormone: defects of transport, metabolism and action. In: De Groot LJ, Chrousos G, Dungan K, Feingold KR, Grossman A, Hershman JM, Koch C, Korbonits M, McLachlan R, New M, Purnell J, Rebar R, Singer F, Vinik A (eds) Endotext [Internet]. MDText.com, South Dartmouth, MA. PMID: 25905294
30. Köhrle J (2015) Selenium and the thyroid. Curr Opin Endocrinol Diabetes Obes 22(5):392–401. https://doi.org/10.1097/MED.0000000000000190. Review. PMID: 26313901
31. Dunn JT, Dunn AD (1999) The importance of thyroglobulin structure for thyroid hormone biosynthesis. Biochimie 81(5):505–509
32. Rokita SE, Adler JM, McTamney PM, Watson JA Jr (2010) Efficient use and recycling of the micronutrient iodide in mammals. Biochimie 92(9):1227–1235. https://doi.org/10.1016/j.biochi.2010.02.013. Epub 2010 Feb 16. Review. PMID: 20167242
33. Wu SY, Green WL, Huang WS, Hays MT, Chopra IJ (2005) Alternate pathways of thyroid hormone metabolism. Thyroid 15(8):943–958. Review. PMID: 16131336
34. Schweizer U, Steegborn C (2015) New insights into the structure and mechanism of iodothyronine deiodinases. J Mol Endocrinol 55(3):R37–R52. https://doi.org/10.1530/JME-15-0156. Epub 2015 Sep 21. Review. PMID: 26390881
35. Meinhold H, Gramm HJ, Meissner W, Zimmermann J, Schwander J, Dennhardt R, Voigt K (1991) Elevated serum diiodotyrosine (DIT) in severe infections and sepsis: DIT, a possible new marker of leukocyte activity. J Clin Endocrinol Metab 72(4):945–953. PMID: 2005222
36. Schweizer U, Johannes J, Bayer D, Braun D (2014) Structure and function of thyroid hormone plasma membrane transporters. Eur Thyroid J 3(3):143–153. https://doi.org/10.1159/000367858. Epub 2014 Sep 10. Review. PMID: 2553889
37. Mondal S, Raja K, Schweizer U, Mugesh G (2016) Chemistry and biology in the biosynthesis and action of thyroid hormones. Angew Chem Int Ed Engl 55(27):7606–7630. https://doi.org/10.1002/anie.201601116. Epub 2016 May 25. Review. PMID: 27226395
38. Groeneweg S, Visser WE, Visser TJ (2017) Disorder of thyroid hormone transport into the tissues. Best Pract Res Clin Endocrinol Metab 31(2):241–253. https://doi.org/10.1016/j.beem.2017.05.001. Epub 2017 May 24
39. Bernal J, Guadaño-Ferraz A, Morte B (2015) Thyroid hormone transporters—functions and clinical implications. Nat Rev Endocrinol 11(7):406–417. https://doi.org/10.1038/nrendo.2015.66. Epub 2015 May 5. Review.
40. Zada D, Blitz E, Appelbaum L (2017) Zebrafish—an emerging model to explore thyroid hormone transporters and psychomotor retardation. Mol Cell Endocrinol. https://doi.org/10.1016/j.mce.2017.03.004. [Epub ahead of print] PMID: 28274736
41. Hoefig CS, Zucchi R, Köhrle J (2016) Thyronamines and derivatives: physiological relevance, pharmacological actions, and future research directions. Thyroid 26(12):1656–1673. Epub 2016 Nov 9. PMID: 27650974
42. Gross J, Pitt-Rivers R (1952) The identification of 3:5:3′-L-triiodothyronine in human plasma. Lancet 1(6705):439–441. PMID: 14898765
43. Hird F Jr, Trikojus VM (1948) Paper partition chromatography with thyroxine and analogues. Aust J Sci 10(6):185–187. PMID: 18875255
44. Sap J, Muñoz A, Damm K, Goldberg Y, Ghysdael J, Leutz A, Beug H, Vennström B (1986) The c-erb-A protein is a high-affinity receptor for thyroid hormone. Nature 324(6098):635–640. PMID: 2879242
45. Di Jeso B, Arvan P (2016) Thyroglobulin from molecular and cellular biology to clinical endocrinology. Endocr Rev 37(1):2–36. https://doi.org/10.1210/er.2015-1090. Epub 2015 Nov 23. Review. PMID: 26595189
46. Braverman LE (1994) Deiodination of thyroid hormones. A 30 year perspective (Berthold Memorial Award Lecture 1994). Exp Clin Endocrinol 102(5):355–363
47. Köhrle J (1999) Local activation and inactivation of thyroid hormones: the deiodinase family. Mol Cell Endocrinol 151(1–2):103–119. Review. PMID: 10411325
48. Gereben B, Zeöld A, Dentice M, Salvatore D, Bianco AC (2008) Activation and inactivation of thyroid hormone by deiodinases: local

action with general consequences. Cell Mol Life Sci 65(4):570–590. Review. PMID: 17989921
49. Peeters RP, Visser TJ (2017) Metabolism of thyroid hormone. In: De Groot LJ, Chrousos G, Dungan K, Feingold KR, Grossman A, Hershman JM, Koch C, Korbonits M, McLachlan R, New M, Purnell J, Rebar R, Singer F, Vinik A (eds) Endotext [Internet]. MDText.com, South Dartmouth, MA. PMID: 25905401
50. Schweizer U, Schlicker C, Braun D, Köhrle J, Steegborn C (2014) Crystal structure of mammalian selenocysteine-dependent iodothyronine deiodinase suggests a peroxiredoxin-like catalytic mechanism. Proc Natl Acad Sci U S A 111(29):10526–10531. https://doi.org/10.1073/pnas.1323873111. Epub 2014 Jul 7. PMID: 25002520
51. Köhrle J (2002) Iodothyronine deiodinases. Methods Enzymol 347:125–167
52. Renko K, Hoefig CS, Hiller F, Schomburg L, Köhrle J (2012) Identification of iopanoic acid as substrate of type 1 deiodinase by a novel nonradioactive iodide-release assay. Endocrinology 153(5):2506–2513. https://doi.org/10.1210/en.2011-1863. Epub 2012 Mar 20. PubMed PMID: 22434082
53. Renko K, Schäche S, Hoefig CS, Welsink T, Schwiebert C, Braun D, Becker NP, Köhrle J, Schomburg L (2015) An improved nonradioactive screening method identifies genistein and xanthohumol as potent inhibitors of iodothyronine deiodinases. Thyroid 25(8):962–968. https://doi.org/10.1089/thy.2015.0058. Epub 2015 Jun 25. PubMed PMID: 25962824
54. Rijntjes E, Scholz PM, Mugesh G, Köhrle J (2013) Se- and s-based thiouracil and methimazole analogues exert different inhibitory mechanisms on type 1 and type 2 deiodinases. Eur Thyroid J 2(4):252–258. https://doi.org/10.1159/000355288. Epub 2013 Nov 27. PubMed PMID: 24783056. PubMed Central PMCID: PMC3923599
55. Dentice M, Salvatore D (2011) Deiodinases: the balance of thyroid hormone: local impact of thyroid hormone inactivation. J Endocrinol 209(3):273–282. https://doi.org/10.1530/JOE-11-0002. Epub 2011 Mar 11. Review. PMID: 21398344
56. Schweizer U, Towell H, Vit A, Rodriguez-Ruiz A, Steegborn C (2017) Structural aspects of thyroid hormone binding to proteins and competitive interactions with natural and synthetic compounds. Mol Cell Endocrinol. https://doi.org/10.1016/j.mce.2017.01.026. [Epub ahead of print] PMID: 28131741
57. Welsh KJ, Soldin SJ (2016) Diagnosis of endocrine disease: how reliable are free thyroid and total T3 hormone assays? Eur J Endocrinol 175(6):R255–R263. Review. PMID: 27737898
58. Spencer CA (2017) Assay of thyroid hormones and related substances. In: De Groot LJ, Chrousos G, Dungan K, Feingold KR, Grossman A, Hershman JM, Koch C, Korbonits M, McLachlan R, New M, Purnell J, Rebar R, Singer F, Vinik A (eds) Endotext [Internet]. MDText.com, South Dartmouth, MA. PMID: 2590533
59. Faix JD (2013) Principles and pitfalls of free hormone measurements. Best Pract Res Clin Endocrinol Metab 27(5):631–645. https://doi.org/10.1016/j.beem.2013.06.007. Epub 2013 Jul 17. Review. PMID: 24094635
60. Engler D, Burger AG (1984) The deiodination of the iodothyronines and of their derivatives in man. Endocr Rev 5(2):151–184. Review. No abstract available. PMID: 6376077
61. Burger A, Reinharz A, Ingbar SH (1975) High-affinity binding of tetraiodothyroacetic acid by a prealbumin in normal rabbit serum. Endocrinology 97(4):919–923
62. Davis PJ, Sudha T, Lin HY, Mousa SA (2015) Thyroid hormone, hormone analogs, and angiogenesis. Compr Physiol 6(1):353–362. https://doi.org/10.1002/cphy.c150011. Review. PMID: 26756636
63. Schmohl KA, Müller AM, Wechselberger A, Rühland S, Salb N, Schwenk N, Heuer H, Carlsen J, Göke B, Nelson PJ, Spitzweg C (2015) Thyroid hormones and tetrac: new regulators of tumour stroma formation via integrin αvβ3. Endocr Relat Cancer 22(6):941–952. https://doi.org/10.1530/ERC-15-0245. Epub 2015 Aug 25
64. Menegay C, Juge C, Burger AG (1989) Pharmacokinetics of 3,5,3′-triiodothyroacetic acid and its effects on serum TSH levels. Acta Endocrinol (Copenh) 121(5):651–658. PMIC: 2588936
65. Kersseboom S, Horn S, Visser WE, Chen J, Friesema EC, Vaurs-Barrière C, Peeters RP, Heuer H, Visser TJ (2014) In vitro and mouse studies supporting therapeutic utility of triiodothyroacetic acid in MCT8 deficiency. Mol Endocrinol 28(12):1961–1970. https://doi.org/10.1210/me.2014-1135. PMID: 25389909
66. Cody V (1988) Thyroid hormone structure-activity relationships: molecular structure of 3,5,3′-triiodothyropropionic acid. Endocr Res 14(2–3):165–176
67. Wood WJ, Geraci T, Nilsen A, DeBarber AE, Scanlan TS (2009) Iodothyronamines are

oxidatively deaminated to iodothyroacetic acids in vivo. Chembiochem 10(2):361–365. https://doi.org/10.1002/cbic.200800607. PMID:19105176

68. Hackenmueller SA, Scanlan TS (2012) Identification and quantification of 3-iodothyronamine metabolites in mouse serum using liquid chromatography-tandem mass spectrometry. J Chromatogr A 1256:89–97. https://doi.org/10.1016/j.chroma.2012.07.052. Epub 2012 Jul 25. PMID: 22885046
69. Richards KH, Schanze N, Monk R, Rijntjes E, Rathmann D, Köhrle J (2017) A validated LC-MS/MS method for cellular thyroid hormone metabolism: uptake and turnover of mono-iodinated thyroid hormone metabolites by PCCL3 thyrocytes. PLoS One 12(8):e0183482. https://doi.org/10.1371/journal.pone.0183482. eCollection 2017. PMID: 28837607
70. Laurino A, De Siena G, Resta F, Masi A, Musilli C, Zucchi R, Raimondi L (2015) 3-iodothyroacetic acid, a metabolite of thyroid hormone, induces itch and reduces threshold to noxious and to painful heat stimuli in mice. Br J Pharmacol 172(7):1859–1868. doi: https://doi.org/10.1111/bph.13032. Epub 2015 Jan 23. PMID: 25439265
71. Schanze N, Jacobi SF, Rijntjes E, Mergler S, Del Olmo M, Hoefig CS, Khajavi N, Lehmphul I, Biebermann H, Mittag J, Köhrle J (2017) 3-iodothyronamine decreases expression of genes involved in iodide metabolism in mouse thyroids and inhibits iodide uptake in PCCL3 thyrocytes. Thyroid 27(1):11–22. https://doi.org/10.1089/thy.2016.0182. Epub 2016 Dec 21. PMID: 27788620
72. Lehmphul I, Hoefig CS, Köhrle J (2017) 3-Iodothyronamine reduces insulin secretion in vitro via a mitochondrial mechanism. Mol Cell Endocrinol. https://doi.org/10.1016/j.mce.2017.07.026. [Epub ahead of print] PMID: 28754352
73. Hoefig CS, Jacobi SF, Warner A, Harder L, Schanze N, Vennström B, Mittag J (2015) 3-iodothyroacetic acid lacks thermoregulatory and cardiovascular effects in vivo. Br J Pharmacol 172(13):3426–3433. https://doi.org/10.1111/bph.13131. Epub 2015 Jun 9. PMID: 25765843
74. Hoefig CS, Wuensch T, Rijntjes E, Lehmphul I, Daniel H, Schweizer U, Mittag J, Köhrle J (2015) Biosynthesis of 3-iodothyronamine from T4 in murine intestinal tissue. Endocrinology 156(11):4356–4364. https://doi.org/10.1210/en.2014-1499. Epub 2015 Sep 8. PMID:26348473
75. Glossmann HH, Lutz OMD (2017) Torpor: the rise and fall of 3-monoiodothyronamine from brain to gut—from gut to brain? Front Endocrinol (Lausanne) 8:118. https://doi.org/10.3389/fendo.2017.00118. PMCID: PMC5450037
76. Dratman MB (1974) On the mechanism of action of thyroxin, an amino acid analog of tyrosine. J Theor Biol 46:255–270
77. Hoefig CS, Renko K, Piehl S, Scanlan TS, Bertoldi M, Opladen T, Hoffmann GF, Klein J, Blankenstein O, Schweizer U, Köhrle J (2012) Does the aromatic L-amino acid decarboxylase contribute to thyronamine biosynthesis? Mol Cell Endocrinol 349(2):195–201. https://doi.org/10.1016/j.mce.2011.10.024. Epub 2011 Oct 28. PMID: 22061622
78. Meyer T, Hesch RD (1983) Triiodothyronamine—a beta-adrenergic metabolite of triiodothyronine? Horm Metab Res 15(12):602–606
79. Cody V, Meyer T, Dohler KD, Hesch RD, Rokos H, Marko M (1984) Molecular structure and biochemical activity of 3,5,3′-triiodothyronamine. Endocr Res 10(2):91–99
80. Scanlan TS, Suchland KL, Hart ME, Chiellini G, Huang Y, Kruzich PJ, Frascarelli S, Crossley DA, Bunzow JR, Ronca-Testoni S, Lin ET, Hatton D, Zucchi R, Grandy DK (2004) 3-Iodothyronamine is an endogenous and rapid-acting derivative of thyroid hormone. Nat Med 10(6):638–642. Epub 2004 May 16. PMID: 15146179
81. Hoefig CS, Köhrle J, Brabant G, Dixit K, Yap B, Strasburger CJ, Wu Z (2011) Evidence for extrathyroidal formation of 3-iodothyronamine in humans as provided by a novel monoclonal antibody-based chemiluminescent serum immunoassay. J Clin Endocrinol Metab 96(6):1864–1872. https://doi.org/10.1210/jc.2010-2680. Epub 2011 Apr 13. PMID: 21490071
82. Galli E, Marchini M, Saba A, Berti S, Tonacchera M, Vitti P, Scanlan TS, Iervasi G, Zucchi R (2012) Detection of 3-iodothyronamine in human patients: a preliminary study. J Clin Endocrinol Metab 97(1):E69–E74. https://doi.org/10.1210/jc.2011-1115. Epub 2011 Oct 26. PMID: 22031514
83. Chiellini G, Nesi G, Digiacomo M, Malvasi R, Espinoza S, Sabatini M, Frascarelli S, Laurino A, Cichero E, Macchia M, Gainetdinov RR, Fossa P, Raimondi L, Zucchi R, Rapposelli S (2015) Design, synthesis, and evaluation of thyronamine analogues as novel potent mouse trace amine associated receptor 1 (mTAAR1) agonists. J Med Chem 58(12):5096–5107. https://

doi.org/10.1021/acs.jmedchem.5b00526. Epub 2015 Jun 5. PMID: 26010728
84. Chiellini G, Bellusci L, Sabatini M, Zucchi R (2017) Thyronamines and analogues—the route from rediscovery to translational research on thyronergic amines. Mol Cell Endocrinol 458:149–155. https://doi.org/10.1016/j.mce.2017.01.002. [Epub ahead of print] PMID: 28069535
85. Khajavi N, Mergler S, Biebermann H (2017) 3-Iodothyronamine, a novel endogenous modulator of transient receptor potential melastatin 8? Front Endocrinol 8:198
86. Visser TJ (1996) Pathways of thyroid hormone metabolism. Acta Med Austriaca 23(1–2):10–16. Review. PMID: 8767510
87. Visser TJ, Rutgers M, de Herder WW, Rooda SJ, Hazenberg MP (1988) Hepatic metabolism, biliary clearance and enterohepatic circulation of thyroid hormone. Acta Med Austriaca 15(Suppl 1):37–39. Review. PMID: 3051833
88. Pietsch CA, Scanlan TS, Anderson RJ (2007) Thyronamines are substrates for human liver sulfotransferases. Endocrinology 148(4):1921–1927. Epub 2007 Jan 4. PMID: 17204552
89. Polk DH, Reviczky A, Wu SY, Huang WS, Fisher DA (1994) Metabolism of sulfoconjugated thyroid hormone derivatives in developing sheep. Am J Phys 266(6 Pt 1):E892–E896
90. van der Spek AH, Fliers E, Boelen A (2017) The classic pathways of thyroid hormone metabolism. Mol Cell Endocrinol. https://doi.org/10.1016/j.mce.2017.01.025. [Epub ahead of print] PMID: 28109953
91. Buitendijk M, Galton VA (2012) Is the kidney a major storage site for thyroxine as thyroxine glucuronide? Thyroid 22(2):187–191. https://doi.org/10.1089/thy.2011.0307. Epub 2011 Dec 16
92. Sarne D (2016) Effects of the environment, chemicals and drugs on thyroid function endotext. In: De Groot LJ, Chrousos G, Dungan K et al (eds) MDText.com, South Dartmouth, MA. Last update: 27 Sept 2016
93. Moreno M, Giacco A, Di Munno C, Goglia F (2017) Direct and rapid effects of 3,5-diiodo-L-thyronine (T2). Mol Cell Endocrinol. https://doi.org/10.1016/j.mce.2017.02.012. [Epub ahead of print] PMID: 28192176
94. Hernandez A (2015) 3,5-diiodo-L-thyronine (t2) in dietary supplements: what are the physiological effects? Endocrinology 156(1):5–7. https://doi.org/10.1210/en.2014-1933. No abstract available. PMID: 25526549
95. Horst C, Rokos H, Seitz HJ (1989) Rapid stimulation of hepatic oxygen consumption by 3,5-di-iodo-L-thyronine. Biochem J 261(3):945–950
96. Goglia F (2015) The effects of 3,5-diiodothyronine on energy balance. Front Physiol 5:528. https://doi.org/10.3389/fphys.2014.00528. eCollection 2014. No abstract available. PMID: 25628573
97. van der Valk F, Hassing C, Visser M, Thakkar P, Mohanan A, Pathak K, Dutt C, Chauthaiwale V, Ackermans M, Nederveen A, Serlie M, Nieuwdorp M, Stroes E (2014) The effect of a diiodothyronine mimetic on insulin sensitivity in male cardiometabolic patients: a double-blind randomized controlled trial. PLoS One 9(2):e86890. https://doi.org/10.1371/journal.pone.0086890
98. Goldberg IJ, Huang LS, Huggins LA, Yu S, Nagareddy PR, Scanlan TS, Ehrenkranz JR (2012) Thyroid hormone reduces cholesterol via a non-LDL receptor-mediated pathway. Endocrinology 153(11):5143–5149. https://doi.org/10.1210/en.2012-1572. Epub 2012 Sep 4. PMID: 22948212
99. Padron AS, Neto RA, Pantaleão TU, de Souza dos Santos MC, Araujo RL, de Andrade BM, da Silva Leandro M, de Castro JP, Ferreira AC, de Carvalho DP (2014) Administration of 3,5-diiodothyronine (3,5-T2) causes central hypothyroidism and stimulates thyroid-sensitive tissues. J Endocrinol 221(3):415–427. https://doi.org/10.1530/JOE-13-0502. Epub 2014 Apr 1
100. Jonas W, Lietzow J, Wohlgemuth F, Hoefig CS, Wiedmer P, Schweizer U, Köhrle J, Schürmann A (2015) 3,5-Diiodo-L-thyronine (3,5-T2) exerts thyromimetic effects on hypothalamus-pituitary-thyroid axis, body composition, and energy metabolism in male diet-induced obese mice. Endocrinology 156(1):389–399. https://doi.org/10.1210/en.2014-1604. PMID: 25322465
101. Lietzow J, Golchert J, Homuth G, Völker U, Jonas W, Köhrle J (2016) 3,5-T2 alters murine genes relevant for xenobiotic, steroid, and thyroid hormone metabolism. J Mol Endocrinol 56(4):311–323. https://doi.org/10.1530/JME-15-0159. Epub 2016 Feb 22. PMID: 26903510
102. Damiano F, Rochira A, Gnoni A, Siculella L (2017) Action of thyroid hormones, T3 and T2, on hepatic fatty acids: differences in metabolic effects and molecular mechanisms. Int J Mol Sci 18(4):pii: E744. https://doi.org/10.3390/ijms18040744

103. da Silva Teixeira S, Filgueira C, Sieglaff DH, Benod C, Villagomez R, Minze LJ, Zhang A, Webb P, Nunes MT (2017) 3,5-diiodothyronine (3,5-T2) reduces blood glucose independently of insulin sensitization in obese mice. Acta Physiol (Oxf) 220(2):238–250. https://doi.org/10.1111/apha.12821. PMID: 27770485
104. Moreno M, Silvestri E, Coppola M, Goldberg IJ, Huang LS, Salzano AM, D'Angelo F, Ehrenkranz JR, Goglia F (2016) 3,5,3′-Triiodo-L-thyronine- and 3,5-diiodo-L-thyronine-affected metabolic pathways in liver of LDL receptor deficient mice. Front Physiol 7:545
105. Orozco A, Lazcano I, Hernández-Puga G, Olvera A (2017) Non-mammalian models reveal the role of alternative ligands for thyroid hormone receptors. Mol Cell Endocrinol 459:59–63. https://doi.org/10.1016/j.mce.2017.03.003. [Epub ahead of print] PMID: 28267601
106. Hernández-Puga G, Mendoza A, León-Del-Río A, Orozco A (2017) Jab1 is a T2-dependent coactivator or a T3-dependent corepressor of TRB1-mediated gene regulation. J Endocrinol 232(3):451–459. https://doi.org/10.1530/JOE-16-0485. Epub 2017 Jan 4. PMID: 28053002
107. Lehmphul I, Brabant G, Wallaschofski H, Ruchala M, Strasburger CJ, Köhrle J, Wu Z (2014) Detection of 3,5-diiodothyronine in sera of patients with altered thyroid status using a new monoclonal antibody-based chemiluminescence immunoassay. Thyroid 24(9):1350–1360. https://doi.org/10.1089/thy.2013.0688. Epub 2014 Aug 1. PMID: 24967815
108. Pietzner M, Lehmphul I, Friedrich N, Schurmann C, Ittermann T, Dörr M, Nauck M, Laqua R, Völker U, Brabant G, Völzke H, Köhrle J, Homuth G, Wallaschofski H (2015) Translating pharmacological findings from hypothyroid rodents to euthyroid humans: is there a functional role of endogenous 3,5-T2? Thyroid 25(2):188–197. https://doi.org/10.1089/thy.2014.0262
109. Pietzner M, Homuth G, Budde K, Lehmphul I, Völker U, Völzke H, Nauck M, Köhrle J, Friedrich N (2015) Urine metabolomics by (1)H-NMR spectroscopy indicates associations between serum 3,5-t2 concentrations and intermediary metabolism in euthyroid humans. Eur Thyroid J 4(Suppl 1):92–100. https://doi.org/10.1159/000381308. Epub 2015 May 28. PMID: 26601079
110. Massolt ET, van der Windt M, Korevaar TI, Kam BL, Burger JW, Franssen GJ, Lehmphul I, Köhrle J, Visser WE, Peeters RP (2016) Thyroid hormone and its metabolites in relation to quality of life in patients treated for differentiated thyroid cancer. Clin Endocrinol 85(5):781–788. https://doi.org/10.1111/cen.13101. Epub 2016 Jun 13. PMID: 27175823
111. Langouche L, Lehmphul I, Perre SV, Köhrle J, Van den Berghe G (2016) Circulating 3-T1AM and 3,5-T2 in critically ill patients: a cross-sectional observational study. Thyroid 26(12):1674–1680. Epub 2016 Oct 25. PMID: 27676423
112. Dietrich JW, Müller P, Schiedat F, Schlömicher M, Strauch J, Chatzitomaris A, Klein HH, Mügge A, Köhrle J, Rijntjes E, Lehmphul I (2015) Nonthyroidal illness syndrome in cardiac illness involves elevated concentrations of 3,5-diiodothyronine and correlates with atrial remodeling. Eur Thyroid J. 4(2):129–137. https://doi.org/10.1159/000381543. Epub 2015 May 23. PMID: 26279999
113. Wu SY, Huang WS, Ho E, Wu ES, Fisher DA (2007) Compound W, a 3,3′-diiodothyronine sulfate cross-reactive substance in serum from pregnant women—a potential marker for fetal thyroid function. Pediatr Res 61(3):307–312. PMID: 17314688
114. Chen D, Yu H, Bao J, Xue W, Xing Y, Zhang L, Green WL, Fisher DA, Wu SY (2012) 3,3′-diiodothyronine sulfate cross-reactive material (compound W) in human newborns. Pediatr Res 72(5):521–524. https://doi.org/10.1038/pr.2012.116. Epub 2012 Aug 20. PMID: 22907618
115. Brabant G, Brabant A, Ranft U, Ocran K, Köhrle J, Hesch RD, von zur Mühlen A (1987) Circadian and pulsatile thyrotropin secretion in euthyroid man under the influence of thyroid hormone and glucocorticoid administration. J Clin Endocrinol Metab 65(1):83–88. PMID: 3584402
116. Russell W, Harrison RF, Smith N, Darzy K, Shalet S, Weetman AP, Ross RJ (2008) Free triiodothyronine has a distinct circadian rhythm that is delayed but parallels thyrotropin levels. J Clin Endocrinol Metab 93(6):2300–2306. https://doi.org/10.1210/jc.2007-2674. Epub 2008 Mar 25. PMID: 18364382
117. Ehrenkranz J, Bach PR, Snow GL, Schneider A, Lee JL, Ilstrup S, Bennett ST, Benvenga S (2015) Circadian and circannual rhythms in thyroid hormones: determining the TSH and free T4 reference intervals based upon time of day, age, and sex. Thyroid 25(8):954–961. https://doi.org/10.1089/thy.2014.0589. PMID: 26061389

118. Richards K, Rijntjes E, Rathmann D, Köhrle J (2017) Avoiding the pitfalls when quantifying thyroid hormones and their metabolites using mass spectrometric methods: the role of quality assurance. Mol Cell Endocrinol 458:44–56. https://doi.org/10.1016/j.mce.2017.01.032

119. Rathmann D, Rijntjes E, Lietzow J, Köhrle J (2015) Quantitative analysis of thyroid hormone metabolites in cell culture samples using LC-MS/MS. Eur Thyroid J 4(Suppl 1):51–58. https://doi.org/10.1159/000430840. Epub 2015 May 28.PMID: 26601073

120. Álvarez E, Madrid Y, Marazuela MD (2017) Comparison of sample preparation strategies for target analysis of total thyroid hormones levels in serum by liquid chromatography-quadrupole time-of-flight-mass spectrometry. Talanta 164:570–579. https://doi.org/10.1016/j.talanta.2016.12.001. Epub 2016 Dec 2. PMID: 28107975

121. Villanger GD, Learner E, Longnecker MP, Ask H, Aase H, Zoeller RT, Knudsen GP, Reichborn-Kjennerud T, Zeiner P, Engel SM (2017) Effects of sample handling and analytical procedures on thyroid hormone concentrations in pregnant women's plasma. Epidemiology 28(3):365–369. https://doi.org/10.1097/EDE.0000000000000606. PMID: 27984425

122. Noyes PD, Lema SC, Roberts SC, Cooper EM, Stapleton HM (2014) Rapid method for the measurement of circulating thyroid hormones in low volumes of teleost fish plasma by LC-ESI/MS/MS. Anal Bioanal Chem 406(3):715–726. https://doi.org/10.1007/s00216-013-7528-3. Epub 2013 Dec 18. PMID:24343452

123. Saba A, Donzelli R, Colligiani D, Raffaelli A, Nannipieri M, Kusmic C, Dos Remedios CG, Simonides WS, Iervasi G, Zucchi R (2014) Quantification of thyroxine and 3,5,3′-triiodo-thyronine in human and animal hearts by a novel liquid chromatography-tandem mass spectrometry method. Horm Metab Res 46(9):628–634. https://doi.org/10.1055/s-0034-1368717. Epub 2014 Mar 3. PMID: 24591048

124. Kiebooms JA, Wauters J, Vanden Bussche J, Vanhaecke L (2014) Validated ultra high performance liquid chromatography-tandem mass spectrometry method for quantitative analysis of total and free thyroid hormones in bovine serum. J Chromatogr A 1345:164–173. https://doi.org/10.1016/j.chroma.2014.04.032. Epub 2014 Apr 18. PMID: 24786658

125. Sakai H, Nagao H, Sakurai M, Okumura T, Nagai Y, Shikuma J, Ito R, Imazu T, Miwa T, Odawara M (2015) Correlation between serum levels of 3,3′,5′-triiodothyronine and thyroid hormones measured by liquid chromatography-tandem mass spectrometry and immunoassay. PLoS One 10(10):e0138864. https://doi.org/10.1371/journal.pone.0138864. PMID: 26426328

126. De Angelis M, Giesert F, Finan B, Clemmensen C, Müller TD, Vogt-Weisenhorn D, Tschöp MH, Schramm KW (2016) Determination of thyroid hormones in mouse tissues by isotope-dilution microflow liquid chromatography-mass spectrometry method. J Chromatogr B Analyt Technol Biomed Life Sci 1033-1034:413–420. https://doi.org/10.1016/j.jchromb.2016.08.037. Epub 2016 Aug 26. PMID: 27649501

127. Bussy U, Chung-Davidson YW, Li K, Fissette SD, Buchinger EG, Li W (2017) A validated LC-MS/MS method for thyroid hormone determination in sea lamprey (Petromyzon marinus) plasma, gill, kidney and liver. J Chromatogr B Analyt Technol Biomed Life Sci 1041-1042:77–84. https://doi.org/10.1016/j.jchromb.2016.12.024. Epub 2016 Dec 19. PMID: 28012382

128. Finan B, Clemmensen C, Zhu Z, Stemmer K, Gauthier K, Müller L, De Angelis M, Moreth K, Neff F, Perez-Tilve D, Fischer K, Lutter D, Sánchez-Garrido MA, Liu P, Tuckermann J, Malehmir M, Healy ME, Weber A, Heikenwalder M, Jastroch M, Kleinert M, Jall S, Brandt S, Flamant F, Schramm KW, Biebermann H, Döring Y, Weber C, Habegger KM, Keuper M, Gelfanov V, Liu F, Köhrle J, Rozman J, Fuchs H, Gailus-Durner V, Hrabě de Angelis M, Hofmann SM, Yang B, Tschöp MH, DiMarchi R, Müller TD (2016) Chemical hybridization of glucagon and thyroid hormone optimizes therapeutic impact for metabolic disease. Cell 167(3):843–857.e14. https://doi.org/10.1016/j.cell.2016.09.014. Epub 2016 Oct 6. PMID: 27720451

Chapter 10

Protocols for Studying How Thyroid Hormone Receptors Trigger Adaptive Thermogenesis in Mice

Karine Gauthier

Abstract

The ability of thyroid hormone T3 to stimulate energy expenditure and regulate different aspects of whole body metabolism has been recognized for over a 100 years. Indeed in 1895 Adolf Magnus Levy was already describing the influence of the thyroid on setting the basal metabolic rate. Now it has been well characterized that the level of circulating T3 is correlated with energy expenditure both in humans and in rodent models (Mullur et al., Physiol Rev 94:355–382, 2014; Silva, Thyroid 5:481–492, 1995). Hypothyroidism and hyperthyroidism are respectively associated with low and high energy expenditure leading to high and low body mass index. Moreover, T3 also reduces the level of cholesterol in the serum, which is one of the key risk factors to develop atherosclerosis (Mullur et al., Physiol Rev 94:355–382, 2014; Hak et al., Ann Intern Med 132:270–278, 2000). Given the worldwide pandemic of obesity and associated metabolic disorders in the last two decades, the metabolic activities of T3 gained a renewed interest. However, one has to keep in mind that despite its desirable actions, T3 cannot be used as a pharmacological agent since it also triggers unacceptable effects including tachycardia, lean mass loss, and osteoporosis. To solve this conundrum, a recent intense effort has been dedicated to understand how T3 fulfills its different activities, looking for the target tissues and the specific T3 receptor (TR) involved. Indeed recently it was shown that injection of T3 in the brain could efficiently trigger thermogenesis (López et al., Nat Med 16:1001–1008, 2010). This questioned the classical paradigm in which most metabolic actions were the result of local T3 action in the different metabolic tissues. To tackle this problem, many new models of transgenic mice carrying selective Knock-out and Knock-in mutation of TR in specific metabolic tissues are currently generated using conditional mutant alleles of *THRA* (Quignodon et al., Mol Endocrinol 21:2350–2360, 2007) and *THRB* (Billon et al., Endocrinology 155:2735–2745, 2014). In parallel some new compounds that allow the targeting of T3 to selective metabolic tissues (Finan et al., Cell 167:843–857, 2016) have also been obtained.

In this chapter we will provide and comment two different protocols that are useful to study the adaptive thermogenesis in response to two physiological stresses: cold exposure and high fat diet feeding. They can also be used to test the thermogenic activity of the new designed compounds.

Key words Adaptive thermogenesis, Mice, Diet-induced obesity, Cold exposure, Thermoneutrality

Michelina Plateroti and Jacques Samarut (eds.), *Thyroid Hormone Nuclear Receptor: Methods and Protocols*, Methods in Molecular Biology, vol. 1801, https://doi.org/10.1007/978-1-4939-7902-8_10,
© Springer Science+Business Media, LLC, part of Springer Nature 2018

1 Adaptive Thermogenesis

1.1 Adaptive Thermogenesis Is Induced by Cold Exposure and High Fat Diet Feeding

Energy expenditure is usually defined as a combination of basal and adaptive thermogenesis [1]. They are respectively defined as heat produced by a resting organism placed at thermoneutrality and the extra heat produced by a combination of physical activity and specific responses triggered to face physiological stressors. Thermoneutrality is the temperature for which the heat needed to maintain temperature is provided by the heat produced as a byproduct of all the inefficient biochemical transformations necessary for life. The two main natural drivers for adaptive thermogenesis are cold exposure that increases the demand of heat production needed to maintain body temperature, and high fat diet to limit weight gain. In both instances the physiological stress is detected by the hypothalamus and triggers the activation of the β-adrenergic system (SNS/β-AR) to increase peripheral heat loss mainly through the activation of BAT and the browning [2] of WAT. Browning is a recently described process during which brown-like adipocytes appear in the WAT depots. They are characterized by high mitochondria content and activity. The β-adrenergic system is at the center of this thermogenic response since concomitant knock out of the 3 β adrenergic receptors (β-AR) leads to hypersensitivity to cold and to diet-induced obesity at thermoneutrality [3]. In both instances, uncoupling is induced. It relies on the uncoupling protein 1 (UCP-1). This protein is expressed in the mitochondrial membranes and decreases the proton gradient necessary for ATP synthase activity. Its overexpression or activation is associated with increased heat generation at the expense of ATP production and as a result in more important energy expenditure. UCP-1 KO mice are hypersensitive to cold [4] and to diet-induced obesity at thermoneutrality.

1.2 Thermoneutrality a Useful Notion to Study Energy Expenditure and Sensitivity to Diet Induced Obesity

Performing experiments at thermoneutrality is critical to study energy expenditure. Maintenance of body temperature is tightly regulated and critical for the animal survival. Mice when reared in animal facilities at 23 °C are chronically cold exposed since thermoneutrality is obtained at 30 °C. The β-adrenergic signaling is thus permanently turned on to efficiently produce heat. Many Knock out models generated to analyze the role of a protein in temperature homeostasis have first been described to be resistant to diet-induced obesity at 23 °C. Further studies showed that it resulted from either a compensatory activation of the SNS/β-AR system or the onset of other mechanisms less efficient than uncoupling to avoid a drop in body temperature. At 30 °C results changed. If the protein is involved in the regulation of adaptive thermogenesis in response to diet, the mice become hypersensitive to obesity. If not the mice models are similar to wild-type. As it is well described in the literature that T3 is central for the thermoregulation it is recommended to determine sensitivity to diet obesity at 30 °C or analyze the results obtained at 23 °C with caution.

1.3 Thyroid Hormone and Adaptive Thermogenesis

T3 signaling is necessary for cold response. Hypothyroid patients and rat [5] are hypersensitive to cold. Hypothyroid mice die from hypothermia within 6 h of cold exposure at 4 °C [6]. In rodents the central role of the BAT in this response is well established. BAT has more recently been identified in human adults and its activity increases during cold exposure [7]. In mice cold triggers local T3 production in the BAT following SNS/β-AR activation [8]. The ensuing local increase in T3 concentration [9] is associated with UCP-1 activation and heat production. T3 status seems to correlate [10] with BAT activity in humans. T3 might regulate BAT activity similarly in both species. The role of T3 signaling in adaptive thermogenesis in response to diet has been less studied. At 23 °C mice that do not express, deiodinase type2 (D2), an enzyme necessary to produce T3 locally, develop obesity similarly to control in response to high fat diet. However, they become hypersensitive at 30 °C and express lower levels of UCP-1 in the BAT [11]. This suggests that a local increase of T3 catalyzed by D2 is important to increase energy expenditure.

Here, we will provide two protocols to study adaptive thermogenesis (1) in response to cold (2) in response to high fat diet.

2 Protocols

2.1 Protocol for Studying Adaptive Thermogenesis in Response to Cold

In this protocol, we aim to assess whether mutant animals can appropriately respond to

**Animals and diets:*

Eight 2- to- 3 month-old mice per group from the same sex to avoid variability introduced by different hormonal levels between males and females. C57black6 is the recommended genetic background to perform metabolic studies. Animals are housed two per cage, to limit stress.

Before the experiment mice were housed under recommended conditions and had *ad libidum* access to chow diet (Harlan Teklad TD7901, Madison WI) and water. The western style diet must be stored at 4 °C, and changed every 3 days when in food dispensers at any temperature.

**Required apparatus:*

A cold (4 °C) ventilated room.

IPTT-300 transponders and detector device (DAS-7007S5) (PLEXX, Netherlands).

A scale.

Isoflurane anesthetizer.

**Procedure:*

2.1.1 Implantation of the Transponders

- Mice are briefly anesthetized with 3% isoflurane to allow implantation of the IPTT device using the provided syringe. Transponders should be subcutaneously located on the side of the animal, under the armpit. The easy detection of temperature and its reproducibility should be tested using the detector.
- A week of recovery is recommended before the experiment.

2.1.2 72 h Cold Exposure

- In the morning food and mice are weighed.
- Body temperature is recorded.
- Mice are switched to 4 °C.
- Every 4 h, food amount, mice weight and body temperature are recorded. If at any time an animal loses 20% of its initial weight, it should be removed from the protocol and restored to room temperature.

2.1.3 Notes

Using transponders is much easier and reliable for measuring body temperature, than other technics such as rectal temperature or thermography. Rectal temperature is a really reliable measurement of temperature; however, rapid temperature changes such as the ones observed during cold exposure cannot be detected. Moreover, it needs repetitive handling of the animals that stresses them and affects the measure itself. Thermography is not usable either since it reflects surface temperature that might not mirror body temperature in a cold environment.

Be careful when implanting the transponders. They can easily cross the peritoneum. If so they won't be immobilized and their detection will be difficult. If you can see them it will help the measurements.

At 4 °C, energy expenditure will be high. Make sure, water bottles do not leak. If wet the mice are less likely to survive.

Usually body temperature is mostly maintained at the expanse of higher food intake.

If body temperature drops under 30 °C even though weight does not drastically change, the mouse will certainly die.

2.2 Protocol for Studying Diet-Induced Obesity at Thermoneutrality

2.2.1 Animals and Diets

Eight 2- to- 3 month-old mice per group from the same sex to avoid variability introduced by different hormonal levels between males and females. C57black6 is the recommended genetic background to study metabolic disorders. Animals are housed two per cage, to limit stress.

Chow (Harlan Teklad TD7901, Madison WI) and western style (Harlan Teklad TD88137) diets were used. Before the experiments mice are housed under recommended conditions and have *ad libidum* access to chow diet and water. The western style diet must be stored at 4 °C, and changed every 3 days when in food dispensers at any temperature.

2.2.2 Required Apparatus

A thermoregulated and ventilated mouse incubator, large enough to hold all cages for the experiment, and can insure a 30 °C inside temperature.

A NMR minispec (Brucker, Germany) or any other device that allows an easy measurement of body composition.

A scale.

2.2.3 Procedure

1. Baseline metabolic parameters are measured at animal facility temperature and under regular diet:

Weight and body composition is measured every week. Mice do not need to be anesthetized to measure body composition with minispec.

Food intake is evaluated by weighing food every day. Despite the presence of two animals per cage, an approximation per day, per animal can be calculated from the average of several consecutive days.

Additional tests can be realized such as glucose tolerance (GTT) and insulin tolerance (ITT) tests.

Eventually indirect calorimetry can be performed to determine Energy Expenditure using metabolic cages according to the manufacturer's recommendation.

Before switching the temperature to 30 °C, blood can be drawn for blood chemistry. To this end a small notch is realized using a scalpel, and blood sampled on a heparin coated tube. After centrifugation (15 mn, 3500 rpm, ($2800 \times g$), 21 °C), plasma is obtained and stored at −80 °C.

2.2.4 Metabolic Parameters and Diet-Induced Obesity at 30 °C

The temperature is elevated to 30 °C. One week of acclimation is observed for the mice to recover.

Half of the mice are switched to the high fat diet, half maintained on chow. Mice put on weight and fat very fast.

The procedure is the same as the one described above for the study at room temperature.

2.2.5 Notes

The western style diet must be stored at 4 °C, and changed every 3 days when in food dispensers at any temperature. It contains a lot of fat, if kept in the dispensers, it rots.

Fed a western diet, the fur of mice becomes greasy and important leaking might favor irritation, wounds, and inflammation. To limit this risk add paper in the cage.

At 30 °C, water bottles need to be filled more frequently.

References

1. Lowell BB, Spiegelman BM (2000) Towards a molecular understanding of adaptive thermogenesis. Nature 404(6778):652–660
2. Contreras C, Nogueiras R, Diéguez C et al (2017) Traveling from the hypothalamus to the adipose tissue: the thermogenic pathway. Redox Biol 12:854–863
3. Jimenez M, Léger B, Canola K et al (2002) Beta(1)/beta(2)/beta(3)-adrenoceptor knockout mice are obese and cold-sensitive but have normal lipolytic responses to fasting. FEBS Lett 530(1–3):37–40
4. Enerback S, Jacobsson A, Simpson EM et al (1997) Mice lacking mitochondrial uncoupling protein are cold-sensitive but not obese. Nature 387(6628):90–94
5. Bianco AC, Silva JE (1987) Intracellular conversion of thyroxine to triiodothyronine is required for the optimal thermogenic function of brown adipose tissue. J Clin Invest 79:295–300
6. Christoffolete MA, Linardi CC, de Jesus L et al (2004) Mice with targeted disruption of the Dio2 gene have cold-induced overexpression of the uncoupling protein 1 gene but fail to increase brown adipose tissue lipogenesis and adaptive thermogenesis. Diabetes 53(3):577–584
7. Virtanen KA, Nuutila P (2011) Brown adipose tissue in humans. Curr Opin Lipidol 22(1):49–54
8. Silva JE, Larsen PR (1983) Adrenergic activation of triiodothyronine production in brown adipose tissue. Nature 305(5936):712–713
9. Silva JE Larsen PR (1985) Potential of brown adipose tissue type II thyroxine 5′-deiodinase as a local and systemic source of triiodothyronine in rats. J Clin Invest 76(6):2296–2305
10. Broeders EP, Vijgen GH, Havekes B et al (2016) Thyroid hormone activates brown adipose tissue and increases non-shivering thermogenesis—a cohort study in a group of thyroid carcinoma patients. PLoS One 11(1):e0145049
11. Castillo M, Hall JA, Correa-Medina M, Ueta C et al (2011) Disruption of thyroid hormone activation in type 2 deiodinase knockout mice causes obesity with glucose intolerance and liver steatosis only at thermoneutrality. Diabetes 60(4):1082–1089

Chapter 11

Pharmacological Inhibition of Lysosomal Activity as a Method For Monitoring Thyroid Hormone-induced Autophagic Flux in Mammalian Cells In Vitro

Jin Zhou, Rohit A. Sinha, Ronny Lesmana, Winifred W. Y. Yau, and Paul M. Yen

Abstract

Autophagy is an evolutionarily conserved intracellular catabolic process that is essential for cellular housekeeping and nutrient homeostasis. Recently, we provided evidence that thyroid hormone (TH) is a major inducer of autophagy in mammalian cells. Here, we describe a method for detecting TH-induced autophagic flux in hepatic, muscle, and brown adipocyte cells using lysosomal inhibitor bafilomycin A1 (BafA1) and conventional Western blot techniques.

Key words Thyroid hormone (TH), Autophagic flux, Bafilomycin A1, HepG2, L6 myoblast, Primary brown preadipocytes, LC3B-II

1 Introduction

Macroautophagy (hereafter referred to as autophagy) was first discovered by De Duve in the late 1960s as a glucagon-induced cellular catabolic process in rat liver [1]. Autophagy involves lysosomal-mediated degradation of cellular fuel reserves to provide energy and amino acids during starvation [2]. It is a highly dynamic, multi-step process involving biogenesis of a double membrane structure known as autophagosome, incorporation of cellular cargoes into the autophagosome, and finally fusion of autophagosome with lysosomes forming autolysosomes for degradation of cellular cargo [3]. The molecular mechanisms of autophagy began to be elucidated shortly after the identification of the various genes involved in this process by the pioneering work of the Ohsumi laboratory [4].

Jin Zhou and Rohit A. Sinha contributed equally to this work.

Michelina Plateroti and Jacques Samarut (eds.), *Thyroid Hormone Nuclear Receptor: Methods and Protocols*, Methods in Molecular Biology, vol. 1801, https://doi.org/10.1007/978-1-4939-7902-8_11,
© Springer Science+Business Media, LLC, part of Springer Nature 2018

Thyroid hormone signaling has an intricate cross-talk with the autophagy pathway to co-regulate metabolic homeostasis [5]. Indeed, TH is required for several metabolic processes such as lipolysis, fat oxidation, gluconeogenesis, cellular differentiation, mitochondrial biogenesis and thermogenesis [6]. We have recently shown that TH stimulated autophagy in hepatic and muscle cells both in vitro and in vivo. Autophagy-mediated lipolysis of fat droplets termed "lipophagy," is crucial for hepatic lipolysis and fatty acid oxidation by TH [7, 8]. TH-mediated autophagy also is important for selective mitochondrial clearance via "mitophagy" [9], and protein aggregate removal via "aggrephagy" [10]. Similarly, TH induction of autophagy is necessary for both mitochondrial activity and biogenesis in muscle cells [11] .

The detection of active autophagy flux is complex and requires validation by complementary molecular techniques. Autophagic flux refers to the entire process of completed autophagy including the delivery of cargo to lysosomes via autophagosomes, subsequent breakdown of cargo components, and release of the resulting macromolecules back into the cytosol. Although classical methods such as electron microscopy can demonstrate the presence of autophagosomes, they only capture a single time point during autophagy and thus cannot serve as a dynamic readout of autophagic flux. Thus, several methods employing different protein markers of autophagy have been developed to analyze autophagic flux [12].

One of the most commonly used markers for detecting autophagy is the visualization of the protein levels of LC3B-II in cells by Western blotting or immunofluorescence. LC3B is initially synthesized in an unprocessed form pro-LC3B, which is converted to LC3B-I by ATG4B. LC3B-I is further modified into the PE-conjugated form LC3B-II upon induction of autophagy (Fig. 1) [13]. The presence of LC3B-II suggests the formation of autophagosomes, and its level correlates with autophagosomal number. However, an increase in LC3B-II alone cannot serve as a reliable marker of the autophagic flux in cells since it is degraded during the latter stages of autophagy (Fig. 1). Thus, it is possible that a lysosomal block also could lead to an increase in LC3B-II. Similarly, SQSTM1/p62 protein abundance is another commonly used marker to measure autophagic flux. p62 serves as a link between LC3B-II and ubiquitinated substrates which become incorporated into the completed autophagosome and are degraded in autolysosomes, so it can serve as a readout of autophagic degradation[14]. However, several transcription factors may increase the transcription of p62 upon autophagy induction, so an increase in synthesis of p62 may complicate the interpretation of p62 protein levels when used alone as a readout of autophagic flux. Thus, a concomitant increase in LC3B-II and a decrease in SQSTM1/p62 suggest increased autophagic flux. However, there

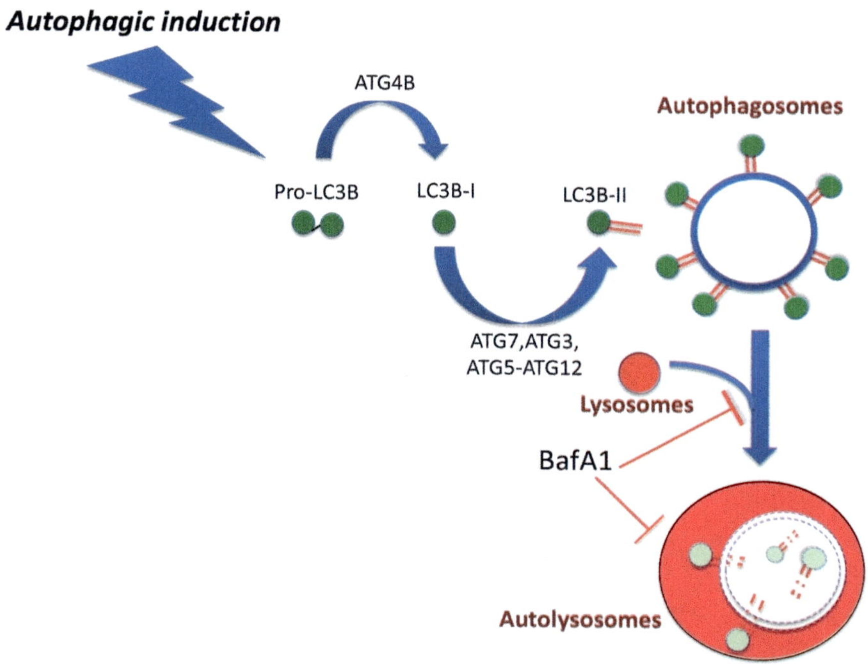

Fig. 1 The turnover of LC3B-II. This schematic sketch shows the dynamics of LC3B-II during the process of autophagy. Upon autophagy induction due to either energy or nutrient depletion, the protease ATG4B cleaves Pro-LC3B to form mature LC3B-I. Concomitantly, LC3B-I that is cytosolic is conjugated to phosphatidylethanolamine to form LC3-phosphatidylethanolamine conjugate (LC3B-II) via other ATG family proteins, which is recruited to autophagosomal membranes. Autophagosomes fuse with lysosomes to form autolysosomes, and intra-autophagosomal components including LC3B-II are degraded by lysosomal hydrolases. Therefore, the levels of LC3B-II detected inside the cells are a reflection of the net balance between the rate at which the autophagosomes are being made and the rate at which they are being degraded during autolysosome formation. Use of BafA1 inhibits both the autophagosome-lysosome fusion as well as lysosomal acidification serves to block the completion of the autophagic flux in cells

are situations such as rapid autophagic flux when both are unchanged or even decreased.

In this chapter, we will describe a widely accepted technique for assessing autophagic flux that relies on measuring the abundance of LC3B-II protein levels in the absence or presence of a lysosomal inhibitor bafilomycin A1 (BafA1) using Western blotting. This method relies on the pharmacological inhibition of either lysosomal-autophagosomal fusion or lysosomal acidification by BafA1 [15]. Besides BafA1, there are a variety of pharmacological agents that can be used to block these processes such as chloroquine, lysosomal protease inhibitors, and NH_4Cl (Fig. 1). Using a lysosomal inhibitor such as BafA1 in both TH-treated and non-treated control cells helps to determine whether TH treatment causes an increase in autophagosome formation or a decrease in autophagic degradation.

2 Materials

2.1 Cell Culture Medium and Reagents

1. DMEM, high glucose, HEPES, no phenol red.
2. Fetal bovine serum (FBS).
3. Horse serum.
4. New born calf serum.
5. 0.3% *Dowex* 1-X10 ion exchange resin.
6. Dowex stripped horse serum.
7. Dowex stripped FBS.
8. Penicillin-Streptomycin (p/s) Solution (100×).
9. DMSO.
10. BafA1.
11. Collagenase.
12. Gentamycin.
13. Isobutylmethylxanthine (IBMX).
14. Dexamethasone.
15. Insulin.
16. Rosiglitazone.
17. Triiodothyronine (T_3).
18. SodiuM bicarbonate.
19. Culture flasks and plates.

2.2 SDS-PAGE and Transfer

1. Resolving gel buffer: 1.5 M Tris–HCl, pH 8.8. Weigh 90.85 g Tris–HCl to cylinder. Add water to a volume of 400 ml. Mix and adjust pH to pH 8.8 with HCl. Make up to 500 ml with water. Store at room temperature.
2. Stacking gel buffer: 0.5 M Tris–HCl, pH 6.8. Weigh 30.3 g Tris–HCl and prepare a 500 ml solution (pH 6.8) as in previous step. Store at room temperature.
3. 30% Acrylamide/Bis Solution, 29:1.
4. Ammonium persulfate: 10 % solution in water.
5. *N,N,N′,N′*-tetramethylethylenediamine (TEMED).
6. 10× Tris/Glycine/SDS running buffer: 10× premixed electrophoresis buffer, contains 25 mM Tris, 192 mM glycine, 0.1% SDS, pH 8.3 following dilution to 1× with water.
7. Transfer buffer: 192 mM glycine, 25 mM Tris, and 10% methanol.
8. 2× Laemmli Sample Buffer (BIO-RAD 1610737).
9. Dithiothreitol (DTT).
10. PVDF transfer membrane.

11. Molecular weight marker.
12. Ponceau S solution.

2.3 Immunoblotting

1. Nonfat milk powder.
2. PBS containing 0.1% Tween 20 (PBST).
3. Blocking buffer: 5% milk in PBST.
4. Primary antibody dilution buffer: 1% BSA in PBST.
5. Secondary antibody dilution buffer: 5% milk in PBST.
6. Anti-LC3B antibody (Cell Signaling, Cat #2775).
7. Anti-β-actin antibody.
8. Mouse anti-rabbit IgG-HRP.
9. Chicken anti-mouse IgG-HRP.
10. Enhanced chemiluminescent substrate ECL™ Select Western Blotting Detection Reagent (GE Healthcare, RPN2235).

3 Methods

Determination of Autophagic Flux Using BafA1 in vitro

We have used the following protocols to study autophagic flux in HepG2-TRβ, L6 myoblasts, and primary brown preadipocytes, respectively. We perform in vitro BafA1 treatments to assess autophagic flux in each cell line of interest. We have found that the efficiency of BafA1-induced lysosomal inhibition varies in different cell types. Therefore, we highly recommend performing a dose response/time course of BafA1 in the cell line of interest in order to identify the optimal dose and duration of BafA1 treatment.

3.1 Cell Culture and BafA1 Treatment

3.1.1 HepG2 Cells Over-Expressing THRα or THRβ

Culture of HepG2 Cells Over-Expressing THRα or THRβ

1. Maintain HepG2 cells with stably transfected human *THRα/B* at 37 °C in DMEM supplemented with 10% FBS as described previously [8].
2. Plate cells in 6- or 12-well plates followed by culture in TH-depleted media (DMEM + 10% Dowex-stripped FBS) [8], for 2 days before adding T_3 (100 nM) for 48 h.
3. For autophagic flux analysis, add 100 nM BafA1 to the culture medium 6 h before protein isolation.

3.1.2 L6 Myoblasts

Culture of L6 Myoblasts

L6 cells are commercially available from American Type Culture Collection (ATCC, Mannassas, VA, USA). Please note that variable handling of this cell line can select for cells with different growth kinetics or poor differentiation. To avoid these problems, please make sure that low passage L6 Stocks are used initially and low-passage cells (<passage 10) for each experiment in order to maintain original cell characteristics.

Maintain L6 cells in L6 proliferation medium (DMEM, 15% FBS, p/s, and 1mM sodium bicarbonate). The cells are grown on Nunc delta surface plastic culture flasks. Cells are cultured at 37 °C in 20 ml of proliferation medium in a humidified incubator with 5% CO_2 for 48 h. Cell density is maintained at approximately 70% density (typically found 72 h after initial plating at 10% density in 75 cm^2 flasks).

1. Remove and discard culture medium.
2. Briefly rinse the cells with Phosphate-Buffered Saline (PBS) × 2 times.
3. Add 2.0 ml of Trypsin EDTA, swirl to cover all surface cells and incubate for 2 min in 37 °C.
4. Check under a microscope whether cells are detached from the flask surface.
5. Resuspend cells with 5 ml fresh warm medium and centrifuge them for 3 min with 1300 × *g*.
6. Discard the supernatant and resuspend pellets using 5 ml medium.
7. Add appropriate aliquots of the cell suspension to new culture vessels.
8. Incubate cultures at 37 °C.

Differentiation and Treatment of L6 Cells

1. Grow cells to 40–50% density in 12- or 24-well plates with proliferation medium.
2. Check growth of cells after 24 h (make sure that cells are attached to the bottom of plate).
3. Change medium to L6 differentiation medium (DMEM, 2% Horse serum, p/s, and 1 mM sodium bicarbonate). Cells typically reach 90–100% confluency after 48 h.
4. Incubate the cells in differentiation medium for 4–7 days.
5. Change medium for every 48 h.
6. Replace L6 differential medium with L6 treatment medium (DMEM, 2% Dowex stripped horse serum, p/s, and 1 mM sodium bicarbonate) after cells are fully differentiated, and culture the cells for 2 days. Cells are then treated in the presence or absence of 10 nM T_3 for 24–48 h [11]. For autophagic flux analysis, L6 cells are co-treated with 30 nM of BafA1 for 24 h in the presence or absence of 100 nM T_3 for 24 h.

3.1.3 Primary Brown Preadipocytes

Primary Culture of Brown Preadipocytes

1. Harvest interscapular brown adipose tissue from five to six mice (2–3 weeks old).
2. Mince tissue using scissors in a 10 cm plate with collagenase (300 μl).

3. Transfer the tissue to a 50 ml tube and add another 2.7 ml of collagenase.
4. Put the tube on a thermal shaker at 37 °C for 30–40 min. Vortex for 10 s every 5 min.
5. Add 10 ml plain DMEM to the digestion mixture and filter through a 100 μm mesh into a 15 ml tube.
6. Centrifuge at 330×*g* at 4 °C for 5 min. Remove the supernatant.
7. Resuspend the pellet with 10 ml of plain DMEM and filter through a 100 μm mesh.
8. Centrifuge at 330×*g* at 4 °C for 5 min. Remove the supernatant.
9. Resuspend the pellet with 10 ml of culture medium (DMEM containing 10% new born calf serum, gentamycin and p/s) and transfer to a new 10 cm culture plate.

Differentiation and Treatment of Brown Adipocytes

1. Seed brown preadipocytes onto 12-well plates for differentiation.
2. Initiate adipogenesis on Day 0 with induction medium (DMEM containing 10% FBS, 0.5 mM IBMX, 1 μM dexamethasone, 172 nM insulin, 1 nM T_3 and 1 μM rosiglitazone).
3. Change medium to culture medium supplemented with 172 nM insulin on Day 2.
4. Change medium to T_3-depleted culture medium on Day 4.
5. On Day 6, mature brown adipocytes are ready for experimental treatment. Incubate cells with T_3-depleted culture medium containing 10 nM T_3 for 24 h. For autophagic flux analysis, add 50 nM BafA1 to the medium 6 h before harvest.

3.2 LC3B SDS-PAGE and Western Blot Analysis

3.2.1 Preparation of Protein Samples from Cultured Cells

Make 1× cell lysis buffer by diluting 2× Laemmli sample buffer with H_2O and 1 M DTT. 100 μl/well is then added to 12-well plates. Of note, T_3 treatment usually does not affect cell number under our experimental conditions. If the cell number changes during the experiment, we recommend lysing the cells with RIPA buffer followed by measurement of protein concentrations, and adjusting the protein concentration to the same levels before mixing the protein samples with 2× Laemmli sample buffer.

1. Prepare 1× cell lysis buffer by adding 2× Laemmli sample buffer, H_2O, 1 M DTT in a volume distribution of 5:4:1.
2. Rinse the cells with PBS. Lyse the cells by adding 100 μl 1× cell lysis buffer. Stir the lysis buffer using pipet tips until the lysis buffer becomes very viscous. Try to transfer the entire contents into 1.5 ml Eppendorf tube.

3. Boil the samples at 100 °C for 10 min. Spin down the samples for 30 s. Cool down the samples to room temperature. Protein samples are now ready for SDS-PAGE or can be stored up to one month at −30 °C.

3.2.2 LC3B SDS-PAGE and Immunoblot Analysis

14% resolving gel is recommended for LC3B SDS-PAGE since it provides good separation of the LC3B-I and LC3B-II bands.

1. Load 10 μl of protein sample using gel loading tip. Run PAGE using 1× SDS running buffer.
2. Soak the PVDF membrane with 100% methanol for approximately 1 min while the gel is running. The methanol can be reused again for this same purpose. Rinse the membrane with distilled water. Soak the membrane in 1× transfer buffer (Transfer buffer: 192 mM glycine, 25 mM Tris, and 10% methanol) in a separate container. Keep the PVDF membranes and filter papers refrigerated.
3. Stop the electrophoresis when the dye front reaches the very bottom of the gel.
4. Transfer onto a PVDF membrane using constant voltage at 25 V for 30 min at room temperature using Trans-Blot® Turbo™ Blotting System (Bio-Rad).
5. Stain the membrane with Ponceau S solution to visualize the transfer efficiency. Ponceau S binding to protein is reversible and can later be washed away in blocking buffer.
6. Rinse the membrane with distilled water to remove excess Ponceau S solution.
7. Block in blocking buffer for 1 h at room temperature on a shaker.
8. Rinse the membrane with 1× PBST.
9. Prepare the rabbit-anti-LC3B and mouse-anti-actin primary antibodies by diluting them 1:1000 in 5 ml of 1% BSA in PBST in a 50 ml Falcon tube.
10. Incubate the membrane with primary antibodies on Roller Mixer (Ratek Laboratory) inside the cold room.
11. Wash the membrane with PBST for 10 min at room temperature ×3 on the shaker.
12. Prepare secondary antibody by diluting it 1:5000 in 5% milk in PBST. Incubate the membrane in secondary antibody solution for 1 h at room temperature on the shaker.
13. Wash the membrane with PBST for 10 min at room temperature ×3 on the shaker.
14. Develop the blots with Enhanced Chemiluminescent Substrate ECL™ Select Blotting Detection Reagent using Gel Doc™ XR and Gel Documentation System (BIO-RAD).

15. Quantify LC3B-II, and actin protein bands using ImageJ and report the data as LC3B-II/actin.

4 Interpretation of Results

The level of cellular LC3B-II at any given point in time is a snapshot based upon the rate of autophagosome biogenesis and degradation after lysosomal fusion. Therefore, the use of lysosomal inhibitors to analyze autophagic flux enables one to ascertain the rate at which autophagosome formation is being induced by a drug/hormone or nutritional state when its degradation is stopped. In other words, any increase in the levels of LC3B-II that occurs in the presence of a lysosomal inhibitor is due to autophagy induction or autophagosome biogenesis prior to, and during, the period when lysosomal activity was blocked. The following are some potential scenarios. In Fig. 2 scenario 1: the relative increase in LC3B-II/ACTB in treated cells plus BafA1 compared to treated cells alone is greater than the increase in LC3B-II/ACTB in control cells (untreated) plus BafA1 compared to control cells alone, suggests that there is an increase in autophagic flux. An increase in LC3B-II levels after lysosomal activity is inhibited by BafA1 can only be explained by increased autophagy induction. In cases, where LC3B-II levels are unchanged or even decreased, an increase in LC3B-II level after BafA1 treatment can be due to high autophagic flux, particularly if p62 levels are decreased without Bafl treatment. On the contrary, in Fig. 2 scenario 2: the relative increase in LC3B-II/ACTB in treated cells plus BafA1 compared to treated

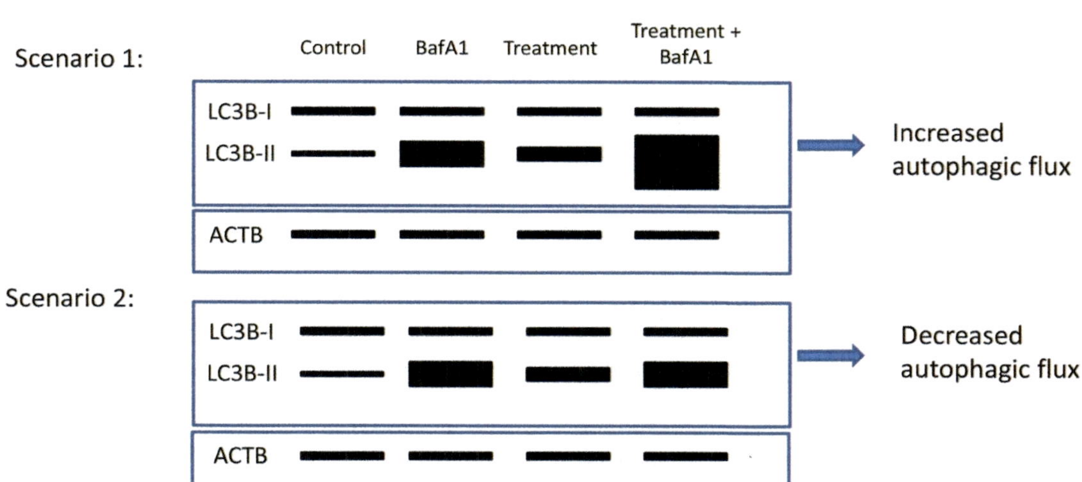

Fig. 2 Interpretation of the LC3B-II levels as a readout of autophagic flux. This figure represents a Western blot result showing the levels of LC3B-I, LC3-II and β-Actin (ACTB) upon probing with anti-LC3B and anti-ACTB antibodies

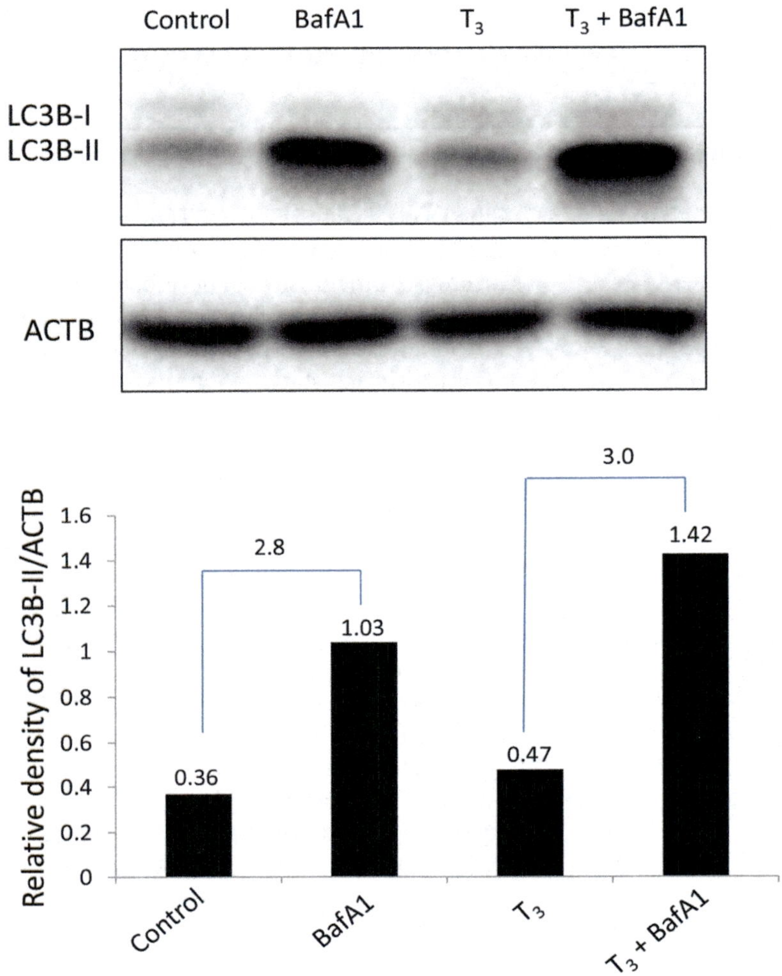

Fig. 3 T_3 increases autophagic flux in vitro. A representative Western blot and densitometric analysis of LC3B-II/ACTB ratio showing increased autophagic flux in T_3 treated HepG2 overexpressing THRβ

cells alone is less than the increase in LC3B-II/ACTB in control cells + BafA1 compared to control cells alone, it is likely that autophagosome accumulation (LC3B-II) observed after treatment is due to inhibition of autophagic degradation rather than increased autophagosome synthesis. In this case, blocking autophagosome-lysosome fusion or lysosomal function decreases net autophagic flux. A real example of scenario 1 is shown in Fig. 3, where T_3 increases autophagic flux as described in Fig. 1.

When interpreting initial autophagy results, it is critical to ascertain whether an increase in LC3B-II is due to increased biosynthesis or degradation of autophagosomes, and reflects a genuine increase in autophagic flux. While an increase in LC3B-II with a concomitant decrease in p62 levels is highly suggestive of increased autophagy, formal confirmation using a lysosomal inhibitor such as BafA1 is necessary to prove autophagic flux. In general, it is highly recommended to monitor autophagy flux by using at least two techniques. Detection of autophagic flux by transfecting tandem mRFP-GFP-LC3 vector followed by fluorescent imaging-based analysis of yellow and red fluorescence in live or fixed cells are widely used techniques [12, 16]. Another probe to detect autophagic flux both in vitro and in vivo has been developed recently. This GFP-LC3-RFP-LC3ΔG probe can be cleaved by endogenous ATG4 protease into equimolar amounts of GFP-LC3 and RFP-LC3ΔG. GFP-LC3 is degraded by autophagic flux, whereas RFP-LC3ΔG remains in the cytosol, serving as an internal control. Thus, autophagic flux can be estimated by calculating the GFP/RFP ratio, which is inversely correlated with autophagic flux [17]. Additionally, increases in lysosomal activity can be monitored by using acridine orange[18] or monodansylcadaverine staining[19].

Acknowledgement

This work was supported by Duke-NUS Medical School Faculty Funds (to P.M.Y.), Singapore National Medical Research Council grant NMRC/CSA/0054 /2013 (to P.M.Y.), NMRC/CIRG/1340/2012 (to P.M.Y.), NMRC/CIRG/1457/2016 (to P.M.Y.), Wellcome Trust/DBT India Alliance Fellowship (IA/I/16/2/502691) to R.A.S.

References

1. Deter RL, Baudhuin P, De Duve C (1967) Participation of lysosomes in cellular autophagy induced in rat liver by glucagon. J Cell Biol 35:C11–C16
2. Boya P, Reggiori F, Codogno P (2013) Emerging regulation and functions of autophagy. Nat Cell Biol 15:713–720
3. Feng Y, He D, Yao Z, Klionsky DJ (2014) The machinery of macroautophagy. Cell Res 24:24–41
4. Tsukada M, Ohsumi Y (1993) Isolation and characterization of autophagy-defective mutants of Saccharomyces cerevisiae. FEBS Lett 333:169–174
5. Sinha RA, Singh BK, Yen PM (2016) Reciprocal crosstalk between autophagic and endocrine signaling in metabolic homeostasis. Endocr Rev 2016:er20161103
6. Brent GA (2012) Mechanisms of thyroid hormone action. J Clin Invest 122:3035–3043
7. Tseng YH, Ke PY, Liao CJ, Wu SM, Chi HC, Tsai CY, Chen CY, Lin YH, Lin KH (2014) Chromosome 19 open reading frame 80 is upregulated by thyroid hormone and modulates autophagy and lipid metabolism. Autophagy 10:20–31
8. Sinha RA, You SH, Zhou J, Siddique MM, Bay BH, Zhu X, Privalsky ML, Cheng SY, Stevens RD, Summers SA, Newgard CB, Lazar MA, Yen PM (2012) Thyroid hormone stimulates hepatic lipid catabolism via activation of autophagy. J Clin Invest 122:2428–2438
9. Sinha RA, Singh BK, Zhou J, Wu Y, Farah BL, Ohba K, Lesmana R, Gooding J, Bay BH, Yen PM (2015) Thyroid hormone induction of mitochondrial activity is coupled to mitophagy via ROS-AMPK-ULK1 signaling. Autophagy 11:1341–1357
10. Chi HC, Chen SL, Tsai CY, Chuang WY, Huang YH, Tsai MM, Wu SM, Sun CP, Yeh

CT, Lin KH (2016) Thyroid hormone suppresses hepatocarcinogenesis via DAPK2 and SQSTM1-dependent selective autophagy. Autophagy 12:2271–2285

11. Lesmana R, Sinha RA, Singh BK, Zhou J, Ohba K, Wu Y, Yau WW, Bay BH, Yen PM (2016) Thyroid hormone stimulation of autophagy is essential for mitochondrial biogenesis and activity in skeletal muscle. Endocrinology 157:23–38
12. Klionsky DJ et al (2016) Guidelines for the use and interpretation of assays for monitoring autophagy (3rd edition). Autophagy 12:1–222
13. Hemelaar J, Lelyveld VS, Kessler BM, Ploegh HL (2003) A single protease, Apg4B, is specific for the autophagy-related ubiquitin-like proteins GATE-16, MAP1-LC3, GABARAP, and Apg8L. J Biol Chem 278:51841–51850
14. Bjorkoy G, Lamark T, Brech A, Outzen H, Perander M, Overvatn A, Stenmark H, Johansen T (2005) p62/SQSTM1 forms protein aggregates degraded by autophagy and has a protective effect on huntingtin-induced cell death. J Cell Biol 171:603–614
15. Mizushima N, Yoshimori T (2007) How to interpret LC3 immunoblotting. Autophagy 3:542–545
16. Kimura S, Noda T, Yoshimori T (2007) Dissection of the autophagosome maturation process by a novel reporter protein, tandem fluorescent-tagged LC3. Autophagy 3:452–460
17. Kaizuka T, Morishita H, Hama Y, Tsukamoto S, Matsui T, Toyota Y, Kodama A, Ishihara T, Mizushima T, Mizushima N (2016) An autophagic flux probe that releases an internal control. Mol Cell 64:835–849
18. Paglin S, Hollister T, Delohery T, Hackett N, McMahill M, Sphicas E, Domingo D, Yahalom J (2001) A novel response of cancer cells to radiation involves autophagy and formation of acidic vesicles. Cancer Res 61:439–444
19. Munafo DB, Colombo MI (2001) A novel assay to study autophagy: regulation of autophagosome vacuole size by amino acid deprivation. J Cell Sci 114:3619–3629

Chapter 12

Analysis of Physiological Responses to Thyroid Hormones and Their Receptors in Bone

J. H. Duncan Bassett and Graham R. Williams

Abstract

Thyroid hormone has profound effects on skeletal development and adult bone maintenance. Here, we review the current literature concerning thyroid hormone action in bone and cartilage in relation to human disease and animal models. We describe state-of-the-art imaging and biomechanical methods used to determine structural and functional parameters in the skeletal phenotyping of mouse models.

Key words Thyroid hormone, Thyroid hormone receptor, Bone, Skeleton, Cartilage, Chondrocyte, Osteoblast, Osteoclast, X-ray microradiography, Micro-CT, Biomechanical testing, Osteoporosis

1 Introduction

1.1 Skeletal Physiology

Four principal cell types are involved in skeletal development and the maintenance of adult bone mass. In early embryogenesis mesenchyme precursors condense and define a template for the future skeleton. These cells differentiate into chondrocytes that proliferate and secrete a matrix containing aggrecan, elastin, and type II collagen to form a cartilage model of the skeletal element [1]. Osteoclasts are polarized multinucleated cells that resorb bone mineral and matrix and are derived from fusion of mononuclear–myeloid precursors of the monocyte-macrophage lineage [2]. Bone-forming osteoblasts originate from mesenchymal stem cells and secrete and mineralize a matrix rich in type I collagen. Following bone formation, osteoblasts differentiate into bone lining cells or become entombed within bone matrix to develop into osteocytes [3]. Osteocytes comprise 95% of bone cells and develop dendritic processes that ramify though networks of canaliculi and sense fluid shear stresses, communicating via gap junctions [4].

Michelina Plateroti and Jacques Samarut (eds.), *Thyroid Hormone Nuclear Receptor: Methods and Protocols*, Methods in Molecular Biology, vol. 1801, https://doi.org/10.1007/978-1-4939-7902-8_12,
© Springer Science+Business Media, LLC, part of Springer Nature 2018

1.2 Development

1.2.1 Intramembranous Ossification

The flat bones of the skull form by intramembranous ossification (Fig. 1). Mesenchyme progenitors condense and differentiate directly to osteoblasts, which secrete an osteoid matrix that mineralizes to shape an ossification center. Surrounding mesenchyme organizes to form the periosteum, and inner surface cells differentiate into lining osteoblasts. Progressive bone formation results in fusion of adjacent ossification centers [5].

1.2.2 Endochondral Ossification

Long bones form on a cartilage scaffold via endochondral ossification (Fig. 2) [1], in which mesenchyme precursors condense and differentiate into chondrocytes to form a cartilage template.

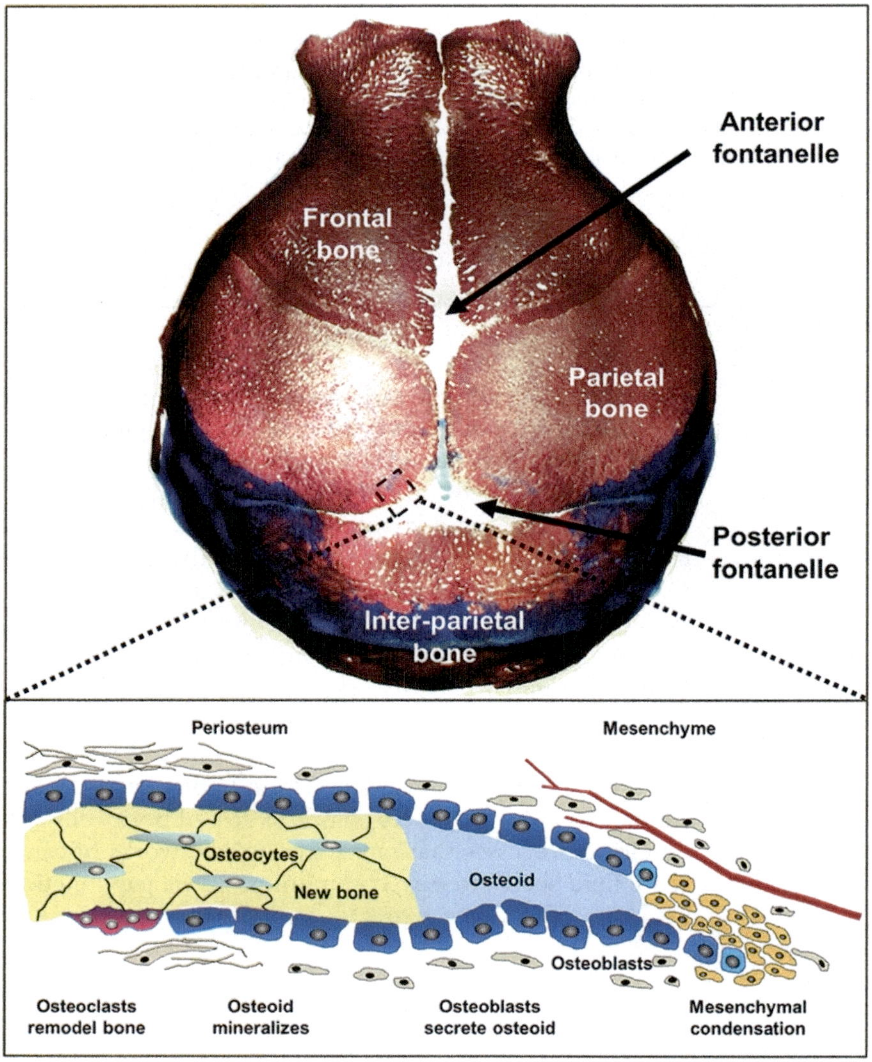

Fig. 1 *Intramembranous ossification.* Skull vault from neonatal mouse at post-natal day 1 stained with alizarin red (bone) and alcian blue (cartilage) showing sutures and fontanelles. Lower panel shows a schematic representation of the cells and processes involved in intramembranous bone formation

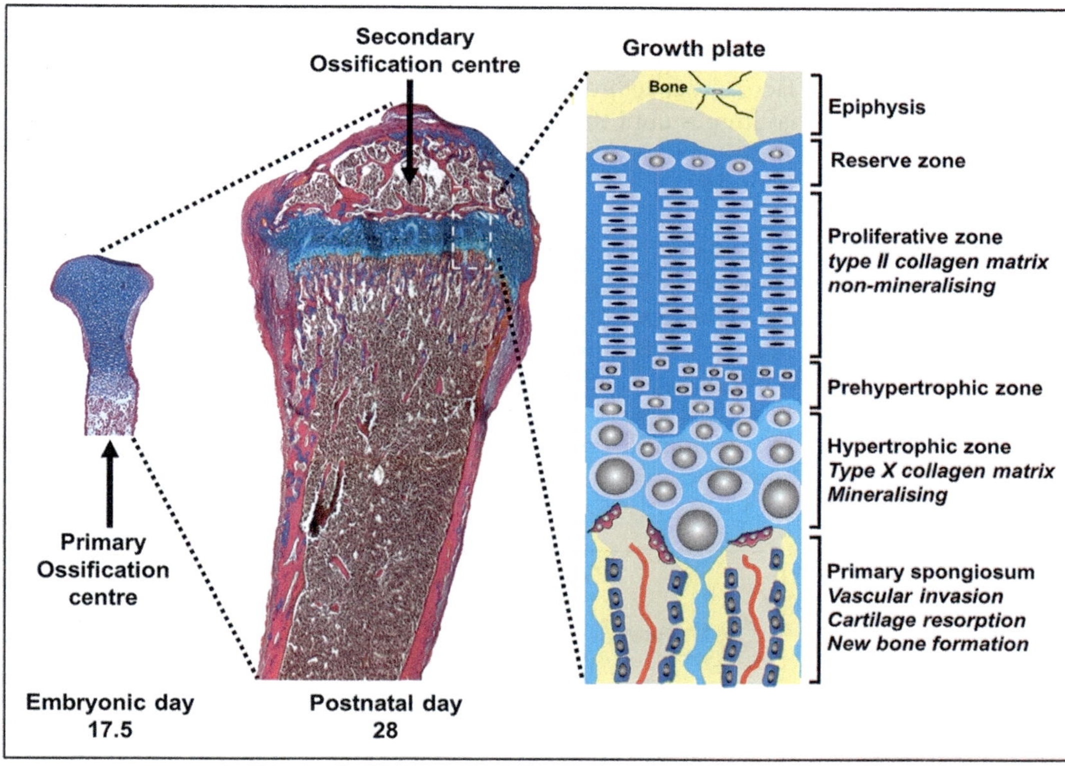

Fig. 2 *Endochondral ossification.* Sections of proximal tibial from mice at embryonic day 17.5 and post-natal day 28 stained with alcian blue (cartilage) and Van Gieson (bone matrix, red) to show primary and secondary ossification centers and the growth plate. On the right is a schematic representation of the growth plate between the epiphysis (above) and primary spongiosum (below) showing the chondrocyte zones

A program of chondrocyte proliferation, differentiation, and apoptosis leads to cartilage mineralization. Migration of osteoblasts via invading blood vessels facilitates replacement of mineralized cartilage with trabecular bone. Mesenchyme precursors in the perichondrium differentiate to osteoblasts and form a cortical bone collar. Secondary ossification centers develop at the ends of long bones and are separated from the primary ossification center by the growth plates.

1.2.3 Growth and Bone Maturation

The growth plates comprise reserve, proliferative, and hypertrophic zones (Fig. 2) [1]. Reserve zone chondrocytes progress to the proliferative zone, form columns, and secrete type II collagen. Hypertrophic chondrocytes secrete type X collagen and increase in volume by tenfold [6, 7] until apoptosis releases angiogenic factors that stimulate vascular invasion, remodeling of calcified cartilage and trabecular bone formation. This ordered process is the mechanism of linear growth [1]. Endochondral ossification is regulated by a local feedback loop involving Indian hedgehog (IHH) and parathyroid hormone-related protein (PTHrP) [1], and other factors including systemic hormones, and cytokines and growth factors that act in a paracrine and autocrine manner. Linear growth continues until the growth plates fuse at puberty, but bone mineralization and consolidation of bone mass continue into early adulthood [8].

1.3 Bone Turnover

Integrity of the adult skeleton is maintained by the bone remodeling cycle [9] (Fig. 3), in which osteoclast and osteoblast activities are orchestrated by osteocytes [10]. Osteocytes secrete transforming growth factor-β (TGFβ) and sclerostin, resulting in basal inhibition of osteoclastogenesis and osteoblastic bone formation. Mechanical load or micro-fractures decrease local TGFβ levels [11], leading to recruitment of osteoclast progenitors. Osteocytes and bone lining cells secrete macrophage colony stimulating factor

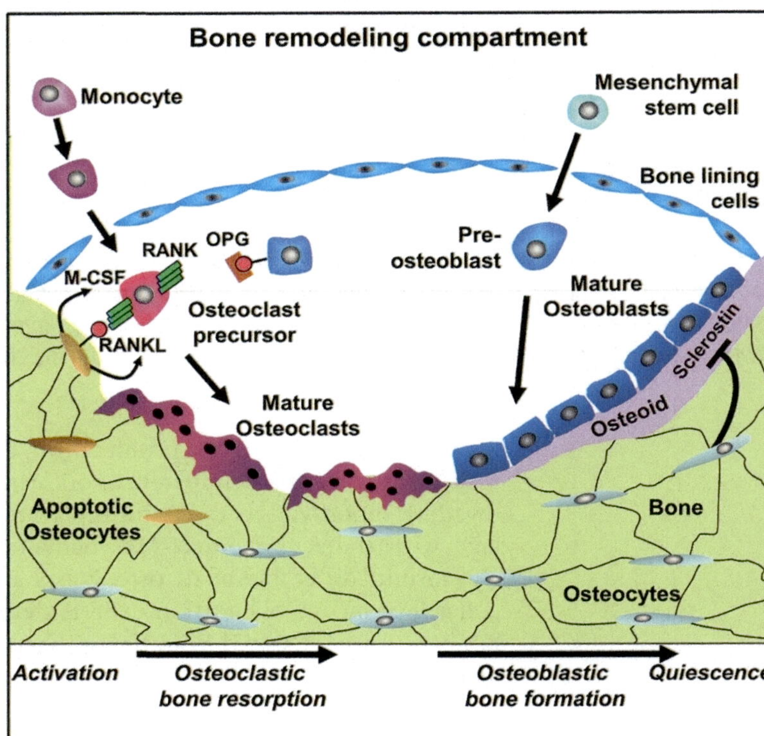

Fig. 3 *Bone remodeling compartment.* The bone remodeling cycle, in which osteoclast and osteoblast activities are orchestrated by osteocytes. Bone remodeling follows changes in mechanical load, micro-fractures, or exposure to systemic or paracrine factors. Monocyte/macrophage precursors differentiate to mature osteoclasts and resorb bone. Osteoclastogenesis is induced by M-CSF and RANKL but inhibited by OPG. Following bone resorption, osteoblast progenitor cells are recruited to the site of bone resorption, where they synthesize osteoid and mineralize new bone to repair the defect

(M-CSF) and receptor activator of nuclear factor κB ligand (RANKL), the cytokines required for osteoclastogenesis [9]. Osteoblasts and bone marrow stromal cells also secrete osteoprotegerin (OPG), a decoy receptor for RANKL that acts as the physiological inhibitor of RANKL signaling [12]. Thus, the RANKL:OPG ratio determines the degree of osteoclast differentiation and bone resorption. During bone resorption, paracrine factors released from degraded bone matrix recruit osteoblasts that initiate new bone formation. Canonical Wnt signaling is the master regulator of osteogenesis [13], and physiological control of the pathway is mediated by the osteocyte, which secretes soluble factors (sclerostin, Dickkopf1-related protein 1 (DKK1), and secreted frizzled related protein 1 (SFRP1)) that inhibit interactions between Wnt ligands and their receptor and co-receptor [10, 14]. Some osteoblasts become embedded within newly formed bone and differentiate to osteocytes. Secretion of Wnt inhibitors by these osteocytes leads to cessation of bone formation and a return to the quiescent state in which osteoblasts become bone-lining cells [10, 14]. Thus, osteocytes are mechano-sensors that control bone turnover through their regulation of osteoclasts via the RANKL/RANK pathway and osteoblasts via modulation of Wnt signaling. This cycle of coupled bone resorption and formation enables the skeleton to repair old or damaged bone, react to changes in mechanical stress, and respond rapidly to the demands of mineral homeostasis.

1.4 Effects of Altered Thyroid Status on the Human Skeleton

1.4.1 Variation Within the Euthyroid Reference Range

Variations in thyroid status within the normal euthyroid range are associated with changes in bone mineral density (BMD), especially in post-menopausal women [15, 16]. Further studies also demonstrated an association with fracture susceptibility [15, 17, 18]. Overall, thyroid status at the upper end of the euthyroid reference range is associated with lower BMD and an increased risk of fracture in post-menopausal women.

1.4.2 Subclinical Thyroid Disease

Studies investigating the relationship between subclinical hypothyroidism, defined as an elevated TSH concentration in the presence or normal levels of T3 and T4, and BMD or fracture have been inconsistent, although a large individual participant meta-analysis found no association between subclinical hypothyroidism and fracture risk [19]. Similarly, studies of subclinical hyperthyroidism, defined as a suppressed TSH concentration in the presence or normal levels of T3 and T4, have been inconsistent. Nevertheless, the recent large individual participant meta-analysis demonstrated an increased risk of fracture in individuals with sub-clinical hyperthyroidism [19].

1.5 Hypothyroidism

1.5.1 Children

Congenital and juvenile acquired hypothyroidism delay bone development and cause growth retardation. In severe cases (Fig. 4) there is post-natal growth arrest with a complex skeletal dysplasia

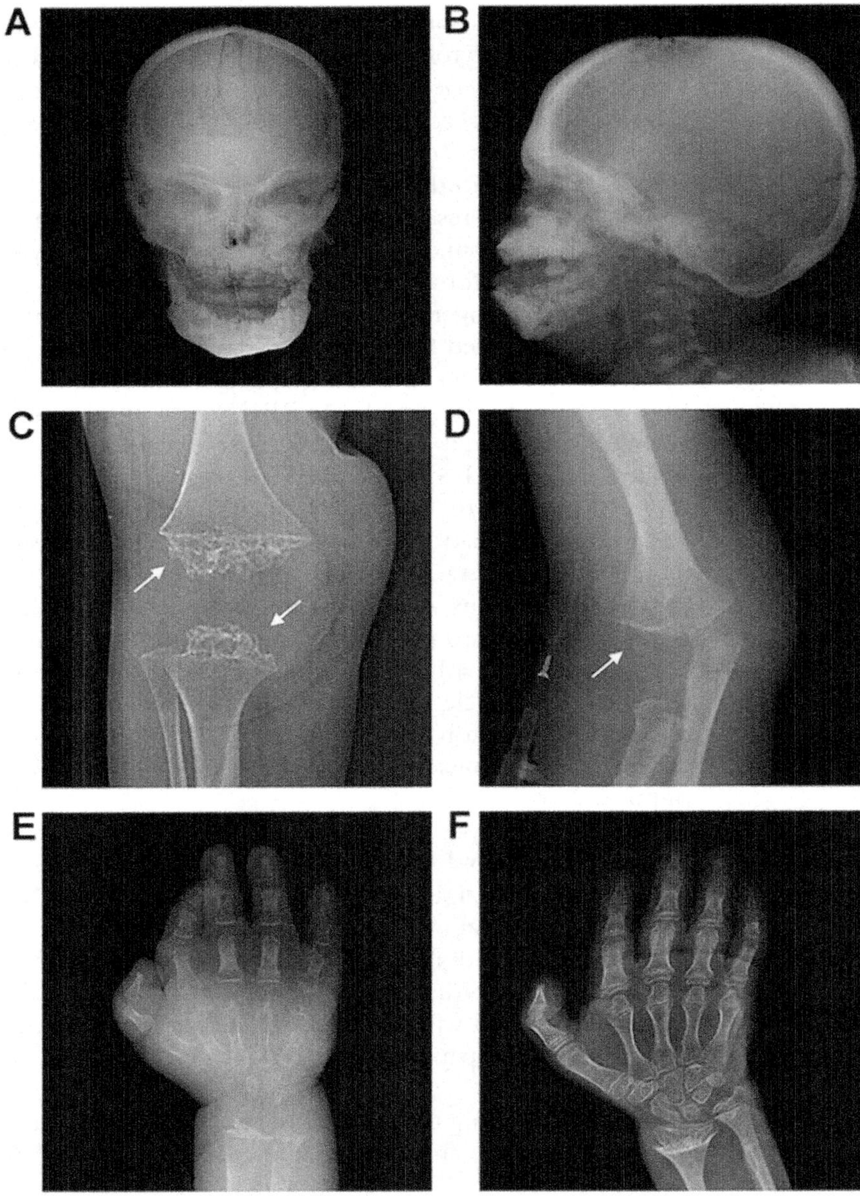

Fig. 4 *Skeletal consequences of severe juvenile hypothyroidism.* X-rays from a 36-year-old female showing effects of untreated congenital hypothyroidism on bone development (images provided by Dr. Jonathan LoPresti, Keck School of Medicine, University of Southern California). (**a**, **b**) Skull X-rays showing patent sutures and fontanelles, delayed tooth eruption and thickening of the skull vault. (**c**) Knee X-ray showing epiphyseal dysgenesis with grossly delayed formation of secondary ossification centers (arrows). (**d**) Elbow X-ray demonstrating gross epiphyseal dysgenesis with failure of mineralization at the proximal ulna and distal humerus (arrow). (**e**) Hand X-ray demonstrating a bone age of 14 months, which represents a delay in maturation of 34 years 10 months. (**f**) Hand X-ray showing the effect of 18 months treatment with T4, which resulted in advancement of bone age by 8 years. This demonstrates classic "catch-up growth" and accelerated skeletal development in response to thyroid hormone replacement

that includes broad flat nasal bridge, hypertelorism, patent fontanelles, scoliosis, vertebral immaturity, absence of ossification centers, and congenital hip dislocation [20]. Thyroid replacement induces rapid "catch up" growth and accelerated skeletal maturation, demonstrating exquisite sensitivity of the developing skeleton to thyroid hormones [21].

1.5.2 Adults

Histomorphometry studies of the effects of hypothyroidism showed a state of low bone turnover with decreased osteoclast and osteoblast activities, resulting in a net increase in mineralization without a major change in bone volume [22]. However, determination of clinical effects of hypothyroidism on bone mass would require long-term follow-up of untreated patients, and data are not available. Studies of the effect of prolonged hypothyroidism on fracture susceptibility are confounded by treatment and study heterogeneity, but data suggest that hypothyroidism alone is unlikely to be related to fracture risk, whereas long-term supraphysiological replacement with thyroid hormones may result in an increased risk of fracture [23, 24].

1.6 Hyperthyroidism

1.6.1 Children

In contrast, juvenile thyrotoxicosis accelerates skeletal development and growth, although advanced bone age results in premature cessation of growth and persistent short stature. In severe cases in young children, craniosynostosis can result from early closure of the cranial sutures [25], while maternal hyperthyroidism is also a risk factor for craniosynostosis [26]. These data further demonstrate the sensitivity of the developing skeleton to thyroid hormones.

1.6.2 Adults

Severe osteoporosis in thyrotoxicosis is rare because of prompt diagnosis and treatment, although undiagnosed hyperthyroidism is an important contributor to bone loss in patients presenting with fracture [27]. Several population studies have investigated the relationship between hyperthyroidism and fracture. Overall, these have demonstrated an association between thyrotoxicosis and low BMD and an increased risk of fracture, especially in postmenopausal women [19].

1.7 Effects of Thyroid Hormone Receptor Mutation on the Human Skeleton

1.7.1 Resistance to Thyroid Hormone (RTH) α)

Heterozygous *THRA* mutations were reported in three families with delayed skeletal development, normal serum TSH, low/normal T4, and high/normal T3 levels (OMIM #614450, RTHα) [28–31]. Overall, a severe skeletal dysplasia has been recognized that indicates impaired T3 action in bone and cartilage. Features include delayed closure of the skull sutures leading to macrocephaly with skull vault thickening, flattened nasal bridge, and hypertelorism. Growth retardation with disproportionate short stature, delayed bone age and tooth eruption, and defective bone mineralization occur in all individuals, while absent hip secondary ossification centers and epiphyseal dysgenesis are present in severe cases. Cortical bone thickening is present in adults. Radiological features consistent with hypothyroidism include immature vertebrae with ossification defects, hypoplasia of the acetabulum, and coxa vara [28–33]. A phenotype-genotype correlation has been identified, with missense mutations associated with a milder phenotype than the more severe abnormalities seen in individuals with *THRA* truncation mutations. Skeletal responses to treatment with T4 have been variable and may reflect severity of the mutation and resulting functional defects of the mutant TRα protein [24].

1.7.2 Resistance to Thyroid Hormone (RTH) β)

RTHβ caused by dominant negative mutations of the *THRB* gene (OMIM #188570) was first described in 1967 [34], and mutations were identified in 1989 [35]. The mutant TRβ disrupts negative feedback of TSH resulting in the characteristic elevation of T4 and T3 and inappropriately normal or increased TSH. The syndrome results in a mixed phenotype of hyperthyroidism and hypothyroidism. Thus, symptoms and signs of thyroid hormone deficiency and excess can coexist. Tissue responses to T3 are dependent upon: severity of the TRβ mutation; genetic background; relative concentrations of TRα, wild-type TRβ and dominant negative mutant TRβ proteins in individual cells; and whether the patient has received prior treatment with anti-thyroid drugs or surgery. Consequently, a wide range of abnormalities are described and interpretation of the skeletal phenotype is complex [24]. Some individuals have been reported with skeletal abnormalities consistent with the consequences of hypothyroidism, including delayed growth and bone age or short stature [36, 37]. Others, however, have abnormalities consistent with hyperthyroidism, including craniosynostosis, advanced bone age, short stature, osteoporosis, and fracture [38]. Findings in human RTHβ come from patients with a range of mutations of varying severity in diverse genetic backgrounds. Well-controlled long-term prospective analysis of large families with specific mutations will be the only way to document the skeletal consequences of human RTHβ.

1.8 T3 Actions in Skeletal Cells

Monocarboxylate transporter-8 (MCT8) and L-amino acid transporter-1 and -2 are expressed in all bone cells, and MCT10 is expressed in cartilage [39–41]. Nevertheless, the role of thyroid hormone transporters in bone is currently unknown. The relative activities of the type 2 and 3 deiodinases (DIO2 and DIO3) determine intracellular availability of T3 in target tissues including bone, whereas DIO1 is not expressed in the skeleton [40, 42]. DIO3 is expressed mostly during intra-uterine development when it protects developing tissues from premature exposure to thyroid hormones, whereas levels of DIO2 rise mainly from birth [24]. Expression of DIO2 in the growth plate regulates the pace of chondrocyte differentiation during early endochondral ossification

[43–45], and its expression in bone-forming osteoblasts controls T3 regulation of adult bone mineralization and strength [46]. Both TRα1 and TRβ1 are expressed in bone and quantitative RT-PCR studies reveal that levels of TRα1 are at least tenfold greater than TRβ1 [47, 48], suggesting TRα1 is the predominant mediator of T3 action in bone [24].

1.8.1 Chondrocytes

Early studies demonstrated thyroid hormones regulate hypertrophic chondrocyte differentiation and vascular invasion of cartilage. In vitro studies of cultured chondrocytes are conflicting due to differences between species, source of chondrocytes, and culture conditions. Overall, data from many laboratories demonstrate T3 stimulates genes that control chondrocyte maturation and cartilage matrix synthesis, mineralization, and degradation. Thus, T3 is essential for coordinated progression of endochondral ossification [24].

1.8.2 Osteoblasts

Although primary osteoblasts and osteoblastic cell lines respond to T3 in vitro, the consequences vary considerably depending on species, anatomical origin of osteoblasts, cell confluence, stage of differentiation, and dose and duration of treatment. Consensus indicates that T3 stimulates osteoblast proliferation and differentiation and bone matrix synthesis, modification and mineralization. Many of these T3 actions involve interactions with bone matrix and local paracrine and autocrine factors via mechanisms that have yet to be determined [24].

1.8.3 Osteoclasts

Thyroid hormone excess results in increased osteoclast activity and bone loss. Osteoclasts express TRα1 and β1 mRNAs but it is not known whether functional receptors are expressed because TR antibodies lack sensitivity to detect endogenous proteins. Studies of mixed cultures containing osteoclast lineage cells and bone marrow stromal cells have been contradictory and it is uncertain whether T3 stimulates osteoclast activity directly. Thus, it is not clear whether stimulatory effects on osteoclastogenesis and bone resorption are secondary responses to direct T3 actions in osteoblasts, osteocytes, stromal cells, or other bone marrow cells [24].

1.9 Effects of Thyroid Hormone Receptor Deletion and Mutation on the Mouse Skeleton

1.9.1 TRα Mutant Mice

Thra encodes the functional T3 receptor, TRα1, as well as an alternative splice variant TRα2 and truncated TRΔα1 and TRΔα2 isoforms of unknown physiological function. Several knockout mice have been generated to identify specific roles for the individual TR isoforms. $TR\alpha1^{-/-}$ and $TR\alpha2^{-/-}$ mice have selective deletion of TRα1 or α2, but retain shorter TRΔα1 and TRΔα2 transcripts. $TR\alpha^{-/-}$ mice lack TRα1 and TRα2 but retain the Δ transcripts, whereas $TR\alpha^{0/0}$ mice represent a complete *Thra* knockout [49–54].

TRα1$^{-/-}$ mice retain normal TRα2 expression and have mild central hypothyroidism. Limited skeletal analysis indicates they have normal weight gain and linear growth [55]. TRα1$^{-/-}$TRβ$^{-/-}$ double-null mice have resistance to thyroid hormone with markedly elevated thyroid hormones and TSH. Juveniles had growth retardation, delayed endochondral ossification and decreased bone mineralization, and adults had reduced BMD [56–58].

TRα2$^{-/-}$ mice have a compensatory increase in TRα1 expression and mild hypothyroidism [54]. Juveniles exhibit normal linear growth, but adults have reduced trabecular BMD and cortical bone mass [54]. Fusion of green fluorescent protein (GFP) to exon 9 of *Thra* in TRα1^{GFP} mice unexpectedly resulted in loss of TRα2 with only a small increase in TRα1 expression [59]. TRα1^{GFP} mice are euthyroid with normal post-natal development and growth, suggesting abnormalities in TRα2$^{-/-}$ mice may result from overexpression of TRα1.

TRα$^{-/-}$ mice are grossly hypothyroid and have severely delayed skeletal development with decreased bone mineralization [50, 60, 61]. TRα$^{-/-}$TRβ$^{-/-}$ double-null mice have marked resistance to thyroid hormone with increased thyroid hormones and TSH, and also display delayed ossification and growth retardation [60, 61].

TRα$^{0/0}$ mice are euthyroid and have less severe skeletal abnormalities than TRα$^{-/-}$ mutants. Juveniles display growth retardation, delayed endochondral ossification, and reduced bone mineral deposition. Although delayed ossification and reduced mineral deposition were observed in juvenile TRα$^{0/0}$ mice, adults had increased bone mass due to low bone turnover and decreased osteoclastic bone resorption [62]. Deletion of both TRs in TRα$^{0/0}$TRβ$^{-/-}$ double-null mice resulted in a more severe phenotype of delayed bone maturation [63].

The critical role for TRα in bone is also seen in mice with dominant negative mutations of *Thra1*. A patient with severe RTH was found to have a frameshift mutation in *THRB*, termed "PV". The mutation results in a dominant-negative mutant TR that cannot bind T3 and acts as a potent antagonist of wild-type TR function [64]. The homologous mutation was introduced into the mouse *Thra* gene to generate TRα1^{PV} mice. TRα1$^{PV/+}$ mice have mild thyroid failure but a severe skeletal phenotype with growth retardation, delayed ossification, impaired chondrocyte differentiation, and reduced mineral deposition during growth [65–67]. Similar to human RTHα, adult TRα1^{PV} mice have skeletal dysplasia with high bone mass and defective bone remodeling [68].

TRα1$^{R384C/+}$ mice, with a less potent dominant-negative mutation of TRα, are euthyroid. They have transient growth retardation with delayed ossification and progressive increase in bone mass leading to osteosclerosis in adults [62]. Brief T3 supplementation during growth, at a dose sufficient to overcome dominant-negative activity of TRα1^{R384C}, ameliorated the adult phenotype [62].

TRα^{AMI} mice have a floxed *Thra* allele (AF2 mutation inducible) and express a dominant-negative mutant TRα1^{L400R} after Cre-mediated recombination. Mice with global expression of TRα1^{L400R} have a similar skeletal phenotype to TRα1$^{PV/+}$ and TRα1$^{R384C/+}$ mice [69], while restricted expression of TRα1^{L400R} in chondrocytes resulted in delayed endochondral ossification, impaired linear growth, and a skull base defect [70].

Overall, the studies demonstrate an essential requirement for TRα-mediated T3 signaling in skeletal development and for the maintenance of adult bone structure and strength [24].

1.9.2 TRβ Mutant Mice

Two independent TRβ$^{-/-}$ mouse strains recapitulate human RTHβ and display similar skeletal phenotypes [61–63, 71, 72]. Juveniles have accelerated endochondral and intramembranous ossification, advanced bone age, increased mineral deposition, and persistent short stature. Increased T3 target gene expression demonstrated enhanced T3 action in skeletal cells resulting from the effects of elevated thyroid hormones mediated by TRα in bone. Thus, the phenotype is characteristic of the skeletal consequences of thyrotoxicosis [25, 73]. Accordingly, adult TRβ$^{-/-}$ mice have osteoporosis with increased osteoclastic bone resorption [62, 71].

TRβ$^{PV/PV}$ mice with the potent dominant-negative PV mutation have gross RTHβ and display similar but more severe skeletal abnormalities than TRβ$^{-/-}$ mice [48, 65, 67, 74].

Overall, skeletal abnormalities in TRβ mutant mice reflect tissue thyroid hormone excess in the skeleton and are consistent with a predominant physiological role for TRα in bone. Thus, the opposing phenotypes in TR mutant mice provide compelling evidence of distinct roles for TRα and TRβ in the skeleton and hypothalamic-pituitary-thyroid (HPT) axis, and can be explained by predominant expression of TRα in bone [47, 48] and TRβ in hypothalamus and pituitary [72, 75]. In TRα mutants, delayed ossification with impaired bone remodeling in juveniles but high bone mass in adults is a consequence of disrupted T3 action in skeletal T3 target cells, whereas accelerated skeletal development and adult osteoporosis in TRβ mutants result from supraphysiological stimulation of skeletal TRα due to disruption of the HPT axis [24, 66].

2 Materials

2.1 Preparation, Fixation and Storage of Skeletal Samples

- Control and experimental mice of the same strain, age and sex.
- Dissecting instruments.
- 20 ml polystyrene tubes.
- 70% ethanol solution.
- 10% neutral buffered formalin (Sigma Aldrich, St Louis, USA).

2.2 Digital X-ray Microradiography

- Cleaned upper limbs, lower limbs and proximal tail vertebrae from experimental mice of the same strain, age and sex, fixed and stored in 70% ethanol at 4 °C.
- Faxitron MX20 variable kV point projection X-ray source and digital image system (Qados, Cross Technologies plc, Berkshire, UK) or Faxitron Ultrafocus variable kV point projection X-ray source and digital image system (RPS Services Ltd., Byfleet, UK).
- Calibration standards: 1 mm diameter steel wire, 1 mm diameter aluminum wire (Hollinbrow Precision Products (UK) Ltd., Shropshire, UK) and 1 mm diameter polyester fiber.
- Mitutoyo CD-6 CP Digital calipers (Mitutoyo, Andover, UK).
- ImageJ software (Download at http://rsb.info.nih.gov/ij/).
- Adobe Photoshop (Adobe Systems Inc., San Jose, CA, USA).

2.3 Micro-CT Analysis

- Cleaned femur and vertebrae from experimental mice of the same strain, age and sex, fixed and stored in 70% ethanol at 4 °C.
- Scanco μCT 50 in vitro Micro-CT scanner with automatic sample changer (Scanco Medical AG, Brüttisellen, Switzerland).
- Sets of 12 sample holders (various sizes).
- Scanning and analysis workstation.
- Reconstruction and analysis blade server.
- 40 TB external RAID storage.

2.4 Biomechanical Testing

- Cleaned long bones or vertebrae from experimental mice of the same strain, age and sex, fixed and stored in 70% ethanol at 4 °C.
- Dissecting instruments including a scalpel with size 22 disposable blades (Swann-Morton, Sheffield, UK)
- 6 cm polystyrene petri dishes
- 70% ethanol solution.
- Aluminum foil.
- Safety glasses.
- Cyano-acrylate glue (Loctite Precision, Henkel Corporation, Westlake USA).
- Instron 5543 load frame with 100 N or 500 N load cells and Bluehills2 software (Instron Limited, High Wycombe, Buckinghamshire, UK).
- Custom mounts for destructive 3-point bend and compression testing of mouse bones (Quality Test Solutions Ltd., High Wycombe, UK).
- Mitutoyo CD-6 CP Digital calipers (Mitutoyo, Andover, UK).

3 Methods

3.1 Preparation and Fixation of Skeletal Samples (Fig. 5)

Skeletal phenotype analysis can be performed at any stage of mouse development or in adulthood. In the pre-natal and early neonatal periods, skeletal "whole mount" staining of the intact skeleton is often performed together with histological analysis of the proximal tibia growth plate (Fig. 5). At later ages upper limbs, lower limbs, and vertebrae are frequently analyzed individually (Fig. 5). Samples may be fixed in 10% neutral buffer formalin (NBF) to preserve tissue morphology or in 70% ethanol to maintain biomechanical properties. The following protocol describes the preparation and fixation of skeletal samples from post natal mice for comprehensive skeletal phenotyping.

3.1.1 Collection

- Sacrifice the animal using an approved method.
- Carefully remove the skin from the carcass (skin can be left on the paws).
- Remove the abdominal organs.
- To remove the upper limbs from the thoracic wall, cut under the scapula with scissors, and then separate the proximal end of the clavicle from the sternum.
- To remove the lower limbs, cut vertically medial to the pelvis and immediately lateral to the sacral vertebrae. This will avoid damaging the femoral head.
- To separate the caudal vertebrae (Ca), remove the skin from tail with scissors, proximal to distal, and then separate the tail from the body at its base. Count down inter-vertebral spaces and detach the four most proximal vertebrae using a scalpel.
- To separate four lumbar vertebrae, use a scalpel cut in an inter-vertebral space just below the ribs. Count down inter-vertebral spaces and detach four lumbar vertebrae.
- Separate the skull at its base using a scalpel and remove the skin posterior to anterior (Skin can be left on the snout).

3.1.2 Fixation and Storage (Fig. 5)

- Place the intact right upper limb into liquid nitrogen and store at −80 °C.
- Place the intact left upper and lower limbs, the 4 proximal caudal vertebrae and 2 distal lumbar vertebrae into 70% ethanol in a 20 ml polystyrene tube and fix for at least 48 h at 4 °C. Use a *pencil* to label the tube indicating the specimen number and fixation method.
- Place the intact right lower limb, the distal caudal vertebrae and 2 proximal lumbar vertebrae and the skull into 10% NBF in a 20 ml polystyrene tube and fix for 12–24 h at 4 °C. Use a pencil to label the tube indicating the specimen number and

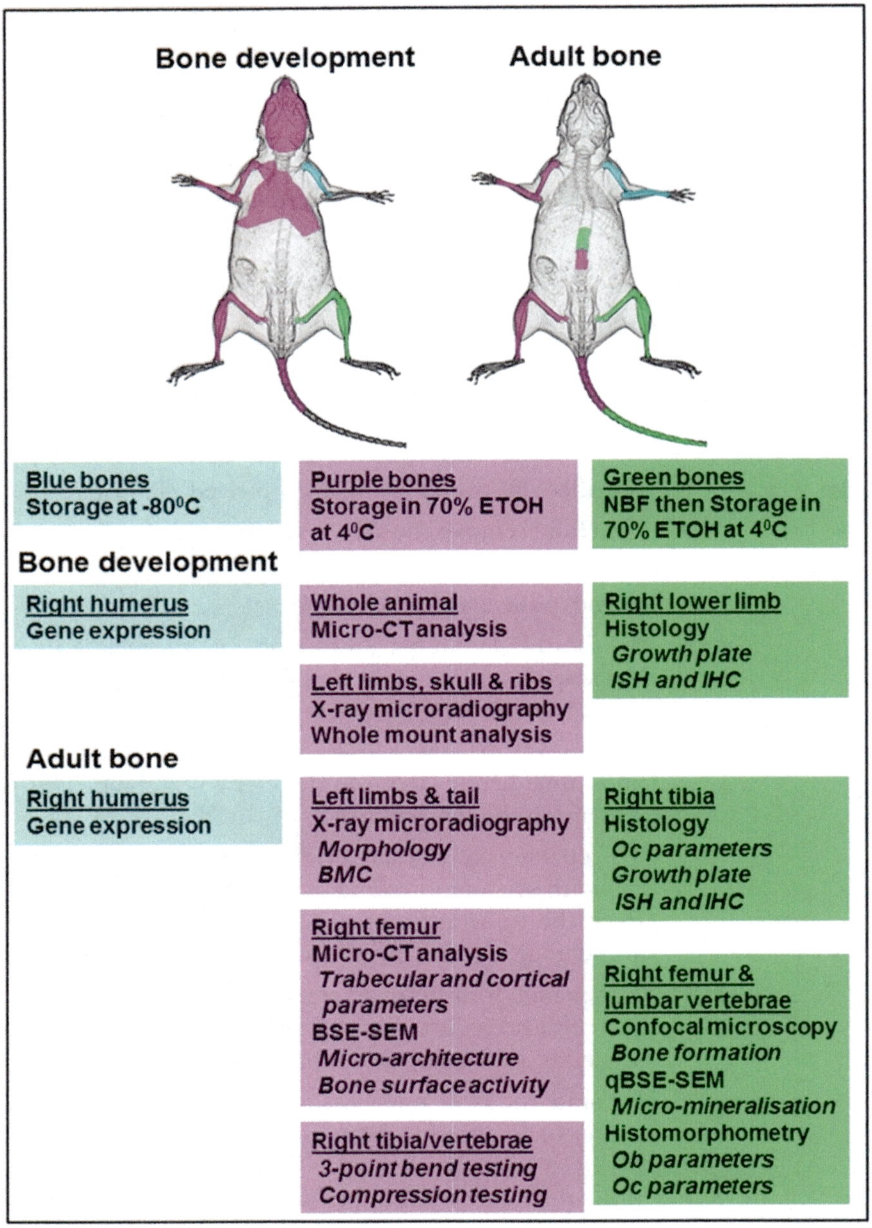

Fig. 5 *Scheme for collection of samples for skeletal phenotype analysis.* Bones in blue should be collected and stored directly at −80 °C. Bones in purple should be collected and fixed directly in 70% ethanol and stored in 70% ethanol at 4 °C. Bones in green should be collected and fixed for 12–24 h in neutral buffered formalin (NBF), transferred to 70% ethanol, and stored in 70% ethanol at 4 °C. Phenotyping techniques for analysis of each sample are listed (BMC = bone mineral content, BSE-SEM = back scattered electron-scanning electron microscopy, ISH = in situ hybridization, IHC = immunohistochemistry, qBSE-SEM = quantitative BSE-SEM, Ob = osteoblast, Oc = osteoclast)

fixation method. After 12–24 h, replace the 10% NBF with 70% ethanol and store at 4 °C.

3.1.3 Sample Preparation

- After fixation dissect the muscle and soft tissues from the bones and joints of the upper and lower limbs. Detach the humerus from the scapula and the paw from the radius and ulna. Detach the femur from the acetabulum being especially careful to avoid damage to the femoral head and then separate the paw from the tibia and fibula.
- Return the dissected bones to their original 20 ml tubes replacing with fresh 70% ethanol and store at 4 °C. Samples can be stored for many months.
- Depending on fixation, samples can be used for histology, histomorphometry, X-ray microradiography, micro-CT, backscattered electron scanning electron microscopy, confocal microscopy, or biomechanical testing (Fig. 5).

3.2 Digital X-ray Microradiography (Fig. 6)

Digital point projection X-ray microradiography is a rapid, cost-effective, and sensitive method to quantify relative bone mineral content and determine morphometric parameters such as bone length in ex-vivo skeletal specimens. The Faxitron MX20 and newer Faxitron Ultrafocus systems can generate 16-bit gray scale images of mouse bones at a sub-10 μm pixel resolution. This protocol describes image acquisition, calibration, montage generation, and analysis [46]. X-ray microradiography imaging can be used at any time after 2 weeks of age. For skeletal phenotype screening, image an upper limb, lower limb, and 4 proximal tail vertebrae.

3.2.1 Image Acquisition

- The Faxitron X-ray source is shielded by a lead lining. X-ray exposure is prevented by a trip switch locking mechanism activated when the source is switched on. It is essential that these safety features are tested at installation and checked during annual maintenance.
- Switch on at least 15 min prior to use to stabilize the X-ray source.
- A 15 s exposure at 26 kV is suitable for most adult mouse bones. Each time the machine is switched on it should be calibrated at this setting, with the sample tray removed.
- Ensure the steel, aluminum, and polyester standards are attached to the under surface of the polycarbonate sample tray at the periphery of the field (Fig. 6a). Place the sample tray on top of the Perspex plastic support and insert into the uppermost level of the cabinet (5× magnification).
- Ensure the Auto Level, Auto Exposure Control, Contrast Assist, and Sharpen Assist toggles are deselected. Then select the "start exposure" button. Ensure all three standards are within the field of view.

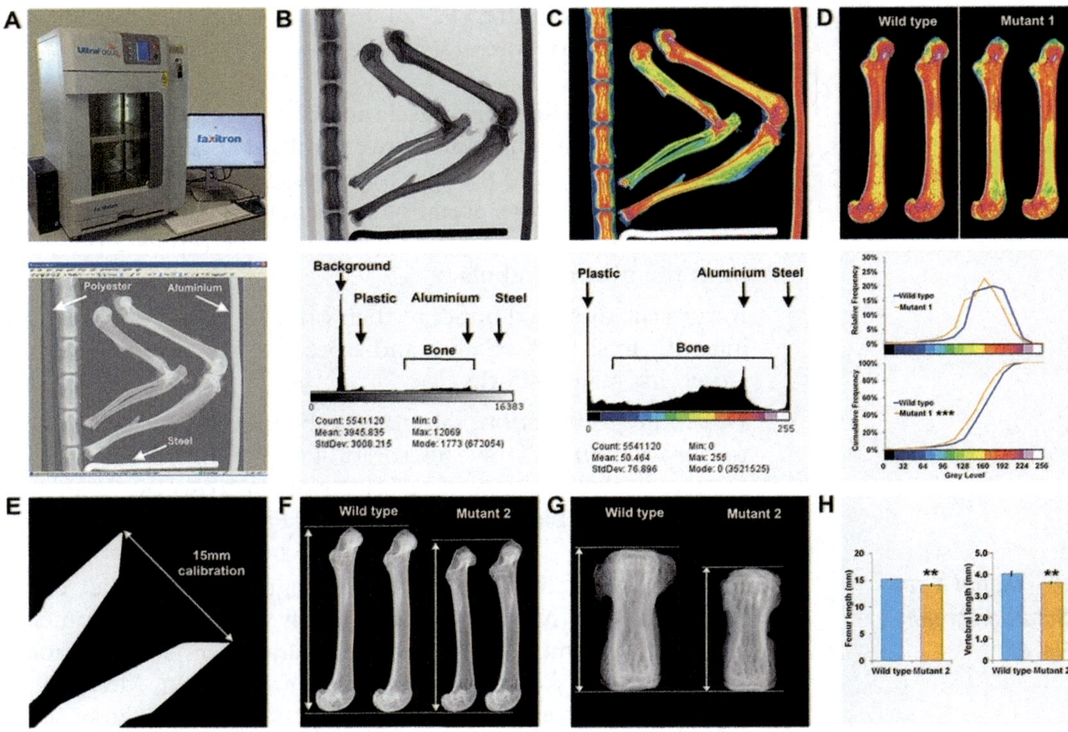

Fig. 6 *X-ray microradiography.* (**a**) Faxitron UltraFocus DXA cabinet with DICOM image below showing the organization of limbs and vertebrae alongside polyester, aluminum and steel calibration standards. (**b**) Inverted DICOM image with histogram below showing gray scale distribution and location of the three standards relative to bone. (**c**) Pseudo-colored 8-bit TIFF image following stretch processing. The histogram below shows the stretched gray scale distribution in relation to 16 color bins. (**d**) Pseudo-color 8-bit TIFF images containing cleaned femur montages. Relative and cumulative frequency plots showing difference in gray scale distributions between wild-type and mutant femurs. (**e**) DICOM image showing calipers set at 15 mm separation for image calibration. (**f**) DICOM image of wild-type and mutant femurs with annotations indicating points used to determine femur length. (**g**) DICOM image of wild-type and mutant caudal vertebrae with annotations indicating points used to determine vertebra length. (**h**) Example graphs showing femur and vertebra lengths in wild-type and mutant mice

- Set the digital micrometer to 15 mm and place it flat on the sample tray. Select the "start exposure" and save the image for future linear calibration.
- To prevent dehydration, remove bones from the 70% ethanol solution only when ready to start imaging. Blot samples briefly and arrange on the sample tray within the imaging area (Fig. 6a) and replace the sample tray in the uppermost slot.
- Select "start exposure" and acquire the image. Check the orientation of bones is correct, taking care to ensure all elements are within the field of view.
- Save the image as a 16 bit DICOM file.
- Once finished, switch the Faxitron off and transfer image data for analysis.

3.2.2 Image Calibration

- Open the DICOM image in ImageJ and press "h" to reveal a histogram showing the distribution of gray scale pixels (Fig. 6b).
- In an Excel spreadsheet, record the minimum and maximum gray level values. Use the ImageJ "Rectangular Selections Tool" to select the polyester standard. Press "h" and record its modal gray level. Similarly, select the steel standard and record its modal value. The aluminum standard acts as an internal reference for the gray scale stretch procedure below.
- Using ImageJ, stretch each of the 16bit DICOM images (16,383 gray levels) to the modal values of the plastic and steel standards. This stretched image is inverted using ImageJ and then converted to an 8 bit TIFF file (256 gray levels). In the stretched image the plastic and steel standards are always assigned gray levels of 0 and 255, respectively (Fig. 6b).
- From "Image" drop down menu select "Lookup Tables" and "16_colors." This applies a pseudo-color scheme in which the 256 gray levels are divided into 16 equal intervals each represented by a different color (Fig. 6c).
- Name and save the modified image and repeat for all samples.

3.2.3 Montage Preparation

- In order to compare samples from different groups of mice it is necessary to generate a single montage for each group.
- Montages can be simply generated using Adobe Photoshop. Open the TIFF files (8 bit indexed-color) of the pseudo-color images to be combined.
- Select the first image and use the "Crop Tool" to select the femur, ensuring the entire bone is included. Using the "Image" and "Image rotation" drop down menu to rotate the image so it is oriented vertically. Increase the canvas size, using a black background, so it is large enough to accommodate all the individual femurs from the group. Rename the montage image and save as a TIFF file without data compression ensuring it remains in indexed-color mode.
- Crop and rotate the femur from each image and paste into the montage positioning appropriately (Fig. 6d). Save the completed montage.
- Prior to analysis, any remaining soft tissue, other bones, and standards must be removed from the montage image. Select the "Pencil Tool" and black for the pencil color then right click and adjust the pencil size to approximately 33 pixel diameter.
- Use the "Zoom Tool" to increase the magnification to at least 200% and then use the pencil to black out any non-femur tissue, being careful not to black out any of the bone itself.

- Select the "Magic Wand Tool" with a 5% tolerance and click adjacent to the first femur. A line should now appear around the bone and if the edges are not clearly highlighted, brush over to black out any remaining soft tissue. Repeat this process for each femur in the montage. Finally, use the "Magic Wand Tool" to select all the femurs and then select the "Paint Bucket Tool" and fill in the background outside the vertebrae with black.
- Name and save the cleaned montage as a new TIFF file (Fig. 6d).
- Repeat steps above to generate a separate montage of each group.
- To create montages of caudal vertebra a slightly different method is used to remove soft tissue from the vertebral images. Crop two equivalent vertebrae from each image and paste them into the group montage. Select a black "Pencil Tool" with a diameter of 13 pixels and zoom the image to at least 300%. Draw a continuous line around the outside of each individual vertebra. Select the "Paint Bucket Tool" and fill in the background outside the vertebrae with black. Name and save the montage as a new TIFF file.

3.2.4 Analysis of Bone Mineral Content

- To determine the relative mineral content of each group, open the cleaned montage file in ImageJ. Under "Plugins," select "Macro," Install, and Open "CustomHistogram."
- Under "Plugins," select "Macro" followed by "CustomHistogram." Select the following values in the dialog box: X Min: 0, X Max: 255, Y Max: auto, Bins: 16 and press OK. Select the "List" button in the displayed histogram and copy the "bin start" and "count" columns and paste into an Excel spreadsheet. This gives the number of pixels in each of 16 equal gray level divisions (bins) and facilitates comparison of gray level distributions between groups (Fig. 6d).
- A statistical comparison of the gray levels cumulative frequency distribution of the different montages is performed using the Kolmogorov-Smirnov test [76].

3.2.5 Analysis of Bone Length

- Determination morphometric parameters is best performed using a 256 gray level version of the montage image. Open the cleaned montage file in ImageJ, select "Lookup Tables" from the "Image" drop down menu and select "Grays." Rename and save the gray-scale montage file.
- To quantify morphometric parameters, it is necessary to calibrate imageJ. To do this open the DICOM image of the digital caliper. Select the "Straight" tool and right click to select the "Straight Line" option. Draw a straight line between and perpendicular to the two surfaces of the digital caliper (Fig. 6e).

Select "Set Scale" from the "Analyze" drop-down menu, type "15" in the "Known distance" box and "mm" in the "Unit of length" box, finally tick the "Global" box and click OK. Select "Set Measurements" from the "Analyze" drop down menu, and deselect all boxes.

- To determine bone length, open the gray-scale montage file in ImageJ and use the "Straight" tool to draw a line from one end of the first bone to the other and then type "m" (Fig. 6f, g). A results window will appear showing the montage name and bone length. Repeat this step for each bone.
- Use the "Open" drop-down menu in ImageJ to open the next gray-scale montage file and repeat step. Finally, copy the results window into an Excel spreadsheet. Calculate the mean bone length and standard deviation for each montage (Fig. 5h).

3.3 Quantitative Micro-computed Tomography (Micro-CT)

Micro-computed tomography is a high resolution method to quantify 3-dimensional trabecular and cortical bone volume and structure, and cortical BMD. The trabecular bone parameters that can be determined include trabecular bone volume, thickness, number and spacing and indices of connectivity. The cortical bone parameters that can be determined include cortical bone volume, thickness, porosity, and mineralization. The protocol describes the ex-vivo analysis of mouse skeletal samples using a Scanco Medical Micro-CT 50 (Fig. 7a). Micro-CT imaging can be performed on any bone but is typically performed on femurs or vertebrae from adult mice aged 12–16 weeks. The protocol describes data acquisition and analysis.

3.3.1 Trabecular Bone Data Acquisition

- Select "machine icon" to turn on the X-ray tube (requires 15 min warm up time) and open the software μCT Tomography V6.3-1.
- Ensure the quality control standard analysis has been performed in the past week (Fig. 7b).
- Carefully disarticulate the hind limbs, separating femur from the tibia.
- Place the first femur in a 6 mm carousel holder distal end downward. Fill the cylinder with 70% ethanol. Dry the tube and seal with Parafilm.
- Follow this procedure for all the femur samples.
- When the X-ray tube is ready the message "do you wish to start scanning?" will be displayed. Select OK.
- Enter sample information for each femur and then select "Save," recording the unique MAE number assigned to each sample.

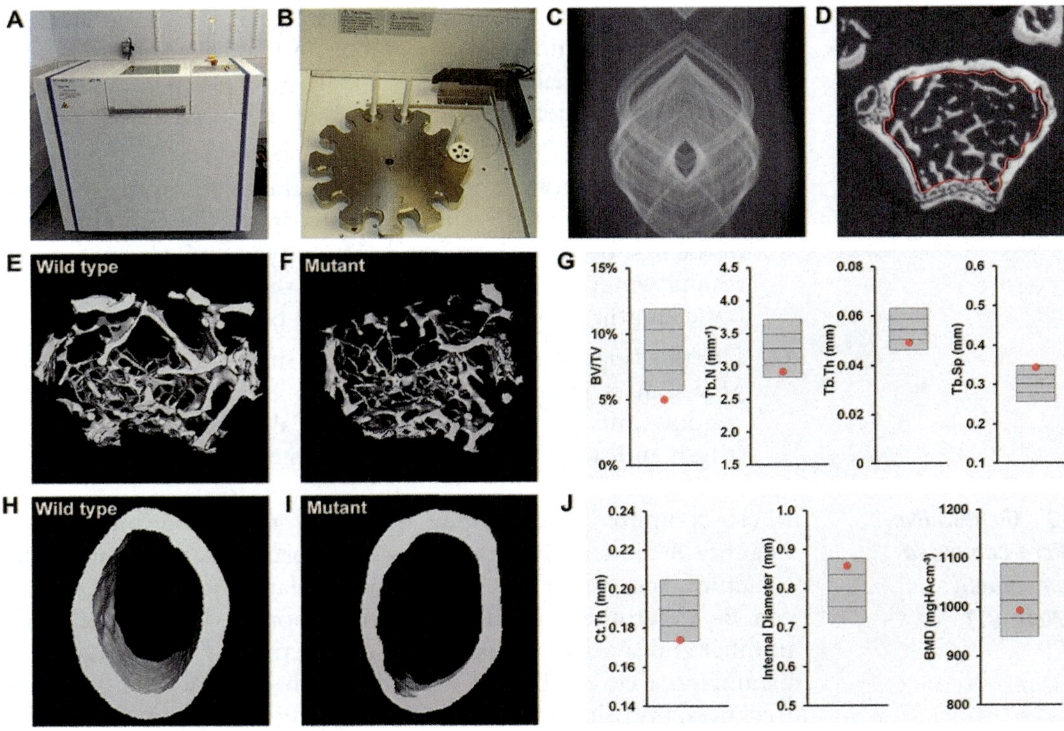

Fig. 7 *Micro-CT.* (**a**) Scanco μCT 50 compact cabinet micro-CT scanner. (**b**) Automatic sample changes with capacity for up to 12 sample holders. (**c**) Sinogram image of raw micro-CT data. (**d**) Image showing transverse cross section of distal femur from three dimensional reconstructed data. Red line shows selection of central trabecular bone compartment as region of interest (ROI) separates from cortical bone. (**e**) Three dimensional reconstruction of trabecular bone compartment ROI from wild-type mouse distal femur. (**f**) Three dimensional reconstruction of trabecular bone compartment ROI from mutant mouse distal femur. (**g**) Graphs showing mean, 1.0SD and 2.0SD values in the wild-type reference range datasets for trabecular bone volume/tissue volume (BV/TV), trabecular number (Tb.N), Trabecular thickness (Tb.Th), and trabecular spacing (Tb.Sp). The mean values for parameters in the mouse mutant are shown in red. (**h**) Reconstruction of mid-shaft transverse section of cortical bone wild-type mouse femur. (**i**) Reconstruction of mid-shaft transverse section of cortical bone mutant mouse femur. (**j**) Graphs showing mean, 1.0SD and 2.0SD values in the wild-type reference range datasets for cortical thickness (Ct.Th), internal cortical diameter, and cortical bone mineral density (BMD). The mean values for parameters in the mouse mutant are shown in red

- Open the loading door, place the tubes into the numbered slots in the carousel, and then shut the door.
- Select "refresh samples" and the scanner will determine the number of samples loaded.
- Assign the samples to the run. Select the carousel number and enter the corresponding MAE number and then select the "trabecular control file." The "trabecular control file" scans a 4 mm region at the base of the 6 mm carousel holder. The trabecular scan is performed at a resolution of 5 μm with a 6.97 mm field of view, 500 projections/180 degrees rotation, 1 s integration time, no averaging, 70 kV, 200 μA, using a 0.5 mm aluminum filter.

- Finally select "add scan."
- Repeat until all samples have been assigned.
- Select "View Scan List" to check the sample list.
- Finally, select "Submit Batch Scans." A sinogram is acquired for each sample (Fig. 7c). The scanning time per femur is approximately 60 min.

3.3.2 Trabecular Bone Data Analysis

- Select "uct_evaluation" and select the images to be analyzed by either selecting the sample from the full list or by typing the MAE number directly.
- Review the reconstructed cross sections to identify the reference point at the base of the growth plate.
- Select the first image of the ROI 100 μm from the growth plate (equivalent to 20 cross sections).
- Use the "Contouring" tool to define the internal cortical surface manually (Fig. 7d). Then use its iterative function to assign the trabecular ROI more accurately. Finally, perform any necessary manual corrections.
- Repeat step every 25 cross sections over the full 1 mm ROI.
- Use the "Morph" tool to iterate between the manually defined ROIs using an "outer value setting" = 500 and "inner value setting" = 50.
- Check the full 1 mm ROI and add additional manually defined contours if required. Careful contouring ensures only trabecular bone is included in the ROI and all cortical bone is excluded.
- Set the following parameters; (a) lower threshold = 180, (b) upper threshold = 1000, (c) gauss sigma = 0.8, (d) gauss support = 1, and (e) component labeling = 10.
- Run "uct_evaluation."
- The evaluation script generates an image of the trabecular ROI (Fig. 7e, f) and calculates the bone volume as a proportion of tissue volume (BV/TV), trabecular number (Tb.N), trabecular thickness (Tb.Th), and trabecular spacing (Tb.Sp) (Fig. 7g).

3.3.3 Cortical Bone Data Acquisition

- Select "machine icon" to turn on the X-ray tube (requires 15 min warm up time) and open the software μCT Tomography V6.3-1.
- Ensure quality control standard analysis has been performed in the past week (Fig. 7b).
- Take a femur sample holder (a plastic 5 ml syringe with the ends removed and six holes, just big enough for a femur, cut across the tube. The six holes are parallel to each other and equally spaced down the tube).

- Select the femur samples to be scanned. Take the first femur and slot it through the top hole in the sample holder (Proximal end first so you do not damage the femoral head). Repeat for the next 5 femur samples. Record sample number, holder number, and location.
- Place the full sample holder into 19 mm carousel holder. Fill the tube with 70% ethanol to cover the bones. Using the carousel holder stand, position the bones parallel to the "Y" axis with the distal end toward you. Carefully place a circular foam pad into the tube to secure the sample holder in alignment. Dry the tube and seal with Parafilm.
- Follow this procedure for all samples.
- Open the loading door, place the tubes into the numbered slots in the carousel, and then shut the door.
- Select "refresh samples" and the scanner will determine the number of samples loaded.
- To add sample details select the carousel position then select "Other" and type in the MAE number.
- Select the "Cortical control file" and then "Scout view." A low resolution 2D X-ray image of the whole 19 mm carousel holder is displayed. Select the 5.5 mm region to be scanned to include the whole femur and press "add scan."
- Repeat steps until details of all femur samples have been entered.
- Select "Start batch scans" from the "Task list"
- The "Cortical control file" then scans each region selected from the scout views. The cortical scan is performed at a resolution of 10 μm, with a 20.5 mm field of view, 500 projections/180 degrees rotation, 1 s integration time, no averaging, 70 kV, 200 μA, 0.5 mm aluminum filter.

3.3.4 Cortical Bone Data Analysis

- Select "uct_evaluation" and enter the MAE number of the femur sample.
- To convert the longitudinal cross sectional images to transverse cross sectional images perform a 90° rotation by selecting "Tasks>Evaluation 3D" and "Reformat to Axial cuts 0 90 90." Define the region to be reformatted using the white box and then select "Start Evaluation."
- Repeat to reformat all the samples to be analyzed.
- Select "uct_evaluation" and enter the MAE number to be analyzed.
- Identify the transverse cross sections that contain the proximal and distal ends of the femur and calculate the cross section that is 56% of this distance.

- The 1.5 mm ROI includes 150 cross sections centered on this point as this is the ideal region to determine cortical bone parameters.
- Use the "Contouring" tool to define the external cortical surface on the top and bottom cross sections of the ROI manually. Then use its iterative function to assign the cortical ROI more accurately.
- Use the "Range" tool to iterate between the two cross sections using an "outer value setting" = 50 and "inner value setting" = 500.
- Set the following parameters; (a) lower threshold = 250, (b) upper threshold = 1000, (c) gauss sigma = 0.8, (d) gauss support = 1, and (e) component labelling = 1.
- Run "uct_evaluation."
- The evaluation script generates an image of the cortical ROI (Fig. 7h, i) and calculates the cortical thickness (Ct.Th), internal diameter (Int.Di), and cortical BMD (Fig. 7j).

3.4 Biomechanical Testing

The biomechanical properties of the mouse skeleton change rapidly with development and are dependent upon bone matrix, mineralization, and skeletal geometry. Destructive 3-point bend and compression testing (Fig. 8) can be used to determine the functional characteristics. Key biomechanical parameters of bone strength and stiffness can be calculated from load displacement curves and skeletal dimensions. This protocol describes 3-point bend testing of the femur, compression testing of caudal vertebrae, and calculation of key biomechanical parameters [46]. Biomechanical analysis can be performed at any age in mice but is technically more challenging before 8 weeks of age.

3.4.1 Femur 3-Point Bend Test

- Attach the 100 N load cell to the Instron 5543 load frame (Fig. 8a).
- Attach the 3-point bend test custom mounts to the base plate of the Instron 5543 load frame and to the 100 N load cell (Fig. 8c).
- Using the digital micrometer, set the separation of the support pins and ensure the loading pin is located centrally. The span must be constant for all groups of femurs analyzed and should be approximately half the total femur length (Subheading 3.2.4).
- Turn on the Inston 5543 load frame and start Bluehills2.
- Select a "flex_load_test" with a constant rate of displacement of 0.03 mm/s and a sample rate of 20 Hz. Raw data logging should include time, extension and load.

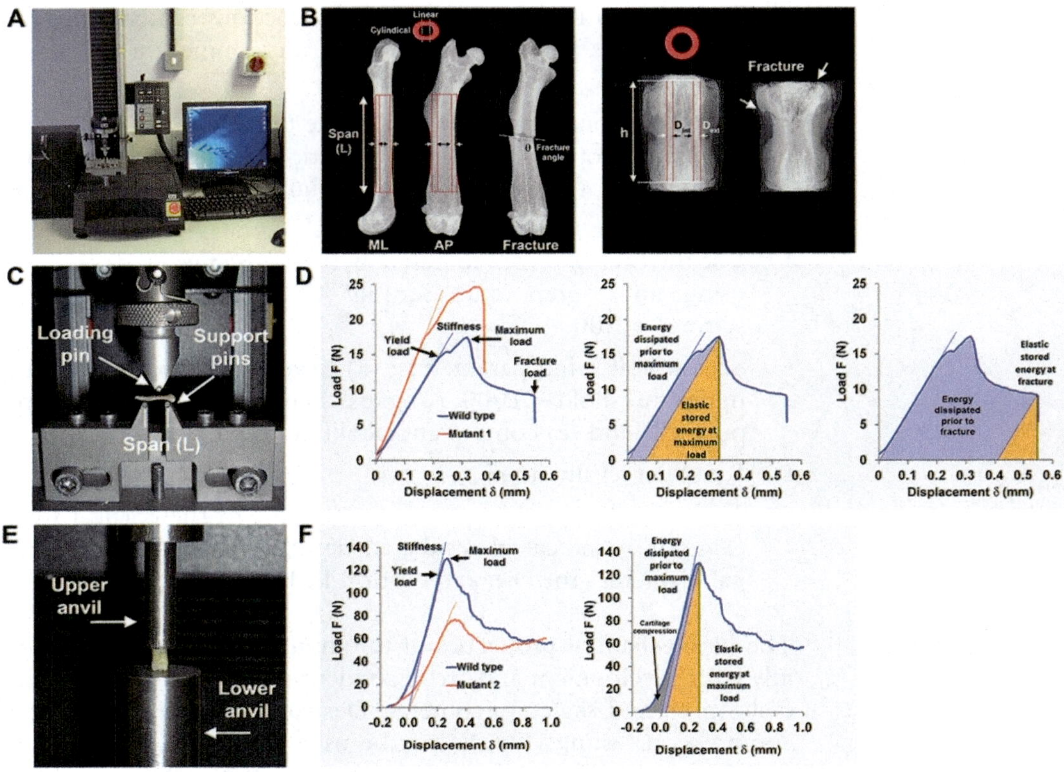

Fig. 8 *Biomechanical testing.* (**a**) Instron 5543 load frame biomechanical testing apparatus. (**b**) X-ray microradiograph images of femur (medio-lateral [ML], antero-posterior [AP] projections and following fracture) and caudal vertebra (before and after fracture) with red lines showing cylindrical tubes and cross-sections from which length, height, and diameter parameters are determined. (**c**) Photograph showing positioning of femur prior to 3-point bend testing. (**d**) Representative load-displacement curves from wild-type and mutant femurs showing yield load, maximum load, fracture load, and gradient of the linear elastic phase (stiffness). Areas under the curve indicating energy dissipated prior to maximum load or fracture (purple) and elastic stored energy at maximum load or fracture (orange). (**e**) Photograph showing positioning of caudal vertebra prior to compression testing. (**f**) Representative compression test curves from wild-type and mutant vertebrae showing yield load, maximum load, fracture load, and gradient of the linear elastic phase (stiffness). Areas under the curve indicating energy dissipated prior to maximum load or fracture (purple) and elastic stored energy at maximum load or fracture (orange). The gray shading represents energy dissipated during cartilage compression before the linear elastic phase of bone compression is reached.

- Ensure the vertical position of the loading pin is appropriate and calibrate the load cell, balance the load, and reset the gauge length.
- Make a screen from aluminum foil around the 3-point bending mount with sufficient opening to allow placement of a femur on the supporting pins. This will retain bone fragments following a brittle fracture.
- Wear safety glasses.

- Select the sample for testing (70% ethanol fixed only). Remove excess 70% ethanol. Separate the femur and tibia carefully at the knee joint using fine scissors. Ensure that all soft tissue has been removed and place the femur in a petri dish containing 70% ethanol.
- Enter the sample information.
- Place the femur carefully and stably onto the supporting pins with the anterior surface upward (Fig. 8c).
- Quickly start the test (samples should be removed from the 70% ethanol for the shortest possible time to prevent desiccation). If the specimen slips or rotates at initial loading stop the test, place the specimen back into 70% ethanol, and repeat.
- Following fracture, return the loading pin to its start position, save data, and replace the bone fragments in the sample tube. The fractured sample can be imaged by Faxitron if required (Fig. 8b).
- Repeat for all samples.
- On completion, end the test and save and export the raw data. An Excel file will be generated for each sample.

3.4.2 Femur 3-Point Bend Test Data Analysis

- Combine all raw data files into one Excel file. For each sample, plot a load-displacement curve. Use the Excel "LINEST" function (least squares method) to determine the trend line for the linear elastic phase of deformation.
- The 3-point bend test load-displacement curve can be subdivided into; (a) a phase of linear elastic deformation ending at the yield point, (b) a phase of plastic deformation including the maximum load, and (c) the point of failure (fracture) (Fig. 8d).
- Use the load-displacement curve and the trend line to determine the load and displacement at yield, maximum load, and fracture (Fig. 8d).
- To calculate stiffness divide the force by the displacement at the yield point (The gradient of the trend line).
- The total work energy (TWE) is the area under the load-displacement curve.
- The elastic stored energy (ESE) is the area of a right-angled triangle with the vertex on the load displacement curve and a hypotenuse with a slope equal to the linear elastic phase of deformation (the trend line) (Fig. 8d).
- The dissipated energy or plastic work (DE) is calculated by subtracting the elastic stored energy from the total work energy (Fig. 8d).

At yield:	$DE_y = 0$	(all work is elastic before yield)
At maximum load:	$DE_m = TWE_m - ESE_m$	
At fracture:	$DE_f = TWE_f - ESE_f$	

- To calculate the fraction of energy dissipated at maximum load and at fracture (plastic work as a fraction of total work) (Fig. 8d) use the formulae:

At maximum load:	$= DE_m/(DE_m + ESE_m)$
At fracture:	$= DE_f/(DE_f + ESE_f)$

3.4.3 Caudal Vertebra Compression Test

- Attach the 500 N load cell to the Instron 5543 load frame.
- Attach the custom mounts for compression testing (Fig. 8e).
- Turn on the Instron 5543 load frame. Start the Bluehills2 software.
- Select a "compression_test" with constant rate of displacement of 0.03 mm/s and a sample rate of 20 Hz. Raw data logging should include time, compressive extension, and compressive load.
- Ensure the vertical position of the upper anvil is appropriate and then calibrate the load cell, balance the load, and reset the gauge length.
- Wear safety glasses.
- Select ethanol-fixed samples for testing. Remove excess 70% ethanol using a paper towel. Dissect caudal vertebrae Ca5 and Ca6 with a scalpel with 22 blade cutting vertically and perpendicular to the long axis of the tail. Place vertebrae in a petri dish containing 70% ethanol.
- Enter sample information.
- Put a small drop of cyano-acrylate glue at the center of the lower anvil (Fig. 8e). With fine forceps pick up Ca5, remove excess ethanol, and carefully place it on the lower anvil ensuring it is aligned vertically (Fig. 8e). Allow 20 s for the glue to bond.
- Start the test. Samples should be removed from the 70% ethanol for as short a period of time as possible to prevent dehydration.
- At the end of the test, return the upper anvil to its start position, save the data, and place the vertebrae in the sample tube. The fractured sample can be imaged by Faxitron if required (Fig. 8b).

- Repeat with Ca6 to test reproducibility.
- Repeat for all tail samples.
- On completion save and export the raw data. An Excel file will be generated for each sample.

3.4.4 Caudal Vertebra Compression Test Data Analysis

- Combine all raw data files into a single Excel file. For each sample plot a load-displacement curve. Use the "LINEST" function (least squares method) to determine the trend line for the linear elastic phase of deformation (Fig. 8f).
- Use the load-displacement curve and the trend line to determine the load and displacement at yield and maximum load (Fig. 8f).
- To calculate stiffness divide the force by the displacement at the yield point (The gradient of the trend line).
- The total work energy (TWE) is the area under the load-displacement curve (AUC). In vertebral compression studies it is important the area representing cartilage compression (ACC) is subtracted from the (Fig. 8f). Thus TWE = AUC − ACC.
- The elastic stored energy (ESE) is the area of a right-angled triangle with the vertex on the load-displacement curve and a hypotenuse with a slope equal to that of the linear elastic phase of deformation (the trend line) (Fig. 8f).
- The dissipated energy (DE) is calculated by subtracting the elastic stored energy from the total work energy.

At yield:	$DE_y = 0$	(all work is elastic before yield)
At maximum load:	$DE_m = TWE_m - ESE_m$	

- To calculate the fraction of energy dissipated at maximum load use the formula: fraction of energy dissipated = $DE_m/(DE_m + ESE_m)$.

4 Notes

1. Skeletal sample preparation and fixation: Care should be taken when using dissection instruments, neutral buffered formalin, and neutral buffered formalin fixed samples. When dissecting samples from juvenile animals particular care should be taken not to break digits when removing skin from the upper and lower limbs or to damage the hip joint when separating the femur from the pelvis. It is also important to ensure that all

sample tubes clearly indicate the method of fixation (ethanol or formalin then ethanol) and are labeled in pencil. When removing muscle and soft tissue form juvenile bones care should be taken to avoid displacing the distal femur and proximal tibia growth plates.

2. Quantitative digital X-ray microradiography: When calibrating the Faxitron it is essential to remove the polycarbonate stage to which the standards are attached. If this is not done negative images of the standards will appear in all subsequent images. When acquiring the Faxitron image it is essential that the Auto Level, Auto Exposure Control, Contrast Assist, and Sharpen Assist toggles are all deselected. If these functions remain selected unwanted stretching of the original DICOM image will occur. When comparing skeletal samples by quantitative Faxitron analysis it is important that skeletal elements are images in a similar orientation. This is achieved by retaining integrity of the knee and elbow joints and imaging the upper and lower limbs at 90% flexion (Fig. 6a). When preparing montages of skeletal elements from one experimental group it is important to maintain the stretched pseudo-colored images as "Indexed color" files rather than RGB or CMYK (Subheading 3.2.2). If the final montage is not Indexed Color, the "custom histogram" analysis in ImageJ will give incorrect results.

3. Quantitative micro-computed tomography: Ethanol or formalin fixed samples may be used for micro-CT analysis. Samples should be dissected to remove all soft tissue, otherwise soft tissue elements may influence the density of the samples and resulting values will not reflect data generated from the standards. Samples should always be scanned in 70% ethanol within the smallest possible sample holder as this will prevent dehydration, which would be detrimental to subsequent techniques. For any study it is preferred that all samples are scanned on the same machine and that a single observer is responsible for drawing the regions of interest.

4. Biomechanical testing: Care should be taken using dissection instruments and safety glasses should be worn. Biomechanical testing should not be performed on neutral buffered formalin fixed bones and it is important to ensure samples do not become dehydrated. The upper and lower pins of the 3-point bend test supports should have a convex profile to minimize cutting of the specimen. Prior to commencing a 3-point bend test it is important to determine a suitable separation between the lower support pins "The span" and approximately half the total length of the shortest femur is recommended. To obtain reproducible results from 3-point bend testing it is important the femur is placed in a consistent stable orientation. This is best achieved with the anterior surface of the femur uppermost but it is advisable to gain experience with test specimens prior to analyzing

critical samples. Similarly, ensuring cuts between caudal vertebrae are perpendicular, through the intervertebral discs and parallel to each other is essential to obtain consistent results for compression testing.

References

1. Kronenberg HM (2003) Developmental regulation of the growth plate. Nature 423(6937): 332–336
2. Zaidi M (2007) Skeletal remodeling in health and disease. Nat Med 13(7):791–801
3. Long F (2012) Building strong bones: molecular regulation of the osteoblast lineage. Nat Rev Mol Cell Biol 13(1):27–38
4. Bonewald LF, Johnson ML (2008) Osteocytes, mechanosensing and Wnt signaling. Bone 42(4):606–615
5. Rice DP, Rice R (2008) Locate, condense, differentiate, grow and confront: developmental mechanisms controlling intramembranous bone and suture formation and function. Front Oral Biol 12:22–40
6. Lassova L, Niu Z, Golden EB, Cohen AJ, Adams SL (2009) Thyroid hormone treatment of cultured chondrocytes mimics in vivo stimulation of collagen X mRNA by increasing BMP 4 expression. J Cell Physiol 219(3):595–605
7. Jacenko O, LuValle PA, Olsen BR (1993) Spondylometaphyseal dysplasia in mice carrying a dominant negative mutation in a matrix protein specific for cartilage-to-bone transition. Nature 365(6441):56–61
8. Bonjour JP, Chevalley T (2014) Pubertal timing, bone acquisition, and risk of fracture throughout life. Endocr Rev 35(5):820–847
9. Raggatt LJ, Partridge NC (2010) Cellular and molecular mechanisms of bone remodeling. J Biol Chem 285(33):25103–25108
10. Bonewald LF (2011) The amazing osteocyte. J Bone Miner Res 26(2):229–238
11. Heino TJ, Hentunen TA, Vaananen HK (2002) Osteocytes inhibit osteoclastic bone resorption through transforming growth factor-beta: enhancement by estrogen. J Cell Biochem 85(1):185–197
12. Tolar J, Teitelbaum SL, Orchard PJ (2004) Osteopetrosis. N Engl J Med 351(27): 2839–2849
13. Baron R, Rawadi G (2007) Wnt signaling and the regulation of bone mass. Curr Osteoporos Rep 5(2):73–80
14. Dallas SL, Prideaux M, Bonewald LF (2013) The osteocyte: an endocrine cell and more. Endocr Rev 34(5):658–690
15. Murphy E, Gluer CC, Reid DM, Felsenberg D, Roux C, Eastell R, Williams GR (2010) Thyroid function within the upper normal range is associated with reduced bone mineral density and an increased risk of nonvertebral fractures in healthy euthyroid postmenopausal women. J Clin Endocrinol Metab 95(7):3173–3181
16. van Rijn LE, Pop VJ, Williams GR (2014) Low bone mineral density is related to high physiological levels of free thyroxine in peri-menopausal women. Eur J Endocrinol 170(3): 461–468
17. Leader A, Ayzenfeld RH, Lishner M, Cohen E, Segev D, Hermoni D (2014) Thyrotropin levels within the lower normal range are associated with an increased risk of hip fractures in euthyroid women, but not men, over the age of 65 years. J Clin Endocrinol Metab 99(8):2665–2673
18. Svare A, Nilsen TI, Asvold BO, Forsmo S, Schei B, Bjoro T, Langhammer A (2013) Does thyroid function influence fracture risk? Prospective data from the HUNT2 study, Norway. Eur J Endocrinol 169(6):845–852
19. Blum MR, Bauer DC, Collet TH, Fink HA, Cappola AR, da Costa BR, Wirth CD, Peeters RP, Asvold BO, den Elzen WP, Luben RN, Imaizumi M, Bremner AP, Gogakos A, Eastell R, Kearney PM, Strotmeyer ES, Wallace ER, Hoff M, Ceresini G, Rivadeneira F, Uitterlinden AG, Stott DJ, Westendorp RG, Khaw KT, Langhammer A, Ferrucci L, Gussekloo J, Williams GR, Walsh JP, Juni P, Aujesky D, Rodondi N, Thyroid Studies C (2015) Subclinical thyroid dysfunction and fracture risk: a meta-analysis. JAMA 313(20):2055–2065
20. Huffmeier U, Tietze HU, Rauch A (2007) Severe skeletal dysplasia caused by undiagnosed hypothyroidism. Eur J Med Genet 50(3): 209–215
21. Rivkees SA, Bode HH, Crawford JD (1988) Long-term growth in juvenile acquired hypothyroidism: the failure to achieve normal adult stature. N Engl J Med 318(10):599–602
22. Eriksen EF, Mosekilde L, Melsen F (1986) Kinetics of trabecular bone resorption and formation in hypothyroidism: evidence for a positive balance per remodeling cycle. Bone 7(2):101–108

23. Abrahamsen B, Jorgensen HL, Laulund AS, Nybo M, Bauer DC, Brix TH, Hegedus L (2015) The excess risk of major osteoporotic fractures in hypothyroidism is driven by cumulative hyperthyroid as opposed to hypothyroid time: an observational register-based time-resolved cohort analysis. J Bone Miner Res 30(5):898–905
24. Bassett JH, Williams GR (2016) Role of thyroid hormones in skeletal development and bone maintenance. Endocr Rev 37(2): 135–187
25. Segni M, Leonardi E, Mazzoncini B, Pucarelli I, Pasquino AM (1999) Special features of Graves' disease in early childhood. Thyroid 9(9):871–877
26. Rasmussen SA, Yazdy MM, Carmichael SL, Jamieson DJ, Canfield MA, Honein MA (2007) Maternal thyroid disease as a risk factor for craniosynostosis. Obstet Gynecol 110(2 Pt 1):369–377
27. Bours SP, van Geel TA, Geusens PP, Janssen MJ, Janzing HM, Hoffland GA, Willems PC, van den Bergh JP (2011) Contributors to secondary osteoporosis and metabolic bone diseases in patients presenting with a clinical fracture. J Clin Endocrinol Metab 96(5): 1360–1367
28. Bochukova E, Schoenmakers N, Agostini M, Schoenmakers E, Rajanayagam O, Keogh JM, Henning E, Reinemund J, Gevers E, Sarri M, Downes K, Offiah A, Albanese A, Halsall D, Schwabe JW, Bain M, Lindley K, Muntoni F, Khadem FV, Dattani M, Farooqi IS, Gurnell M, Chatterjee K (2012) A mutation in the thyroid hormone receptor alpha gene. N Engl J Med 366(3):243–249
29. Moran C, Schoenmakers N, Agostini M, Schoenmakers E, Offiah A, Kydd A, Kahaly G, Mohr-Kahaly S, Rajanayagam O, Lyons G, Wareham N, Halsall D, Dattani M, Hughes S, Gurnell M, Park SM, Chatterjee K (2013) An adult female with resistance to thyroid hormone mediated by defective thyroid hormone receptor alpha. J Clin Endocrinol Metab 98(11):4254–4261
30. van Mullem A, van Heerebeek R, Chrysis D, Visser E, Medici M, Andrikoula M, Tsatsoulis A, Peeters R, Visser TJ (2012) Clinical phenotype and mutant TRalpha1. N Engl J Med 366(15):1451–1453
31. van Mullem AA, Chrysis D, Eythimiadou A, Chroni E, Tsatsoulis A, de Rijke YB, Visser WE, Visser TJ, Peeters RP (2013) Clinical phenotype of a new type of thyroid hormone resistance caused by a mutation of the TRalpha1 receptor: consequences of LT4 treatment. J Clin Endocrinol Metab 98(7):3029–3038
32. Tylki-Szymanska A, Acuna-Hidalgo R, Krajewska-Walasek M, Lecka-Ambroziak A, Steehouwer M, Gilissen C, Brunner HG, Jurecka A, Rozdzynska-Swiatkowska A, Hoischen A, Chrzanowska KH (2015) Thyroid hormone resistance syndrome due to mutations in the thyroid hormone receptor alpha gene (THRA). J Med Genet 52(5):312–316
33. Moran C, Agostini M, Visser WE, Schoenmakers E, Schoenmakers N, Offiah AC, Poole K, Rajanayagam O, Lyons G, Halsall D, Gurnell M, Chrysis D, Efthymiadou A, Buchanan C, Aylwin S, Chatterjee KK (2014) Resistance to thyroid hormone caused by a mutation in thyroid hormone receptor (TR) alpha1 and TRalpha2: clinical, biochemical, and genetic analyses of three related patients. Lancet Diabetes Endocrinol 2(8):619–626
34. Refetoff S, DeWind LT, DeGroot LJ (1967) Familial syndrome combining deaf-mutism, stuppled epiphyses, goiter and abnormally high PBI: possible target organ refractoriness to thyroid hormone. J Clin Endocrinol Metab 27(2):279–294
35. Sakurai A, Takeda K, Ain K, Ceccarelli P, Nakai A, Seino S, Bell GI, Refetoff S, DeGroot LJ (1989) Generalized resistance to thyroid hormone associated with a mutation in the ligand-binding domain of the human thyroid hormone receptor beta. Proc Natl Acad Sci U S A 86(22):8977–8981
36. Weiss RE, Refetoff S (2000) Resistance to thyroid hormone. Rev Endocr Metab Disord 1(1–2):97–108
37. Refetoff S, Dumitrescu AM (2007) Syndromes of reduced sensitivity to thyroid hormone: genetic defects in hormone receptors, cell transporters and deiodination. Best Pract Res Clin Endocrinol Metab 21(2):277–305
38. Weiss RE, Dumitrescu A, Refetoff S (2010) Approach to the patient with resistance to thyroid hormone and pregnancy. J Clin Endocrinol Metab 95(7):3094–3102
39. Capelo LP, Beber EH, Fonseca TL, Gouveia CH (2009) The monocarboxylate transporter 8 and L-type amino acid transporters 1 and 2 are expressed in mouse skeletons and in osteoblastic MC3T3-E1 cells. Thyroid 19(2): 171–180
40. Williams AJ, Robson H, Kester MH, van Leeuwen JP, Shalet SM, Visser TJ, Williams GR (2008) Iodothyronine deiodinase enzyme activities in bone. Bone 43(1):126–134
41. Abe S, Namba N, Abe M, Fujiwara M, Aikawa T, Kogo M, Ozono K (2012) Monocarboxylate transporter 10 functions as a thyroid hormone transporter in chondrocytes. Endocrinology 153(8):4049–4058

42. Waung JA, Bassett JH, Williams GR (2012) Thyroid hormone metabolism in skeletal development and adult bone maintenance. Trends Endocrinol Metab 23(4):155–162
43. Capelo LP, Beber EH, Huang SA, Zorn TM, Bianco AC, Gouveia CH (2008) Deiodinase-mediated thyroid hormone inactivation minimizes thyroid hormone signaling in the early development of fetal skeleton. Bone 43(5): 921–930
44. Gouveia CH, Christoffolete MA, Zaitune CR, Dora JM, Harney JW, Maia AL, Bianco AC (2005) Type 2 iodothyronine selenodeiodinase is expressed throughout the mouse skeleton and in the MC3T3-E1 mouse osteoblastic cell line during differentiation. Endocrinology 146(1):195–200
45. Dentice M, Bandyopadhyay A, Gereben B, Callebaut I, Christoffolete MA, Kim BW, Nissim S, Mornon JP, Zavacki AM, Zeold A, Capelo LP, Curcio-Morelli C, Ribeiro R, Harney JW, Tabin CJ, Bianco AC (2005) The Hedgehog-inducible ubiquitin ligase subunit WSB-1 modulates thyroid hormone activation and PTHrP secretion in the developing growth plate. Nat Cell Biol 7(7):698–705
46. Bassett JH, Boyde A, Howell PG, Bassett RH, Galliford TM, Archanco M, Evans H, Lawson MA, Croucher P, St Germain DL, Galton VA, Williams GR (2010) Optimal bone strength and mineralization requires the type 2 iodothyronine deiodinase in osteoblasts. Proc Natl Acad Sci U S A 107(16):7604–7609
47. Bookout AL, Jeong Y, Downes M, Yu RT, Evans RM, Mangelsdorf DJ (2006) Anatomical profiling of nuclear receptor expression reveals a hierarchical transcriptional network. Cell 126(4):789–799
48. O'Shea PJ, Harvey CB, Suzuki H, Kaneshige M, Kaneshige K, Cheng SY, Williams GR (2003) A thyrotoxic skeletal phenotype of advanced bone formation in mice with resistance to thyroid hormone. Mol Endocrinol 17(7):1410–1424
49. Bassett JH, Williams GR (2009) The skeletal phenotypes of TRalpha and TRbeta mutant mice. J Mol Endocrinol 42(4):269–282
50. Chassande O, Fraichard A, Gauthier K, Flamant F, Legrand C, Savatier P, Laudet V, Samarut J (1997) Identification of transcripts initiated from an internal promoter in the c-erbA alpha locus that encode inhibitors of retinoic acid receptor-alpha and triiodothyronine receptor activities. Mol Endocrinol 11(9):1278–1290
51. Flamant F, Gauthier K (2012) Thyroid hormone receptors: the challenge of elucidating isotype-specific functions and cell-specific response. Biochim Biophys Acta 1830(7): 3900–3907
52. Flamant F, Samarut J (2003) Thyroid hormone receptors: lessons from knockout and knock-in mutant mice. Trends Endocrinol Metab 14(2):85–90
53. O'Shea PJ, Williams GR (2002) Insight into the physiological actions of thyroid hormone receptors from genetically modified mice. J Endocrinol 175(3):553–570
54. Salto C, Kindblom JM, Johansson C, Wang Z, Gullberg H, Nordstrom K, Mansen A, Ohlsson C, Thoren P, Forrest D, Vennstrom B (2001) Ablation of TRalpha2 and a concomitant overexpression of alpha1 yields a mixed hypo- and hyperthyroid phenotype in mice. Mol Endocrinol 15(12):2115–2128
55. Wikstrom L, Johansson C, Salto C, Barlow C, Campos Barros A, Baas F, Forrest D, Thoren P, Vennstrom B (1998) Abnormal heart rate and body temperature in mice lacking thyroid hormone receptor alpha 1. EMBO J 17(2): 455–461
56. Gothe S, Wang Z, Ng L, Kindblom JM, Barros AC, Ohlsson C, Vennstrom B, Forrest D (1999) Mice devoid of all known thyroid hormone receptors are viable but exhibit disorders of the pituitary-thyroid axis, growth, and bone maturation. Genes Dev 13(10): 1329–1341
57. Kindblom JM, Gevers EF, Skrtic SM, Lindberg MK, Gothe S, Tornell J, Vennstrom B, Ohlsson C (2005) Increased adipogenesis in bone marrow but decreased bone mineral density in mice devoid of thyroid hormone receptors. Bone 36(4):607–616
58. Kindblom JM, Gothe S, Forrest D, Tornell J, Vennstrom B, Ohlsson C (2001) GH substitution reverses the growth phenotype but not the defective ossification in thyroid hormone receptor alpha 1−/−beta−/− mice. J Endocrinol 171(1):15–22
59. Wallis K, Dudazy S, van Hogerlinden M, Nordstrom K, Mittag J, Vennstrom B (2010) The thyroid hormone receptor alpha1 protein is expressed in embryonic postmitotic neurons and persists in most adult neurons. Mol Endocrinol 24(10):1904–1916
60. Fraichard A, Chassande O, Plateroti M, Roux JP, Trouillas J, Dehay C, Legrand C, Gauthier K, Kedinger M, Malaval L, Rousset B, Samarut J (1997) The T3R alpha gene encoding a thyroid hormone receptor is essential for postnatal development and thyroid hormone production. EMBO J 16(14):4412–4420
61. Gauthier K, Chassande O, Plateroti M, Roux JP, Legrand C, Pain B, Rousset B, Weiss R,

Trouillas J, Samarut J (1999) Different functions for the thyroid hormone receptors TRalpha and TRbeta in the control of thyroid hormone production and post-natal development. EMBO J 18(3):623–631

62. Bassett JH, Nordstrom K, Boyde A, Howell PG, Kelly S, Vennstrom B, Williams GR (2007) Thyroid status during skeletal development determines adult bone structure and mineralization. Mol Endocrinol 21(8):1893–1904
63. Gauthier K, Plateroti M, Harvey CB, Williams GR, Weiss RE, Refetoff S, Willott JF, Sundin V, Roux JP, Malaval L, Hara M, Samarut J, Chassande O (2001) Genetic analysis reveals different functions for the products of the thyroid hormone receptor alpha locus. Mol Cell Biol 21(14):4748–4760
64. Parrilla R, Mixson AJ, McPherson JA, McClaskey JH, Weintraub BD (1991) Characterization of seven novel mutations of the c-erbA beta gene in unrelated kindreds with generalized thyroid hormone resistance. Evidence for two "hot spot" regions of the ligand binding domain. J Clin Invest 88(6): 2123–2130
65. Kaneshige M, Suzuki H, Kaneshige K, Cheng J, Wimbrow H, Barlow C, Willingham MC, Cheng S (2001) A targeted dominant negative mutation of the thyroid hormone alpha 1 receptor causes increased mortality, infertility, and dwarfism in mice. Proc Natl Acad Sci U S A 98(26):15095–15100
66. O'Shea PJ, Bassett JH, Cheng SY, Williams GR (2006) Characterization of skeletal phenotypes of TRalpha1 and TRbeta mutant mice: implications for tissue thyroid status and T3 target gene expression. Nucl Recept Signal 4:e011
67. O'Shea PJ, Bassett JH, Sriskantharajah S, Ying H, Cheng SY, Williams GR (2005) Contrasting skeletal phenotypes in mice with an identical mutation targeted to thyroid hormone receptor alpha1 or beta. Mol Endocrinol 19(12): 3045–3059
68. Bassett JH, Boyde A, Zikmund T, Evans H, Croucher PI, Zhu X, Park JW, Cheng SY, Williams GR (2014) Thyroid hormone receptor alpha mutation causes a severe and thyroxine-resistant skeletal dysplasia in female mice. Endocrinology 155(9):3699–3712
69. Quignodon L, Vincent S, Winter H, Samarut J, Flamant F (2007) A point mutation in the activation function 2 domain of thyroid hormone receptor alpha1 expressed after CRE-mediated recombination partially recapitulates hypothyroidism. Mol Endocrinol 21(10): 2350–2360
70. Desjardin C, Charles C, Benoist-Lasselin C, Riviere J, Gilles M, Chassande O, Morgenthaler C, Laloe D, Lecardonnel J, Flamant F, Legeai-Mallet L, Schibler L (2014) Chondrocytes play a major role in the stimulation of bone growth by thyroid hormone. Endocrinology 155(8): 3123–3135
71. Bassett JH, O'Shea PJ, Sriskantharajah S, Rabier B, Boyde A, Howell PG, Weiss RE, Roux JP, Malaval L, Clement-Lacroix P, Samarut J, Chassande O, Williams GR (2007) Thyroid hormone excess rather than thyrotropin deficiency induces osteoporosis in hyperthyroidism. Mol Endocrinol 21(5): 1095–1107
72. Forrest D, Hanebuth E, Smeyne RJ, Everds N, Stewart CL, Wehner JM, Curran T (1996) Recessive resistance to thyroid hormone in mice lacking thyroid hormone receptor beta: evidence for tissue-specific modulation of receptor function. EMBO J 15(12): 3006–3015
73. Bassett JH, Williams GR (2008) Critical role of the hypothalamic-pituitary-thyroid axis in bone. Bone 43(3):418–426
74. Kaneshige M, Kaneshige K, Zhu X, Dace A, Garrett L, Carter TA, Kazlauskaite R, Pankratz DG, Wynshaw-Boris A, Refetoff S, Weintraub B, Willingham MC, Barlow C, Cheng S (2000) Mice with a targeted mutation in the thyroid hormone beta receptor gene exhibit impaired growth and resistance to thyroid hormone. Proc Natl Acad Sci U S A 97(24): 13209–13214
75. Abel ED, Ahima RS, Boers ME, Elmquist JK, Wondisford FE (2001) Critical role for thyroid hormone receptor beta2 in the regulation of paraventricular thyrotropin-releasing hormone neurons. J Clin Invest 107(8):1017–1023
76. Demidenko E (1994) Kolmogorov-Smirnov Test for image comparison. In: Laganá A, Gavrilova ML, Kumar V, Mun Y, CJK T, Gervasi O (eds) Computational science and its applications – ICCSA 2004. Lecture notes in computer science, vol 3046. Springer, Berlin, pp 933–939

Chapter 13

Evaluating the Hypothalamic-Pituitary-Thyroid (HPT) Axis in Mice

F. E. Wondisford

Abstract

As the genome of experimental animals has become easier to manipulate, a number of mouse models have been developed to understand in vivo thyroid hormone action. A major site of thyroid hormone action is the HPT axis. While several methods are available that provide a detailed understanding of the HPT axis in mice, many authors choose to include only cursory data about this axis, which can lead to erroneous conclusions about in vivo thyroid hormone action. A standard protocol is proposed to evaluate the HPT axis in mice.

Key words HPT axis, TSH and thyroid hormone (TH) levels, TRH testing, Hypothyroidism, T_3 suppression test, TSH subunit mRNA levels, T_3 regulation of hypothalamic TRH

1 Introduction

One of the fundamental measures of thyroid hormone action in humans and animals is serum TSH. Human serum TSH measurements have progressively improved over the past four decades, and serum TSH is widely believed to be the best single test of the thyroid axis functioning in humans. By extension using serum TSH to evaluate the HPT axis in mouse models of disordered thyroid hormone (TH) action such as when the TH receptor (THR) is deleted or mutated is now standard in the field [1–4]. However, mouse TSH measurements have never been proven to be as sensitive nor specific as human TSH measurements. Another way to assess the HPT axis is by measurements of mouse thyroid hormone levels, but these measurements may also be unreliable due to significant inbreed strain differences and the lack of established serum free TH kits. The American Thyroid Association went so far as to recommend estimating free hormone levels using radiolabeled TH and a resin binding assay, even though the quantity of serum needed for this test would be impractical in mice [5]. Finally, a number of investigators continue to report isolated TSH subunit and TRH mRNA levels as

Michelina Plateroti and Jacques Samarut (eds.), *Thyroid Hormone Nuclear Receptor: Methods and Protocols*,
Methods in Molecular Biology, vol. 1801, https://doi.org/10.1007/978-1-4939-7902-8_13,
© Springer Science+Business Media, LLC, part of Springer Nature 2018

indicators of thyroid hormone status even though no study has convincingly shown they can be used for this purpose. Given these uncertainties, a protocol is outlined that dynamically tests the HPT axis and gives definitive information on thyroid status in mice.

2 Materials

2.1 TSH and Thyroid Hormone (TH) Levels

ELISA kits (total T_3 and T_4 and free T_4) were purchased from Alpha Diagnostic International. Mouse serum TSH assay kits were obtained from Millipore (Rat Pituitary Panel Multiplex Assay, Luminex).

2.2 TRH Testing

TRH (Innovative Research of America) pellets were purchased for use as subcutaneous implants.

2.3 Hypothyroidism

Low-iodine diet containing 0.15% PTU was obtained from Harlan Tekland). Methimazole (MMI) was purchased from Sigma Corp.

2.4 T_3 Suppression Test

L-T_3 was purchased from Sigma Corp.

2.5 TSH Subunit mRNA Levels

Total RNA was extracted by standard methodology (TRIzol Reagent; Life Technologies, Invitrogen). RT was carried out using the Access RT-PCR system (Promega). Real-time RT-PCR analyses were performed in a fluorescent temperature cycler (MyiQ Single-Color Real-Time PCR Detection System; Bio-Rad) using SYBR Green I (Bio-Rad) according to the recommendations of the manufacturer.

2.6 T_3 Regulation of TRH

Formalin and Superfrost Plus slides were purchased from Fisher Scientific. Sections were cut using a Leica SM 2000R sliding microtome (Leica Microsystems, Bannockburn, IL). An in vitro transcription system was purchased from Promega Corp. Kodak BioMax-MR film was purchased from Eastman Kodak Co. Images were captured and quantitated using Image Pro Plus software and an Olympus BH2-RFCA microscope (Olympus America Inc.) equipped with a Sony DXC-960MD color analog video camera (Sony Corp.).

3 Methods

3.1 TSH and Thyroid Hormone (TH) Levels

Serum TH levels were measured according to the manufacturer's instruction. These kits were specifically designed to measure hormone levels in mice or rats where applicable (*see* **Note 1**).

3.2 TRH Testing

After 14 days of LoI/PTU and MMI treatment, placebo pellets or pellets containing 10 mg of TRH (Innovative Research of America) were implanted subcutaneously. Blood samples were collected from the tail vein 0, 5, and 10 days after pellet implantation, and TSH was assayed as described above (*see* **Note 2**).

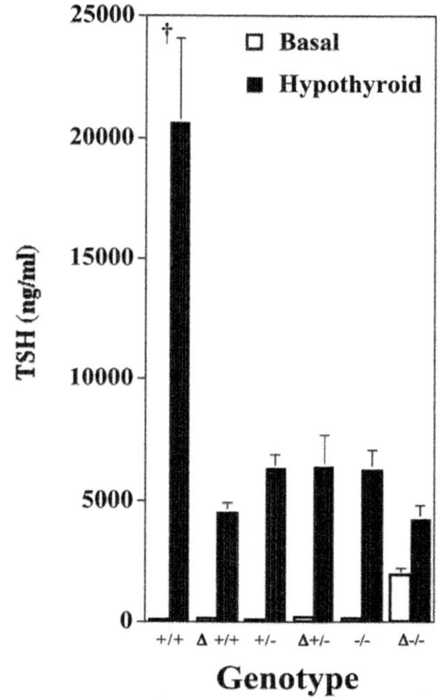

Fig. 1 *Serum TSH response to hypothyroidism.* TSH concentrations were measured before (basal, open bars) and 4 week after mice were rendered hypothyroid by ^{131}I administration (hypothyroid, closed bar). Wt, heterozygous (+/−), or homozygous (−/−) *Thrb* KO mice were compared to littermates harboring a pituitary-expressing *Thrb* transgene containing the Δ337T mutation (from Ref. 6)

3.3 Hypothyroidism

TH deficiency can be induced in two ways: (1) by administering 150 μCi of ^{131}I by i.p. injection to mice who have been placed on a low iodine diet for 8 days or (2) with a low-iodine diet containing 0.15% PTU (LoI/PTU; Harlan Tekland) and 0.05% methimazole (MMI, Sigma) in water for 3–4 weeks. Hypothyroidism was confirmed by measuring the TH and TSH levels (Fig. 1) (*see* **Note 3**).

3.4 T_3 Suppression Test

After 5 weeks of treatment to induce hypothyroidism, mice received either vehicle or daily i.p. injections of a low (0.2 μg per 100 g body weight per day), medium (0.5 μg per 100 g body weight per day), or high (1.0 μg per 100 g body weight per day) dose of L-T_3 (Sigma) for 5 days each. The LoI/PTU diet and MMI in water were given throughout the L-T_3 treatment period. The animals were killed 24 h after the last injection of L-T_3 and TH and TSH levels measured (Fig. 2) [7, 8].

3.5 TSH Subunit mRNA Analysis

For quantitative real-time reverse transcriptase PCR (real-time RT-PCR) analysis, reverse transcription (RT) was carried out on 2 μg of total pituitary RNA. RT-PCR conditions: cycled 40 times at 95 °C for 15 s, 60 °C for 30 s, and 72 °C for 30 s. Primers used for the amplification of cDNAs of interest: 5′-GTGTATGGG CTGTTGCTTCTCC-3′ and 5′-GCACTCCGTATGATTCTCC

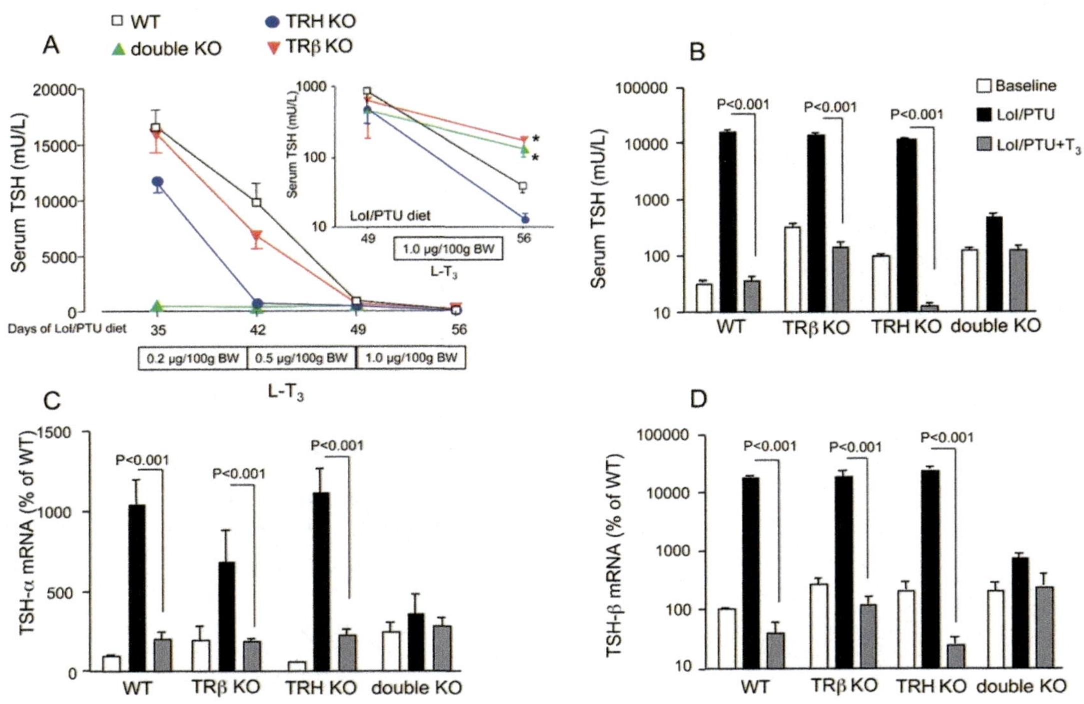

Fig. 2 *Effect of thyroid hormone deficiency and excess on regulation of the HPT axis.* (**a**) Serum TSH levels were sequentially determined in WT and KO mice after a Lol/PTU diet + MMI water for at least 1 month followed by treatment with escalating L-T3 doses (0.2, 0.5, and 1 μg/100 g of body weight/day) for 7 days each. (**b**) Serum TSH at baseline, after Lol/PTU diet, and after the highest dose of L-T3 treatment. (**c**, **d**) Linear and log scales, respectively, of TSH-α and TSH-β subunit mRNA levels in the anterior pituitary (from Ref. 7)

ACTCTG-3′ for the TSH α-subunit; 5′-TCTCGCCGTCCTCC TCTCCGTGCTT-3′ and 5′-AGTTGGTTCTGACAGCCTCGT G-3′ for the TSH β-subunit; and 5′-CGGCTACCACATCCAAGG AA-3′ and 5′-GCTGGAATTACCGCGGCT-3′ for the 18 S ribosomal subunit. Relative mRNA levels (2ΔCt) were determined by comparing the PCR cycle threshold (Ct) between groups. The purity of the PCR products was checked by analyzing the melting curves. Each sample is measured in duplicate, and each experiment is repeated at least three times (Fig. 2). All the results were expressed relative to WT expression considered as 100% (*see* **Note 4**).

3.6 T_3 Regulation of TRH

In situ hybridization histochemistry was performed following a previously described protocol [6]. Mice were anesthetized and perfused intracardially following protocol with 10% phosphate-buffered formalin (Fisher Scientific). Brains were removed to a 10% sucrose solution in 10% phosphate-buffered formalin overnight at 4 °C with gentle shaking to be sectioned the next day. Coronal sections 30 μm thick were collected free floating in cold phosphate-buffered saline treated with diethylpyrocarbonate and mounted on slides. After drying, the slides were treated with a 0.001% proteinase K solution. Hybridization was carried out overnight at 65 °C on a slide warmer using 1×10^7 cpm of probe/ml of hybridization solution.

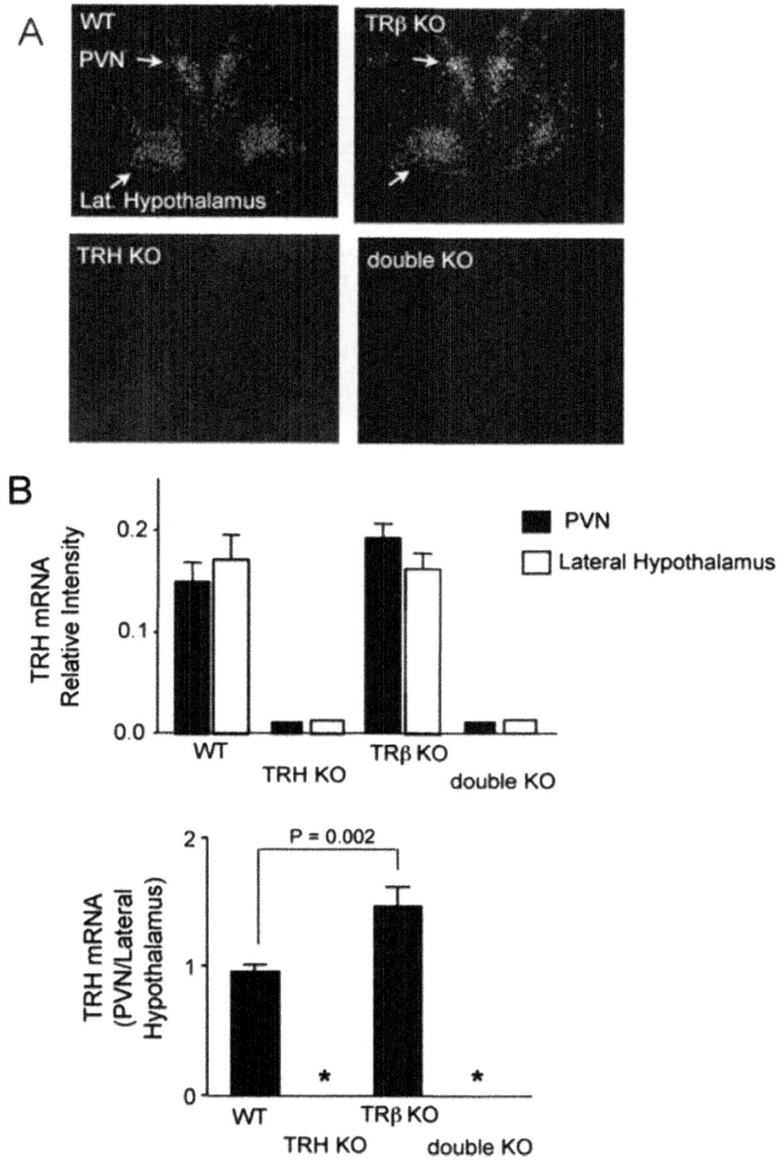

Fig. 3 *TRH mRNA level in KO mice.* (**a**) Dark field photomicrographs showing TRH mRNA. (**b**) Relative pre-proTRH mRNA expression (from Ref. 7)

To prepare the riboprobe used in hybridization, bases 272–629 of exon 3 of the mouse thyrotropin-releasing hormone gene (NCBI accession number NM_009426) were PCR amplified from WT mouse genomic DNA and cloned into pGEMT. Transcription was completed using 1 μg of DNA template with an in vitro transcription system, which included ^{35}S-UTP. Post-hybridization washes and film signal detection of sections followed 3 day exposure. Images of sections after development were captured and quantitated. After subtracting background measurements, values for the paraventricular nucleus (PVN) and for the lateral hypothalamus (LH) of each animal were calculated. TRH expression ratios (PVN/LH) for each animal were calculated (Fig. 3, *see* **Note 5**).

4 Notes

1. To ensure that the measurement of TH is reliable, it is suggested that a diluted sample also be measured in the same assay. It is important that a non-rodent species of antibody is used.
2. In cases where overall anterior pituitary function may be suppressed, TRH administration can be helpful in determining the set-point of the HPT axis and indirectly the extent of T_3 suppression. TRH testing is rarely performed in humans due to a heavy reliance on serum TSH measurements. With less certainty in TSH measurements in rodents, TRH testing should still be performed in many cases.
    - An important consideration within the HPT axis is whether the presence of THRB affects ligand-independent activation of TSH subunit gene expression. Whereas ligand-dependent suppression of serum TSH levels is normal in TR-β+/− mice, ligand-independent activation of TSH production by hypothyroidism is markedly impaired. The presence of the Δ337T pituitary transgene also appears to impair both T_3-dependent and independent THRB function. It remains probable that the unoccupied THRB may dictate the level at which TSH is regulated in the HPT axis.
3. Dynamic regulation of the HPT axis remains the gold-standard. Hypothyroidism induced by either radioiodine or PTU/MMI treatment probes the response of the HPT axis to a lack of negative feedback, while T_3 replacement assesses the sensitivity of the axis to T_3 suppression. The importance of this approach is illustrated in Fig. 2, which demonstrates that a lack of T_3 suppression induced by KO of *Thrb* in mice can be masked by the lack of ligand-independent activation from TRH (measured during hypothyroidism) in mice lacking both *Trh* and *Thrb* (double KO).
4. TRH expression within the paraventricular nucleus (PVN) is regulated by T_3, but TRH expression in the lateral hypothalamus is not regulated. Many studies report hypothalamic TRH mRNA levels without regard for these two discrete cell populations. For this reason total hypothalamic mRNA levels for TRH are not useful in assessing the status of the HPT axis. Moreover, the effect of hypothyroidism on TRH mRNA levels is quite modest compared to that observed on either serum TSH or TSH subunit mRNA levels. In fact, the increase in PVN TRH mRNA levels found in hypothyroidism is only statistically significant when corrected for TRH expression in the LH (Fig. 3).

Acknowledgment

These studies were supported by the National Institutes of Health (DK49126).

References

1. Chiamolera MI, Wondisford FE (2009) Minireview: Thyrotropin-releasing hormone and the thyroid hormone feedback mechanism. Endocrinology 150(3):1091–1096
2. Costa-e-Sousa RH, Hollenberg AN (2012) Minireview: The neural regulation of the hypothalamic-pituitary-thyroid axis. Endocrinology 153(9):4128–4135
3. Ortiga-Carvalho TM, Sidhaye AR, Wondisford FE (2014) Thyroid hormone receptors and resistance to thyroid hormone disorders. Nat Rev Endocrinol 10(10):582–591
4. Ortiga-Carvalho TM, Chiamolera MI, Pazos-Moura CC, Wondisford FE (2016) Hypothalamus-pituitary-thyroid axis. Compr Physiol 6(3):1387–1428
5. Bianco AC, Anderson G, Forrest D, Galton VA, Gereben B, Kim BW, Kopp PA, Liao XH, Obregon MJ, Peeters RP, Refetoff S, Sharlin DS, Simonides WS, Weiss RE, Williams GR (2014) American Thyroid Association Guide to investigating thyroid hormone economy and action in rodent and cell models. Thyroid 1:88–168
6. Abel ED, Moura EG, Ahima RS, Campos-Barros A, Pazos-Moura CC, Boers ME, Kaulbach HC, Forrest D, Wondisford FE (2003) Dominant inhibition of thyroid hormone action selectively in the pituitary of thyroid hormone receptor-β null mice abolishes the regulation of thyrotropin by thyroid hormone. Mol Endocrinol 17(9):1767–1776
7. Nikrodhanond AA, Ortiga-Carvalho TM, Shibusawa N, Hashimoto K, Liao XH, Refetoff S, Yamada M, Mori M, Wondisford FE (2006) Dominant role of thyrotropin-releasing hormone in the hypothalamic-pituitary-thyroid axis. J Biol Chem 281(8):5000–5007
8. Machado DS, Sabet A, Santiago LA, Sidhaye AR, Chiamolera MI, Ortiga-Carvalho TM, Wondisford FE (2009) A thyroid hormone receptor mutation that dissociates thyroid hormone regulation of gene expression in vivo. Proc Natl Acad Sci U S A 106(23):9441–9446

Chapter 14

Thyroid Hormone Action: The p43 Mitochondrial Pathway

Chantal Wrutniak-Cabello, François Casas, and Gérard Cabello

Abstract

The possibility that several pathways are involved in the multiplicity of thyroid hormone physiological influences led to searches for the occurrence of T3 extra nuclear receptors. The existence of a direct T3 mitochondrial pathway is now well established. The demonstration that TRα1 mRNA encodes not only a nuclear thyroid hormone receptor but also two proteins imported into mitochondria with molecular masses of 43 and 28 kDa has provided new clues to understand the pleiotropic influence of iodinated hormones.

The use of a T3 photo affinity label derivative (T3-PAL) allowed detecting two mitochondrial T3 binding proteins. In association with western blots using antibodies raised against the T3 nuclear receptor TRα1, mitochondrial T3 receptors were identified as truncated TRα1 forms. Import and in organello transcription experiments performed in isolated mitochondria led to the conclusion that p43 is a transcription factor of the mitochondrial genome, inducing changes in the mitochondrial/nuclear crosstalk. In vitro experiments indicated that this T3 mitochondrial pathway affects cell differentiation, apoptosis, and transformation. Generation of transgenic mice demonstrated the involvement of this mitochondrial pathway in the determination of muscle phenotype, glucose metabolism, and thermogenesis.

Key words Mitochondrial T3 receptor, T3 photo affinity label derivative, p43 Mitochondrial import, In organello mitochondrial transcription assay, Mitochondrial/nuclear crosstalk, Animal models

1 Introduction

The involvement of thyroid hormone in a great number of complex physiological events involved in development, growth, cell metabolism, or thermogenesis introduced, for a long time, a major difficulty to understand the molecular basis of these influences. Indeed, two possibilities had to be considered. So different effects could be explained by only one general pathway controlling the expression of a large panel of genes, or, alternatively, multiple mechanisms induced by a diversity of receptors could be at the origin of the broad spectrum of thyroid hormone influences. This last possibility was supported by the detection of specific triiodothyronine (T3) binding sites localized in the plasma membrane,

Michelina Plateroti and Jacques Samarut (eds.), *Thyroid Hormone Nuclear Receptor: Methods and Protocols*, Methods in Molecular Biology, vol. 1801, https://doi.org/10.1007/978-1-4939-7902-8_14,
© Springer Science+Business Media, LLC, part of Springer Nature 2018

inside mitochondria, or in the nucleus (for review [1]). First, the pioneering study of Tata et al. [2] brought the first data indicating that thyroid hormone clearly affects the expression of a number of nuclear genes in agreement with the detection of specific T3 binding sites localized in the nuclei. More recently, two genes encoding two different T3 nuclear receptors were simultaneously identified [3, 4], acting as T3-dependent transcription factors after binding to specific binding sequences occurring in gene promoters called T3 responsive elements (TRE) [5–8].

In parallel, very fast influences of T3 on cell 2-deoxyglucose and calcium uptakes [9] involving an adenylate cyclase activation [10] were described and related to putative membrane receptors displaying a high affinity for T3 [11]. These data were more recently substantiated by the demonstration that the αVβ3 integrin binds T3 and T4 with a very high affinity and stimulates in minutes phosphorylation pathways increasing the activity of several transcription factors [12, 13]. In addition, after T3 binding, a 30 kDa TRα1 membrane protein has been shown to activate protein kinase GII, Src, ERK, and Akt signaling [14], thus reinforcing the possibility that T3 could exert a part of its influences through membrane receptors.

In the following chapters, we will focus on another thyroid hormone pathway, the direct mitochondrial T3 pathway, in order to provide an up to date review on this particular aspect of thyroid hormone activity.

1.1 Identification of the Direct Mitochondrial T3 Pathway

Sterling's team reported that T3 increases ATP production, mitochondrial respiration, and ATP/ADP exchanger activity after a very short latency period in rats or in isolated mitochondria, independently of the presence of protein synthesis inhibitors [15, 16]. More recently, Enriquez et al. [17] convincingly reported that T3 directly stimulates the expression of the organelle genome by inducing changes in the protein occupancy of the D-loop considered as the promoter of mitochondrial genome transcription and increases the ratio mRNA/rRNA. In agreement with these data suggesting the occurrence of a direct mitochondrial pathway, the occurrence of specific T3 binding sites in mitochondria has been reported by several teams [18–20].

For the identification of T3 binding sites, we have tested on highly purified mitochondria, the T3 derivative T3-PAL previously used to successfully characterize the molecular mass of nuclear receptors [21]. This approach allowed us to detect two different specific binding sites with molecular masses of 43 and 28 kDa, localized in the organelle [22]. Although the 43 kDa protein was located inside the mitochondrial matrix, the 28 kDa protein was specifically detected in the inner membrane, in agreement with the T3 binding site previously reported by Sterling and Milch [19].

Interestingly, Bigler and Eisenman have reported that, in addition to the T3 nuclear receptor, three truncated forms of TRα1 proteins were synthesized through the use of alternative initiation sites of translation in the TRα1 transcript [23, 24]. Among them, two displayed molecular masses similar to that described for the mitochondrial T3 binding proteins. Overexpression of these two truncated TRα1 proteins in CV1 cells demonstrated that they are specifically addressed to mitochondria. In addition, expression of the 43 kDa protein stimulates mitochondrial activity and induces an astonishing increase in mitochondriogenesis, well-known T3 influences [22]. Moreover, two different antibodies allowed us to detect a 43 kDa protein located in the mitochondrial matrix of highly purified mitochondria, and an inner membrane 28 kDa protein. In addition, these antibodies immunoprecipitated a 43 kDa protein bound to T3-PAL, thus demonstrating that T3 binding proteins are identical to that recognized by TRα1 antibodies [22].

Subsequently to this study, other members of the nuclear receptor superfamily, such as glucocorticoid, estrogen, vitamin D, RXRα, or PPARγ receptors in the organelle [25–29], were detected in mitochondria. These observations clearly suggest that the mitochondrial localization of TRα1 truncated forms has probably to be expanded to other members of this superfamily.

When compared to the nuclear TRα1 receptor, p43 is essentially deleted from the first nuclear localization sequence located in the A/B domain [30, 31]. However, it possesses the DNA binding domain and the second NLS of TRα1 occurring in the hinge region and the multifunctional sequence, including the ligand binding domain. The more important deletion occurring in the p28 N-terminal domain additionally suppresses the DNA binding domain and the second NLS (Fig. 1). More surprising is the observation that p43 and p28 are devoid of the reported classical mitochondrial import signals (MIS) [32–35], thus leading to acute studies using in organello import experiments with the aim of characterizing atypical MIS.

As shown in Fig. 2, only the C-terminal part of the T3 mitochondrial receptors is involved in their mitochondrial import. In contrast to the N-terminal part, fusion of this large sequence to a cytosolic protein (EGFP) induced its mitochondrial import. Moreover, fusion of EGFP to helices 5 or 10/11 occurring in TRα1 also induced its organelle addressing. In agreement with these data, a simultaneous deletion of these two helical structures abolishes p43 or p28 mitochondrial import. These results led to the conclusion that two different sequences located in the C-terminal part of these proteins independently induce the mitochondrial import of p43 and p28 [30].

The presence of a DNA binding domain in p43 explains that, in EMSA experiments, p43 binds to canonical or previously identified natural TRE sequences [22, 26]. In addition, five TRE-

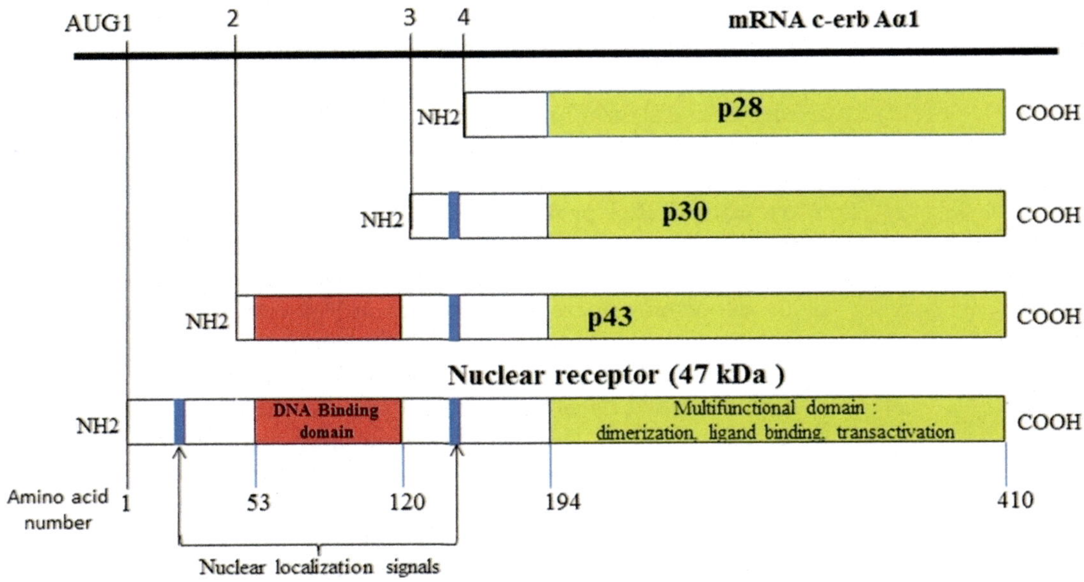

Fig. 1 Functional structure of p43, p30, and p28 TRα1 proteins relatively to the nuclear TRα1 receptor

like sequences, recognized by p43, were detected in the mitochondrial genome, two of them being located in the D-loop, considered the DNA sequence inducing transcription in the organelle [36, 37]. Furthermore, in isolated mitochondria, in the T3 presence, p43 induces a potent stimulation of mitochondrial gene expression, similar to that observed in the same experiment for TFAM, a well-known mitochondrial transcription factor (Fig. 2 [36]). Noteworthy, this influence occurs in less than 5 min, with characteristics similar to that demonstrated for the direct T3 transcriptional activity associated with an increased mRNA/rRNA ratio observed by Enriquez et al. [17]. As expected, this influence was associated with a stimulation of protein synthesis inside the organelle [36]. Last, these influences were abrogated after the deletion of the p43 DNA binding domain, thus clearly suggesting that binding to the organelle DNA is a necessary event. In this field, the identification of two other truncated forms of RXRα or PPARγ inside the organelle, interacting with p43 and increasing its transcriptional activity, suggests that, as shown in the nucleus, the mitochondrial T3 receptor could act through heterodimerical complexes [28, 29].

Besides this transcriptional activity, p43 displays also a very specific influence through another pathway. Chocron et al. [38] reported that T3 stimulates fatty acid oxidation in CV-1 cells considered devoid of endogenous TR receptors, only after expression of exogenous p43. They also reported that this action is related to a physical interaction with the trifunctional protein, which stimulates fatty acid oxidation in minutes. This result clearly suggests

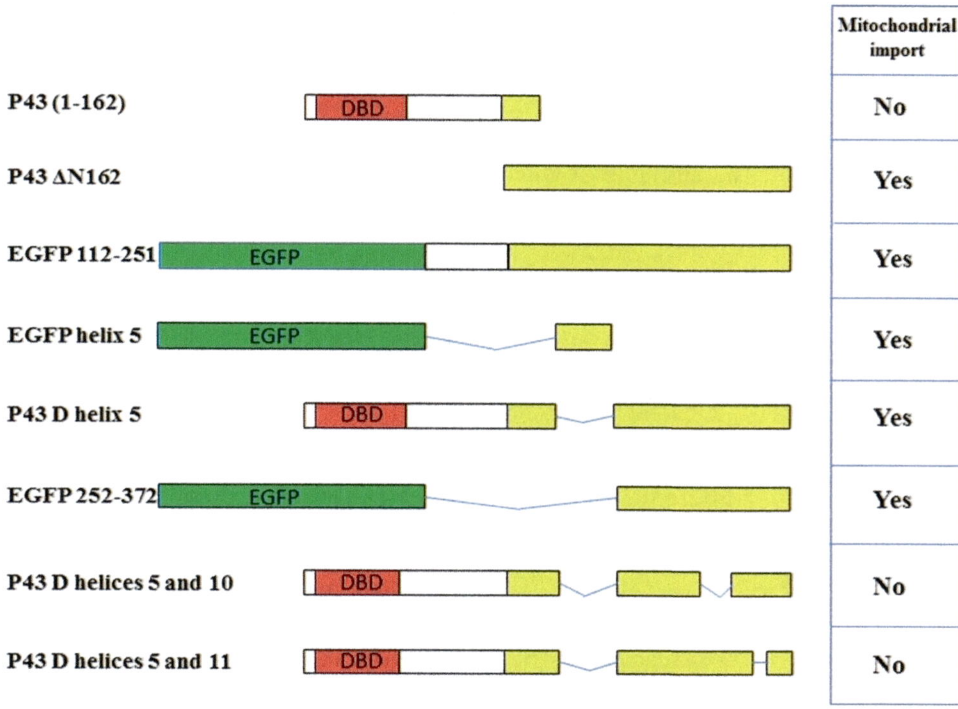

Fig. 2 Characterization of the sequences involved in the mitochondrial import of p43. (**a**) The N-terminal p43 binding domain does not induce a mitochondrial import. (**b**) The p43 ligand binding domain is imported into the organelle and (**c**) drives the mitochondrial import of EGFP, a cytosolic protein. (**d**) Helix 5 of TRα1 is sufficient to drive the mitochondrial import of EGFP, but (**e**) its deletion does not abrogate p43 organelle import. (**f**) Sequence 252-372 contains another mitochondrial import signal driving EGFP into mitochondria. (**g** and **h**) a simultaneous deletion of helices 5 and 10 or 5 and 11 of TRα1 abrogates the mitochondrial import of p43 (*DBD* DNA binding domain)

that p43 could be at the origin of the very fast influence of T3 reported at the mitochondrial level [15, 16].

1.2 Physiological Importance of the Mitochondrial T3 Pathway

Important influences of the p43 pathway could be expected, as changes in mitochondrial activity do not only affect the energetic metabolism, but could also influence the expression of a number of nuclear genes through the mitochondrial/nuclear crosstalk [39]. Indeed, changes in mitochondrial activity affect cell signaling. The organelle contains an important pool of calcium imported according to the mitochondrial membrane potential. Calcium uptake and release occurs independently in relation to the level of mitochondrial activity, thus affecting calcium pulses [40] and consequently the expression of genes regulated by this pathway [41, 42]. In addition, mitochondria are a major site of reactive oxygen species (ROS) production [43] depending on the organelle activity. When released in the cytosol, these oxidative molecules influence the activity of transcription factors such as AP-1 or NFκB [44, 45].

Furthermore, mitochondrial metabolism influence the amount of the fatty acid pool, generating PPAR agonists, or of long chain fatty acids, affecting the expression of genes involved in lipid metabolism or AP2 [46–48]. Interestingly, p43 expression importantly increases ROS production [49] and clearly affects calcium signaling in *Xenopus* eggs [50], thus influencing the mitochondrial/nuclear crosstalk and its nuclear target genes.

1.2.1 Influence of p43 at the Cell Level

One of the best examples of the importance of the p43 pathway at the cell level probably concerns myoblast differentiation. In these myogenic cells, p43 overexpression represses c-Myc expression [51], a proto-oncogene efficiently suppressing the irreversible cell cycle arrest needed to allow terminal differentiation [52–54]. Furthermore, it stimulates the expression and the myogenic activity of myogenin, a crucial transcription factor inducing myoblast differentiation [51, 55]. Consequently, these changes in gene expression lead to the irreversible arrest of myogenic cells proliferation and the induction of terminal differentiation, and consequently a clear stimulation of myoblast differentiation. These data could be expanded to other cells as Kaneko et al. [56] and Cordeau-Lossouarn et al. [57] have reported that mitochondrial activity also influences erythrocytes and neuron differentiation. Furthermore, p43 upregulates slow myosin heavy chain (MyHC-I) expression through stimulation of calcineurin expression, a calcium phosphatase not only involved in the stimulation of myoblast differentiation [58], but also upregulating of MyHC-I expression, a slow twitch contractile myosin [59].

More surprising is the observation that in human dermal fibroblasts, a stimulation of p43 expression, through the induction of an oxidative stress, increases c-Jun and c-Fos proto-oncogenes expression and inhibits the expression of tumor suppressor genes such as p53, $p21^{WAF1}$ and Rb, inducing cell transformation. Furthermore, the induction of the expression of the myogenic factor Myf 5 induces a defective myogenic phenotype in these transformed cells, leading to the development of a typical rhabdomyosarcoma after injection in nude mice [49].

Last, Saelim et al. [60] also reported that, in CV1 cells and in *Xenopus* oocytes, a specific activation of the p43 pathway inhibits apoptotic processes mediated by cytochrome release.

Overall, these data establish that the p43 pathway influences major physiological events at the cell level, such as differentiation, apoptosis, and transformation, clearly suggesting that developmental processes influenced by thyroid hormone probably involve a mitochondrial component.

1.2.2 In Vivo Influence of the Mitochondrial T3 Pathway

The in vivo influence of the p43 pathway has been studied by generating mice overexpressing the mitochondrial receptor specifically in skeletal muscle, or by abrogating p43 expression.

In agreement with in vitro data, in young mice (2 months old), p43 overexpression in skeletal muscle stimulates mitochondrial activity, mitochondriogenesis, and ROS production [61]. These changes in the organelle activity increase PGC-1α expression, considered a major factor increasing mitochondrial biogenesis [62], and consequently the expression of several nuclear genes involved in this process [63], probably through the mitochondrial/nuclear crosstalk. However, this initial stimulation of mitochondrial activity is associated with a permanent oxidative stress inducing, during ageing, muscle atrophy [61].

Another striking observation, also in agreement with in vitro data, concerns the changes in muscle phenotype. P43 overexpression upregulated MyHC-I expression in soleus, an oxidative slow twitch muscle and MyHC-IIa and MyHC-IIx in gastrocnenimus, an oxido-glycolytic fast twitch muscle [63]. Even in these muscles, p43 overexpression reduces the expression of the fastest myosin isoform, indicating that activation of the p43 pathway induces a general slower/more oxidative phenotype in skeletal muscle mass.

As shown by Pessemesse et al. [64], the general depletion of p43 induces exactly the opposite changes than that described for p43 overexpression. $P43^{-/-}$ mice displays a decreased mitochondrial activity, associated with changes in MyHC isoforms expression in different muscles exactly reverse than that occurring in p43 overexpressing mice. In addition, this shift toward a fastest/glycolytic muscle phenotype is associated with an increased muscle mass, as the size of glycolytic fibers is more important than that of oxidative ones.

As previously shown in $TR\alpha^{0/0}$ mice, depleted of all isoforms encoded by the *Thra* gene [65], depletion of the mitochondrial T3 receptor resulted in a lower body weight than in control animals, a reduced fat deposit and a higher metabolic rate [64, 66].

Recently, it was shown that p43 depletion in mice induced a major defect in insulin secretion both in vivo and in isolated pancreatic islets, and a loss of glucose-stimulated insulin secretion [67]. A down-regulation of the glucose carrier Glut 2 and Kir6.2, a key component of the ATP sensitive channel 2, found in pancreatic islets of $p43^{-/-}$ mice, probably explains this defect in insulin secretion. In addition, $p43^{-/-}$ mice displayed a better insulin sensitivity in skeletal muscle to maintain a normal glycaemia in basal condition. However, a high fat/high sucrose diet, or aging, elicited in these animals a hyperglycemia in fasted state, glucose intolerance and insulin-resistance several features of type-2 diabetes [66, 67].

Several other in vivo p43 influences have also been reported. For instance, body temperature was increased in p43 overexpressing mice, and decreased in p43 null mice placed in a normal environment [63, 68]. In addition, p43 expression reduced T3 circulating levels in association with a down regulation of deiodinase 2 [63], whereas plasma T4 and T3 levels were higher in p43 depleted mice than in control ones [67]. Although unexplained, these changes in thyroid hormone levels could be involved in some aspects of the phenotypes observed in transgenic animals.

1.3 Conclusions

Characterization of mitochondrial hormone receptors as products of the mRNA encoding the T3 nuclear receptor TRα1 leads to the unexpected conclusion that only one messenger is at the origin of two thyroid hormone pathways, initiated in two different compartments. In addition, p30, another product of the TRα1 mRNA, is located in the cellular membrane and induces additional mechanisms. This confers to the *Thra* gene a particular importance in the induction of T3 influences, as the *Thrb* gene encoding TRβ receptors does not possess internal start sites allowing the synthesis of truncated isoforms. However, in particular species such as birds and *Xenopus*, a specific TRβ mRNA encodes a short TRβ0 isoform, displaying a strong homology with p43 (86%), imported into mitochondria [36] and acting similarly inside the organelle [38], at least by stimulating fatty acid oxidation. Another interesting conclusion is that, by increasing mitochondrial activity, p43 influences the mitochondrial crosstalk, thus considerably enlarging the number of T3 target genes, when compared to the number of genes directly targeted by T3 nuclear receptors. This explains, at least partly, the large panel of physiological events regulated by thyroid hormone.

Overall, it appears that the occurrence of these two pathways allows us to control a large spectrum of physiological influences through regulation of the expression of an important panel of genes, but also induces a fine-tuning of thyroid hormone action. Interestingly, as other nuclear receptors have also been detected in mitochondria, these conclusions could be possibly extended to other hormones shown to act through a nuclear pathway.

2 Materials

2.1 Preparation of Mitochondria

1. Triton WR 1339 solution: 200 mg/mL in 0.9% NaCl.
2. Sucrose gradient: 1–2 M sucrose containing 1 mM EDTA, 0.1% BSA and 10 mM Tris–HCl, pH 7.5.
3. Protein assay dye reagent concentrate (Bio-Rad, Germany).
4. Bovine serum albumin (BSA).

2.2 T3 Photo Affinity Labeling

1. Glassware hemolysis tube TK75.
2. *p*-Nitrophenyl-2-diazo-3,3,3-trifluoropropionate (PNP-DTP).
3. Anhydrous dimethyl formamide (DMF).

4. 3,3′,5-Triiodo-L-thyronine sodium salt (Sigma Aldrich) (L-T3).
5. L-3,5,3′-[^{125}I]-triiodothyronine (L-[^{125}I]T3) (specific activity: >3000 μCi/μg, NEN).
6. K5 silica gel columns: preparation in 0.5 cm × 7 cm Pasteur pipet.
7. Glacial acetic acid: ethyl acetate mixture: 1:9, v/v.
8. Extraction buffer: 20 mM Tris–HCl, pH 7.8, 0.25 M sucrose, 1 mM $MgCl_2$, 0.4 M KCl, 1 μL/mL aprotinin, 25 mM 2-mercaptoethanol.
9. Sephadex G-25 (Sigma-Aldrich).
10. Gamma counter.
11. Tissue culture plate, 6-well.
12. UV lamp 254 nm.
13. Kodak XAR-5 films.

2.3 In Organello Mitochondrial Transcription Assays

1. Expression plasmids: pSG5-Δ1 plasmid constructed by insertion of the 1.1 kb EcoR1 fragment of the pF1Δmet1 plasmid (provided by Bigler and Eisenman, [33]) encoding p43 into the EcoR1 site of pSG5; pSG5-mt-TFA plasmid constructed by insertion of the 1.3 kb BamH1/BglII fragment of the pQE9-hmtTFA encoding hTFAM (provided by R. Wiesner) into the BamH1/BglII site of pSG5.
2. p43, TFAM proteins: produced by using the TNT®-coupled transcription/translation rabbit reticulocyte lysate system (Promega, Madison, WI, USA), according to the manufacturer's instructions.
3. DNA probes: constructed by specific digestion of the pST41 plasmid, which contains the mouse mitochondrial DNA, and purification of the specific insert by using the Qiaex II kit (Qiagen). Digestion by *Hinc*II gives a 1,494-pb insert encoding 12S and 16S rRNA. Specific DNA probes for the mouse mitochondrial cytochrome *b* (Cytb), cytochrome *c* oxidase subunit III (COX III), and cytochrome *c* oxidase subunit I (COX I) are generated by PCR from the pST41 plasmid with the following primers: Cytb (5'-CCTACCTGCC CCATCCAACAT and 3'-ATGGTTAGAGTCCTTAATAGC), COX III (5'-CAAACTCATGCATATCACATA and 3'-ATC TGCATTAGACTGAAAAGG), and COX I (5'-AATCGTT GATTATTCTCAACC and 3'-GTGTAAGCTCCTTGGTT GGAT).
4. Incubation buffer: 25 mM sucrose, 75 mM sorbitol, 100 mM KCl, 10 mM K_2HPO_4, 0.05 mM EDTA, 5 mM $MgCl_2$, 1 mM

ATP, 10 mM glutamate, 2.5 mM malate, 1 mg/mL BSA, 10 mM Tris–HCl, pH 7.4.

5. Wash buffer: 10% glycerol, 1 mM Tris–HCl pH 6.8, 0.15 mM $MgCl_2$.
6. Lyse buffer: 10 mM Tris–HCl, pH 7.4, 150 mM NaCl, 1 mM EDTA, 2% SDS, 100 μg pronase.
7. 1× methylmercury gel-running buffer: 50 mM boric acid, 5 mM $Na_2B_4O_7 \cdot 10H_2O$, 10 mM Na_2SO_4, pH 8.1.
8. Hybridation buffer: 0.5 M phosphate buffer pH 7.0, 1 mM EDTA, 7% SDS, 1% BSA.
9. [α-^{32}P]dCTP-labeled DNA probes: preparation by using Megaprime™ DNA labeling system (Amersham) according to the manufacturer's instructions. Removal of labeled nucleotides from DNA labeling reactions by using MicroSpin™ G-50 Columns (Amersham).
10. 20× SSC buffer: 3 M NaCl, 0.3 M sodium citrate, pH 6.5.
11. SSC wash buffer: 25 mM phosphate buffer, 0.25% SDS, 2, 0.6 or 0.1× SSC (dilute 20× SSC).
12. Kodak XAR-5 films.

2.4 Mitochondrial Import Assay

1. Expression plasmids: pSG5-THRß1 encoding human TRß1; pSG5-THRß0 encoding chicken TRß0; pSG5-Δ1 encoding p43; pSG5-FE6 encoding chicken TRα1; pSG5-mt-TFA encoding human TFAM.
2. Proteins (TRß1, TRß0, TRα1, p43 and TFAM) are produced by using the TNT® -coupled transcription/translation rabbit reticulocyte lysate system (Promega, Madison, WI, USA),) in presence of trans-[^{35}S]-methionine (ICN) according to the manufacturer's instructions.
3. Import buffer: 220 mM mannitol, 70 mM sucrose, 20 mM sodium succinate, 1 mM EDTA, 5 mM NADH, 1 mM DTT, 10 mM HEPES-KOH, pH 7.4.
4. Proteinase K.
5. Phenylmethylsulfonylfluoride (PMSF).
6. NP40 lyse buffer: 150 mM NaCl, 1.0% NP-40, 50 mM Tris–HCl, pH 8.0.
7. Apyrase.
8. Digitonine.
9. Fluoryl cyanide *m*-chlorophenylhydrazone (FCCP).
10. Autoradiography films.

3 Methods

The general methodology presented in this review is described in the original papers. Only methods which have led to important data for the topic are described in detail.

All protocols are performed using freshly purified isolated rat liver mitochondria (*see* **Note 1**).

3.1 Preparation of Mitochondria

1. Inject intraperitoneally a solution of Triton WR 1339 (75 mg/100 g of body weight) into rats 4 days before euthanasia (*see* **Note 2**).
2. Sacrifice animals after a 16 h period of food deprivation to reduce cellular stocks of lipids and glycogen.
3. Prepare rat liver mitochondria by classical differential centrifugations and purify by isopycnic centrifugation on linear gradient ranging from 35% to 57.5% sucrose (1–2 M), according to Fleischer and Kervina [69].
4. Determine the protein concentration by using the Bio-Rad protein assay and BSA as standard.

3.2 T3 Photo Affinity Labeling

3.2.1 Synthesis and Purification of the L-T3 Photo affinity Label Derivative

Affinity labeling techniques in conjunction with crosslinking have been used for the identification and characterization of T3 nuclear receptors. Here, the T3 photo affinity label derivative (T3-PAL) was used to identify T3 binding proteins located in mitochondria. Synthesis of [^{125}I]-T3-PAL and protein binding were performed according to Horowitz and Samuels [70].

1. Add 1 mg of *p*-nitrophenyl-2-diazo-3,3,3-trifluoropropionate (PNP-DTP) dissolved in 0.2 mL of DMF to a 12 mm × 75 mm tube containing dried L-T3 or L-[^{125}I]T3. The molar ratio of the PNP-DTP and L-T3 is at least 1000/1. Incubate for 18–22 h in the dark at room temperature (*see* **Note 3**).
2. Separate the unreacted hormone and L-[^{125}I]-T3-PAL by chromatography through two K5 silica gel columns pre-equilibrated with the glacial acetic acid: ethyl acetate mixture. (a) drop the preparation on the column and wait until complete infiltration, (b) eluate with the glacial acetic acid: ethyl acetate mixture collecting fractions of approximately 0.3 mL, (c) count 1 μL aliquots of each fraction in a γ counter and pool the fractions containing the peak (*see* **Note 4**), (d) dry the pooled material using a speed vacuum concentrator to a volume of approximately 0.1 mL, (e) re-chromatograph the concentred material on the second column. Dry completely. The yield of the L-[^{125}I]-T3-PAL is estimated around 50%.
3. Dissolve the dried material in a small volume of anhydrous DMF (*see* **Notes 5** and **6**).

3.2.2 Photo Affinity Labeling

1. Prepare the mitochondrial extract freshly by incubating mitochondria for 20 min at 4 °C in the extraction buffer. Centrifuge at 2000 × *g* for 15 min at 4 °C and use the supernatant for the photo affinity labeling reaction.
2. Adjust the pH of the [^{125}I]-T3-PAL derivate to 7.5 by adding 1 M Tris–HCl, pH 8.2 (final concentration about 20 mM Tris). Incubate at room temperature in the dark for 15 min.
3. Add to the mitochondrial extract (final DMF concentration should be lower than 1%), incubate in the dark for 3 h on ice. The concentration of the [^{125}I]-T3-PAL derivate is between 1 and 5 nM. Competition experiments are performed in parallel with 1000-fold molar excess cold hormone.
4. Drop 0.5 mL of sample on an ice-cold 2.5 mL Sephadex G-25 column, pre-equilibrated with extraction buffer (*see* **Note 7**). Wait until complete infiltration.
5. Eluate with 1.5 mL of extraction buffer in a single fraction.
6. Apply 0.75 mL of eluent par well in a six-well tissue culture plate on ice. Irradiate at 254 nm for 2 min at a distance of 2 cm.
7. Pool the samples and precipitate with two volume of 80% ethanol at −20 °C for 15 h. Centrifuge at 3000 × *g* for 20 min, wash twice with 70% ethanol to remove salt and L-[^{125}I]-T3-PAL which is not covalently bound. Dissolve the samples in 10 mM sodium phosphate buffer, pH 7.4, containing 1% SDS (w/v).
8. Apply identical amounts of protein to a 10% SDS-PAGE gel electrophoresis. Dry the gel and autoradiography at −80 °C using X-ray film with an intensifying screen.

3.3 In Organello Mitochondrial Transcription Assays

The in vitro mitochondrial transcription assay is performed as described previously [71].

1. Prepare purified liver mitochondria.
2. Resuspend the pellet of mitochondria (2 mg/mL) in 0.5 mL of incubation buffer.
3. Incubate at 37 °C for 60 min 5 μL of programmed rabbit reticulocyte lysate or unprogrammed lysate as control in the presence or not of T3 (10^{-8} M).
4. Stop the reaction by cooling on ice.
5. Pellet the mitochondria at 12,000 × *g* for 1 min and wash twice with 1 mL of wash buffer.
6. Solubilize the mitochondria in 500 μL of lyse buffer for 15 min at 37 °C.

7. Extract the nucleic acid according to the popular single-step method of Chomczynski and Sacchi [72].
8. Run samples in a 1.4% agarose-5 mM deionized methylmercuric hydroxide slab gel prepared in 1× methylmercury gel-running buffer at 80 V for 4 h (*see* **Note 8**).
9. Transfer the RNA onto nylon membrane. Fix the RNA under UV using a Biolink apparatus (0.12 J/cm^2).
10. Pre-hybridize the membranes at 65 °C for 30 min in the hybridation buffer.
11. Add [^{32}P]dCTP-labeled DNA probes (2 × 10^6 cpm/mL) and incubate during 24 h. Thereafter, wash the membranes at 65 °C in 2× SSC wash buffer for 5 min, 0.6× SSC wash buffer two times for 20 min and 0.1× SSC wash buffer for 5 min.
12. Expose the membranes to X-ray films and analyze by densitometric scanning. Figure 3 shows an example of northern blots performed after in organello transcription in the presence of unprogrammed reticulocyte lysate or lysate expressing p43 or TFAM, in the presence or absence of T3.

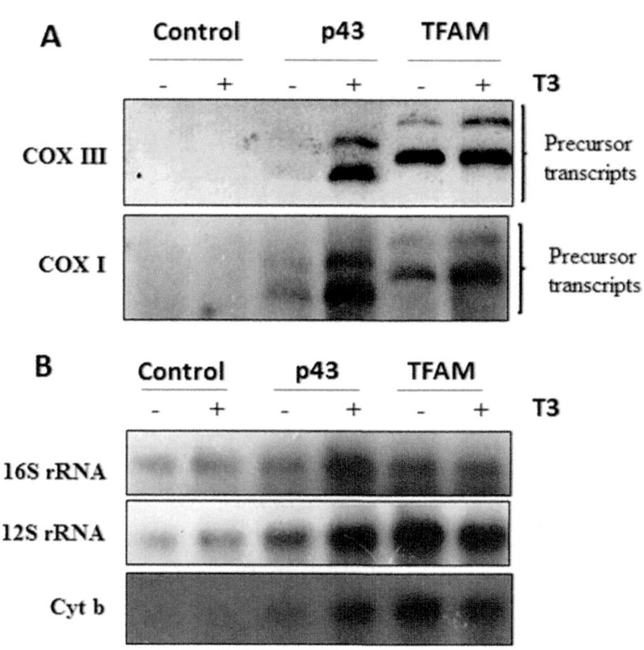

Fig. 3 Typical northern blot performed after in organello mitochondrial transcription in the presence of unprogrammed reticulocyte lysate or lysate expressing p43 or TFAM, in the presence or absence of 10^{-8} M T3. Precursors (**a**) and mature transcripts (**b**) are detected by hybridization with the indicated mitochondrial probes after 60 min of the in organello transcription assay [36]

3.4 Mitochondrial Import Assay

The in vitro mitochondrial import assay is performed according to the method of Komiya and Mihara [73] with minor modifications.

1. Prepare purified liver mitochondria.
2. Resuspend the pellet of mitochondria (2 mg/mL) in 0.1 mL of the import buffer (*see* **Note 9**). To study the internalization modalities of p43 different treatments are also used (*see* **Note 10**).
3. Start the import reaction by adding 10 μL of programmed rabbit reticulocyte lysate containing [^{35}S] methionine-labeled proteins. Incubate at 30 °C for 45 min. Stop the import by cooling in ice.
4. Add 50 μg/mL proteinase K for 10 min at 4 °C (*see* **Note 11**) and then inhibit the reaction by adding 2 mM PMSF.
5. Centrifuge at 12,000 × *g* for 5 min and wash the pellet of mitochondria once with 1 mL of the import buffer.
6. Lyse the pellet of mitochondria in 10 μL of NP40 lyse buffer. Treat an aliquot by proteinase K (*see* **Note 12**).
7. To assess the presence of p43 in the mitochondrial matrix, matrix extracts are prepared after p43 import by an osmotic shock procedure as described by Goglia et al. [18]. (a) After proteinase K treatment, incubate the mitochondria in 100 μL of 8 mM phosphate buffer pH 7.4 for 30 min at 4 °C followed by three cycles of freezing/thawing. (b) Centrifuge at 20,000 × *g* for 30 min at 4 °C. The supernatant contains the matrix proteins.
8. Separate imported proteins treated and untreated by proteinase K by 12% SDS-PAGE gel electrophoresis (*see* **Note 13**), dry the gel, expose to X-ray film, and analyze by laser densitometric scanning. Figure 4 shows the principle of a mitochondrial import experiment and typical autoradiograms of [^{35}S] methionine-labeled proteins after import assay.

4 Notes

1. The mitochondrial preparation must be pure to avoid contamination by nuclear receptors. This can be controlled using specific markers of the different cell organelles (nucleus, microsomes, lysosomes, cell membrane): measurement of enzymatic activities or western blotting with specific antibodies. The mitochondria must remain intact for in organello transcription and import experiments.
2. Triton WR 1339 is a nonionic detergent. It is taken up selectively and accumulates inside the lysosomes, decreasing their

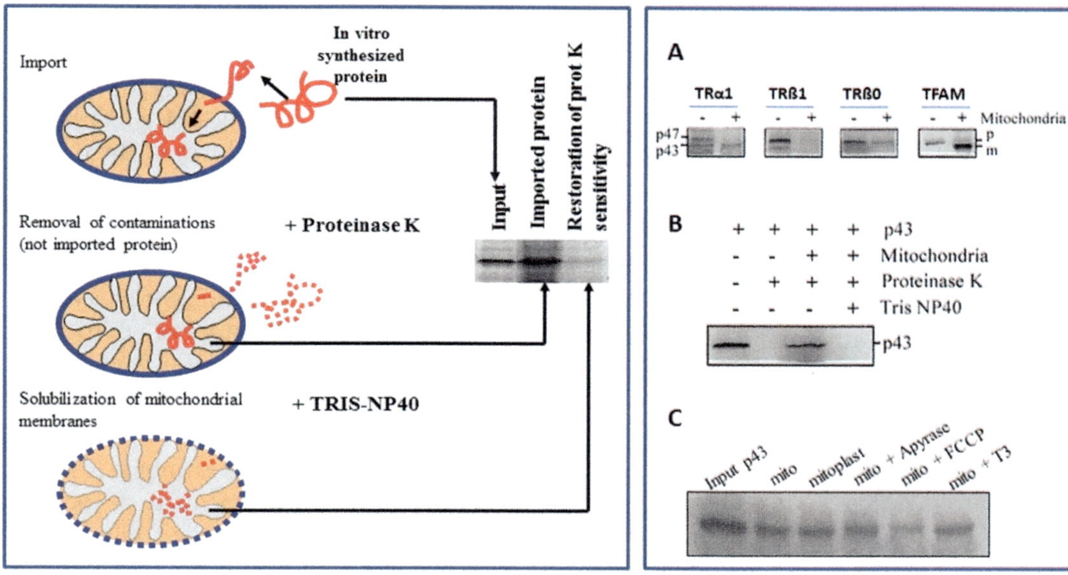

Fig. 4 Principle of a mitochondrial import experiment and typical autoradiograms of [^{35}S] methionine-labeled proteins after the import assay. (**a**) Mitochondrial import of full-length TRα1, TRß1, TRß0, and TFAM. In contrast to full-length nuclear receptors p47 and TRß1, the truncated forms p43 and TRß0 are imported into mitochondria. As a positive control, the mitochondrial transcription factor TFAM is imported after cleavage (*p* precursor protein, *m* mature protein), (**b**) Sensitivity of p43 to proteinase K before import assay, protection from proteolysis after import and restoration of the sensitivity to proteinase K after solubilization of mitochondria by Tris-NP40; (**c**) Import of p43 into mitochondria by an unusual pathway, without cleavage during import; no effect of depletion of mitochondrial ATP stores by apyrase, reduction of mitochondrial membrane potential by FCCP, or exogenous T3 addition [36]

density. Lysosomes and mitochondria can then readily be separated by isopycnic gradient centrifugation.

3. As the reaction is light-sensitive, all steps of synthesis and purification of [^{125}I]-T3-PAL are carried out in a dark room with an inactinic lamp. Glass tubes should be used to avoid absorption of thyroid hormone to plastic.
4. L-[^{125}I]-T3 remains at the origin while L-[^{125}I]-T3-PAL migrates with an RF of greater than 0.8.

   The peak containing the L-[^{125}I]-T3-PAL derivate is eluted usually between fractions 4 and 6.
5. This is important because the final concentration of DMF in the photo labeling reaction should be less than 1%.
6. T3-PAL derivate may be stored in the dark at room temperature if used within 24 h or at 4 °C up to a week.
7. This step removes unbound PAL derivate.
8. Methylmercuric hydroxide is a reversible denaturing agent that completely destroys all secondary structure in RNA and DNA molecules. Electrophoresis through 1.4% agarose-5 mM

CH_3HgOH slab gels has been used for the analysis of the mitochondrial transcription products with high degree of resolution. As methylmercuric hydroxide is uncharged, it is not added to running buffer. Caution: CH_3HgOH is extremely toxic and should always be handled with gloves in the chemical hood.

9. The ATP levels present in reticulocyte lysate are usually sufficient for the import reaction
10. In order to study the internalization modalities of p43 different treatments are also used: (a) use of mitoplast (mitochondria without the outer membrane), (b) depletion of energy by apyrase treatment, (c) depletion of mitochondrial membrane potential using an ionophore as FCCP.
    (a) Mitoplasts are prepared by digitonin treatment as previously described by Ragan et al. [74]: incubate mitochondria in the import buffer with 0.15 mg digitonin/mg of mitochondrial protein for 15 min at 4 °C, centrifuge at 12,000 × *g* for 5 min, and wash the pellet of mitoplast once with 1 mL of the import buffer and use for the import.
    (b) Apyrase treatment: incubate mitochondria and reticulocyte lysate with 20 U/mL apyrase in the import buffer for 30 min at 30 °C, centrifuge the mitochondria at 12,000 × *g* for 5 min, and wash once with 1 mL of the import buffer before use for the import.
    (c) FCCP treatment: incubate mitochondria with 1 mM FCCP in the import buffer at 0 °C for 5 min, centrifuge the mitochondria at 12,000 × *g* for 5 min, wash once with 1 mL of the import buffer before use for the import.
11. After import, contaminations by non-imported proteins are prevented by a proteinase K treatment and only the imported proteins are protected from lysis by mitochondrial membranes.
12. To prove that imported proteins are sensitive to proteinase K and was only protected from the enzymatic proteolysis by mitochondrial membrane, thus demonstrating a true mitochondrial import.
13. Ten percent of the amount of reticulocyte lysate added to mitochondria for the import experiments are loaded in the control lane.

References

1. Wrutniak-Cabello C, Casas F, Rochard P et al (2001) Effets non génomiques des hormones thyroïdiennes. In: Leclère J, Orgiazzi RB, Schlienger JL, Wémeau JL (eds) La thyroïde, des concepts à la pratique clinique, 2nd edn. Elvesier, Amsterdam, p 162
2. Tata JR, Ernster L, Lindberg O et al (1963) The action of thyroid hormones at the cell level. Biochem J 86:408–428
3. Sap J, Muñoz A, Damm K et al (1986) The c-erb-A protein is a high-affinity receptor for thyroid hormone. Nature 324:635–640

4. Weinberger C, Thompson CC, Ong ES et al (1986) The c-erb-A gene encodes a thyroid hormone receptor. Nature 324:641–646
5. Lazar MA (1993) Thyroid hormone receptors: multiple forms, multiple possibilities. Endocr Rev 14:184–193
6. Suen CS, Yen PM, Chin WW (1994) In vitro transcriptional studies of the roles of the thyroid hormone (T3) response elements and minimal promoters in T3-stimulated gene transcription. J Biol Chem 269:1314–1322
7. Glass CK (1994) Differential recognition of target genes by nuclear receptor monomers, dimers and heterodimers. Endocr Rev 15:391–407
8. Harvey CB, Williams GR (2002) Mechanism of thyroid hormone action. Thyroid 12:441–446
9. Segal J, Gordon A (1977) The effects of actinomycin D, puromycin, cycloheximide and hydroxyurea on 3′,5,3-triiodo-L-thyronine stimulated 2-deoxy-D-glucose uptake in chick embryo heart cells in vitro. Endocrinology 101:150–156
10. Segal J (1989) A rapid, extranuclear effect of 3,5,3′-triiodothyronine on sugar uptake by several tissues in the rat in vivo. Evidence for a physiological role for the thyroid hormone action at the level of the plasma membrane. Endocrinology 124:2755–2764
11. Segal J, Ingbar SH (1982) Specific binding sites for triiodothyronine in the plasma membrane of rat thymocytes. Correlation with biochemical responses. J Clin Invest 70:919–926
12. Bergh JJ, Lin HY, Lansing L et al (2005) Integrin αvβ3 contains a cell surface receptor site for thyroid hormone that is linked to activation of mitogen-activated protein kinase and induction of angiogenesis. Endocrinology 146:2864–2871
13. Davis PJ, Davis FB, Cody V (2005) Membrane receptors mediating thyroid hormone action. Trends Endocrinol Metab 16:429–435
14. Kalyanaraman H, Schwappacher R, Joshua J et al (2014) Nongenomic thyroid hormone signaling occurs through a plasma membrane-localized receptor. Sci Signal 7:ra48
15. Sterling K, Brenner MA, Sakurada T (1980) Rapid effect of triiodothyronine on the mitochondrial pathway in rat liver in vivo. Science 210:340–342
16. Sterling K, Brenner MA (1995) Thyroid hormone action: effect of triiodothyronine on mitochondrial adenine nucleotide translocase in vivo and in vitro. Metabolism 34:193–199
17. Enriquez JA, Fernandez-Silva P, Garrido-Pérez N et al (1999) Direct regulation of mitochondrial RNA synthesis by thyroid hormone. Mol Cell Biol 19:657–670
18. Goglia F, Torresani J, Bugli P et al (1981) In vitro binding of triiodothyronine to rat liver mitochondria. Pflugers Arch Eur J Physiol 390:120–124
19. Sterling K, Milch PO (1975) Thyroid hormone binding by a component of mitochondrial membrane. Proc Natl Acad Sci U S A 72:3225–3229
20. Hashizume K, Ichikawa K (1982) Localization of 3,5,3′-L-triiodothyronine receptor in rat kidney mitochondrial membranes. Biochem Biophys Res Commun 106:920–926
21. Pascual A, Casanova J, Samuels HH (1982) Photoaffinity labeling of thyroid hormone nuclear receptors in intact cells. J Biol Chem 257:9640–9647
22. Wrutniak C, Cassar-Malek I, Marchal S et al (1995) A 43-kDa protein related to c-Erb A alpha 1 is located in the mitochondrial matrix of rat liver. J Biol Chem 270:16347–16354
23. Bigler J, Eisenman RN (1988) c-erbA encodes multiple proteins in chicken erythroid cells. Mol Cell Biol 8:4155–4161
24. Bigler J, Hokanson W, Eisenman RN (1992) Thyroid hormone receptor transcriptional activity is potentially autoregulated by truncated forms of the receptor. Mol Cell Biol 12:2406–2417
25. Scheller K, Sekeris CE, Krohne G et al (2000) Localization of glucocorticoid hormone receptors in mitochondria of human cells. Eur J Cell Biol 79:299–307
26. Yager JD, Chen JQ (2007) Mitochondrial estrogen receptors—new insights into specific functions. Trends Endocrinol Metab 18:89–91
27. Silvagno F, De Vivo E, Attanasio A et al (2010) Mitochondrial localization of vitamin D receptor in human platelets and differentiated megakaryocytes. PLoS One 5:e8670
28. Casas F, Domenjoud L, Rochard P et al (2000) A 45 kDa protein related to PPARgamma2, induced by peroxisome proliferators, is located in the mitochondrial matrix. FEBS Lett 478:4–8
29. Casas F, Daury L, Grandemange S et al (2003) Endocrine regulation of mitochondrial activity: involvement of truncated RXRalpha and c-Erb Aalpha1 proteins. FASEB J 17:426–436
30. Carazo A, Levin J, Casas F et al (2012) Protein sequences involved in the mitochondrial import of the 3,5,3′-L-triiodothyronine receptor p43. J Cell Physiol 227:3768–3777
31. Mavinakere MS, Powers JM, Subramanian KS et al (2012) Multiple novel signals mediate thy-

roid hormone receptor nuclear import and export. J Biol Chem 287:31280–31297

32. Von Heijne G, Steppuhn J, Herrmann RG (1989) Domain structure of mitochondrial and chloroplast targeting peptides. Eur J Biochem 180:535–545
33. Schatz G, Dobberstein B (1996) Common principles of protein translocation across membranes. Science 271:1519–1526
34. Neupert W (1997) Protein import into mitochondria. Annu Rev Biochem 66:863–917
35. Pfanner N, Geissler A (2001) Versatility of the mitochondrial protein import machinery. Nat Rev Mol Cell Biol 2:339–349
36. Casas F, Rochard P, Rodier A et al (1999) A variant form of the nuclear triiodothyronine receptor c-ErbAalpha1 plays a direct role in regulation of mitochondrial RNA synthesis. Mol Cell Biol 19:7913–7924
37. Wrutniak C, Rochard P, Casas F et al (1998) Physiological importance of the T3 mitochondrial pathway. Ann Acad Sci N Y 839:93–100
38. Chocron ES, Sayre NL, Holstein D et al (2012) The trifunctional protein mediates thyroid hormone receptor-dependent stimulation of mitochondria metabolism. Mol Endocrinol 26:1117–1128
39. Poyton RO, McEwen JE (1996) Crosstalk between nuclear and mitochondrial genomes. Annu Rev Biochem 65:563–607
40. Pizzo P, Drago I, Filadi R et al (2012) Mitochondrial Ca^{2+} homeostasis: mechanism, role, and tissue specificities. Pflugers Arch 464: 3–17
41. Brookes PS, Yoon Y, Robotham JL et al (2004) Calcium, ATP, and ROS: a mitochondrial love-hate triangle. Am J Physiol Cell Physiol 287:C817–C833
42. Biswas G, Guha M, Avadhani NG (2005) Mitochondria-to-nucleus stress signaling in mammalian cells: nature of nuclear gene targets, transcription regulation, and induced resistance to apoptosis. Gene 354:132–139
43. Poyton RO, Ball KA, Castello R (2009) Mitochondrial generation of free radicals and hypoxic signaling. Trends Endocrinol Metab 20:332–340
44. Mazière C, Conte MA, Degonville J et al (1999) Cellular enrichment with polyunsaturated fatty acids induces an oxidative stress and activates the transcription factors AP1 and NFkappaB. Bioch Biophys Res Commun 265:116–122
45. Dalton TP, Shertzer HG, Puga A (1999) Regulation of gene expression by reactive oxygen. Annu Rev Pharmacol Toxicol 39:67–101
46. Georgiadi A, Kersten S (2012) Mechanisms of gene regulation by fatty acids. Adv Nutr 3:127–134
47. Gremlich S, Bonny C, Waeber G et al (1997) Fatty acids decrease IDX-1 expression in rat pancreatic islets and reduce GLUT2, glucokinase, insulin, and somatostatin levels. J Biol Chem 272:30261–30269
48. Ailhaud GP, Abumrad N, Amri EZ et al (1994) A new look at fatty acids as signal-transducing molecules. World Rev Nutr Diet 75:35–45
49. Grandemange S, Seyer P, Carazo A et al (2005) Stimulation of mitochondrial activity by p43 overexpression induces human dermal fibroblast transformation. Cancer Res 65:4282–4291
50. Saelim N, John LM, Wu J et al (2004) Non transcriptional modulation of intracellular Ca2+ signaling by ligand stimulated thyroid hormone receptor. J Cell Biol 167:915–924
51. Seyer P, Grandemange S, Busson M et al (2006) Mitochondrial activity regulates myoblast differentiation by control of c-Myc expression. J Cell Physiol 207:75–86
52. Miner JH, Wold BJ (1991) c-myc inhibition of MyoD and myogenin-initiated myogenic differentiation. Mol Cell Biol 11:2842–2851
53. Crescenzi M, Crouch DH, Tato F (1994) Transformation by myc prevents fusion but not biochemical differentiation of C2C12 myoblasts: mechanisms of phenotypic correction in mixed culture with normal cells. J Cell Biol 125:1137–1145
54. Cabello G, Casas F, Wrutniak-Cabello C (2010) Transcription factors and muscle differentiation. In: Giordano A, Galderesi U (eds) Cell cycle regulation and differentiation in cardiovascular and neuronal system. Springer, New York. https://doi.org/10.1007/978-1-60327-153-0_3
55. Rochard P, Rodier A, Casas F et al (2000) Mitochondrial activity is involved in the regulation of myoblast differentiation through myogenin expression and activity of myogenic factors. J Biol Chem 275:2733–2744
56. Kaneko T, Watanabe T, Oishi M (1988) Effect of mitochondrial protein synthesis inhibitors on erythroid differentiation of mouse erythroleukemia (friend) cells. Mol Cell Biol 8:3311–3315
57. Cordeau-Lossouarn L, Vayssière JL, Larcher JC et al (1991) Mitochondrial maturation during neuronal differentiation in vivo and in vitro. Biol Cell 71:57–65
58. Seyer P, Grandemange S, Rochard P et al (2011) P43-dependent mitochondrial activity regulates myoblast differentiation and slow

myosin isoform expression by control of Calcineurin expression. Exp Cell Res 317: 2059–2071

59. Chin ER, Olson EN, Richardson JA et al (1998) A calcineurin-dependent transcriptional pathway controls skeletal muscle fiber type. Genes Dev 12:2499–2509
60. Saelim N, Holstein D, Chocron ES et al (2007) Inhibition of apoptotic potency by ligand stimulated thyroid hormone receptors located in mitochondria. Apoptosis 12:1781–1794
61. Casas F, Pessemesse L, Grandemange S et al (2009) Overexpression of the mitochondrial T3 receptor induces skeletal muscle atrophy during aging. PLoS One 4:e5631
62. Fernandez-Marcos PJ, Auwerx J (2011) Regulation of PGC-1α, a nodal regulator of mitochondrial biogenesis. Am J Clin Nutr 93:884S–890S
63. Casas F, Pessemesse L, Grandemange S et al (2008) Overexpression of the mitochondrial T3 receptor p43 induces a shift in skeletal muscle fiber types. PLoS One 3:e2501
64. Pessemesse L, Schlernitzauer A, Sar C et al (2012) Depletion of the p43 mitochondrial T3 receptor in mice affects skeletal muscle development and activity. FASEB J 26: 748–756
65. Pelletier P, Gauthier K, Sideleva O et al (2008) Mice lacking the thyroid hormone receptor-alpha gene spend more energy in thermogenesis, burn more fat, and are less sensitive to high-fat diet-induced obesity. Endocrinology 149:6471–6486
66. Bertrand C, Blanchet E, Pessemesse L et al (2013) Mice lacking the p43 mitochondrial T3 receptor become glucose intolerant and insulin resistant during aging. PLoS One 8:e75111
67. Blanchet E, Bertrand C, Annicotte JS et al (2012) The mitochondrial T3 receptor p43 regulates insulin secretion and glucose homeostasis. FASEB J 26:40–50
68. Bertrand-Gaday C, Pessemesse L, Cabello G et al (2016) Temperature homeostasis in mice lacking the p43 mitochondrial T3 receptor. FEBS Lett 590:982–991
69. Fleischer S, Kervina M (1974) Subcellular fractionation of rat liver. Methods Enzymol 31:6–41
70. Horowitz ZD, Samuels HH (1988) Thyroid receptor. In: Gronemeyer H (ed) Affinity labelling and cloning of steroid and thyroid hormone receptors. Ellis Horwood Ltd, Chichester, pp p79–p83
71. Ostronoff LK, Izquierdo JM, Enríquez JA, Montoya J, Cuezva JM (1996) Transient activation of mitochondrial translation regulates the expression of the mitochondrial genome during mammalian mitochondrial differentiation. Biochem J 316:183–191
72. Chomczynski P, Sacchi N (1987) Single-step method of RNA isolation by acid guanidinium thiocyanate-phenol-chloroform extraction. Anal Biochem 162:156–159
73. Komiya T, Mihara K (1996) Protein import into mammalian mitochondria. Characterization of the intermediates along the import pathway of the precursor into the matrix. J Biol chem 271:22105–22110
74. Ragan CI, Wilson MT, Darley-Usmar VM et al (1987) Sub-fractionation of mitochondria and isolation of the proteins of oxidative phosphorylation. In: Darley-Usmar VM, Rickwood D, Wilson MT (eds) Mitochondria, a practical approach. IRL Press, Washington, DC, pp 79–112

Chapter 15

Evaluating Thyroid Disrupting Chemicals In Vivo Using *Xenopus laevis*

Bilal B. Mughal, Barbara A. Demeneix, and Jean-Baptiste Fini

Abstract

Using in vivo animal model systems for chemical screening can permit evaluation of the signaling pathways implicated. *Xenopus laevis* is an ideal model organism to test thyroid axis disruption as thyroid hormones are highly conserved across vertebrates. Here, we describe a high-throughput assay using non-feeding embryonic stage transgenic *X. laevis* (TH/bZip) to screen for thyroid disrupting chemicals using a 3 day exposure protocol. We further describe a protocol to detect endocrine disruption of thyroid axis by the analysis of gene expression using wild-type *X. laevis*.

Key words *Xenopus laevis*, Thyroid signaling, Endocrine disruption, Thyroid disruption, XETA assay, Neurodevelopment, Gene expression, Behavior analysis

1 Introduction

The number of chemicals being introduced to our environment has increased dramatically since the 1930s, from one million tons to 400 million today annually.[1] The number of different chemicals has also increased to about 90,000. Many of these chemicals have been identified as endocrine disrupting compounds (EDCs). These include synthetic chemicals used as lubricants/solvents and their by-products such as the polychlorinated biphenyls (PCBs) and polybrominated biphenyls (PBBs). Others include plastics and plasticizers such as bisphenol A (BPA) and the phthalates, pesticides, and fungicides including methoxychlor, chlorpyrifos, dichlorodiphenyltrichloroethane (DDT) and vinclozolin, and pharmaceutical agents diethylstilbestrol (DES). Interestingly, certain natural chemicals such as phytoestrogens found in infant formula, can also act as endocrine disruptors [1, 2].

[1] European commission study 2006. "Environment fact sheet: REACH - a new chemicals policy for the EU". Consulted on 21st July 2017 - http://ec.europa.eu/environment/chemicals/reach/pdf/reach.pdf

Michelina Plateroti and Jacques Samarut (eds.), *Thyroid Hormone Nuclear Receptor: Methods and Protocols*, Methods in Molecular Biology, vol. 1801, https://doi.org/10.1007/978-1-4939-7902-8_15,
© Springer Science+Business Media, LLC, part of Springer Nature 2018

Thyroid hormones (TH) play essential roles during early neurodevelopment. Furthermore, in the last few decades, several studies have reported an association between mild maternal thyroid dysfunction and impaired cognitive function outcome in the offspring [3–11]. Due to the critical role of thyroid hormones during pregnancy, raising concern for the impact of EDCs on thyroid signaling. It is therefore necessary to develop high-throughput methods to assess endocrine disrupting potential of the multiple chemicals that are currently on the market.

An advantage of in vivo chemical screening over in vitro cellular assays is that it provides a full spectrum of physiological impacts exerted by a given chemical. Here, we describe a protocol for a high-through put chemical screening to detect endocrine disruption of thyroid axis, using transgenic *Xenopus laevis* embryos. This test is known as the Xenopus Embryonic Thyroid Assay or XETA [12] and has recently been validated at the level of the OECD. Further, using gene expression and behavior analysis, exposed *X. laevis* tadpoles allow for an in-depth look at the possible action on neurodevelopment by the chemical of interest. *Xenopus laevis* is an ideal model organism to test thyroid axis disruption in vivo as thyroid hormones (THs) are highly conserved across vertebrates and orchestrate tadpole metamorphosis. In particular, stage NF45 [13] is most apt for in vivo screening as at this stage the tadpoles possess all the main elements of thyroid hormone signaling (thyroid receptors, deiodinases, transporters) and are metabolically competent [14, 15]. Combination with early embryonic exposure from NF1 to NF41 can provide alternate insight into the early developmental and epigenetic changes brought on by thyroid disrupting chemicals. Taken together, these methods allow for the detection of subtle changes in TH signaling by EDCs that either activate or inhibit TH signaling in vivo.

2 Materials

2.1 Animal Husbandry

1. *Xenopus laevis* mature male and female (Wild type and transgenic TH/bZIP-GFP).
2. Marc's Modified Ringer's (MMR): 0.1 M NaCl, 2.0 mM KCl, 1 mM $MgSO_4$, 2 mM $CaCl_2$, 5 mM HEPES, balanced to pH 7.4.
3. De-jellying solution: 2% Cysteine-hydrochloride in 0.1× MMR. Adjust to pH 7.8–8 with NaOH.
4. Evian™ water 75 cL in glass bottle (or other mineral water with reproducible quality).
5. Human chorionic gonadotropin (hCG).
6. Needles 0.4 × 20 mm.

7. Syringe 1 mL.
8. Glass aquarium (50 L).
9. Incubator (16–30 °C).

2.2 Chemical Exposure

1. 3,3′,5-Triiodo-L-thyronine sodium salt (T_3). Stock solution at 10^{-2}M should be prepared in 30% NaOH 1N, 70% ddH_2o
2. Chemical of interest.
3. Dimethyl sulfoxide (DMSO) or solvent for chemical of interest.
4. 50 mL polypropylen capped tubes (Greiner bio-one).
5. Micro centrifuge tubes 0.2–1.7 mL.
6. Transparent flat 6-well plates (TPP Switzerland).

2.3 XETA Assay

1. Ethyl 3-aminobenzoate methane sulfonate salt (MS-222).
2. Sodium bicarbonate.
3. 96-Well black, conic well plate (Greiner bio-one).
4. Fluorescent microscope equipped with 25× objective and long pass GFP filter.
5. ImageJ—Software.

2.4 Brain Dissections and RNA Extraction

1. Dissecting tools.
2. RNaseZAP™.
3. Micro centrifuge tubes 0.2–1.7 mL.
4. Liquid nitrogen.
5. Centrifuge.
6. RNA extraction kit.
7. RNA quantity and quality assessor: e.g., NanoDrop ThermoScientific & BioAnalyzer (Agilent™).

2.5 Real Time-Quantitative Polymerase Chain Reaction (RT-qPCR)

1. cDNA synthesis kit.
2. PCR machine.
3. PCR tube Strips 0.2 mL.
4. Power SYBR Master Mix.
5. qPCR primers for target genes (Table 1).
6. 384-Well plate for RT-qPCR.
7. RT-qPCR machine: e.g., QuantStudio 6 flex QPCR machine.

2.6 Behavior Analysis

1. Locomotor analysis machine (e.g., DanioVision (Noldus)).
2. 24-Well plate transparent (TPP, Switzerland).

Table 1
Primer list for gene expression analysis using RT-qPCR: Primer list and sequences for the *Xenopus laevis* thyroid hormone signalling components and two reference genes, ef1α1 and odc1

Gene and primer: (F)orward, (R)everse	Sequence
thra F	CGCCTTGGTCTCTTCGGAT
thra R	CCCATACATTGGCTGTTCTTTCTTT
thrb F	AAGAGTGGTTGATTTTGCCAAAA
thrb R	AGGGACATGATCTCCATACAACAG
dio1 F	CAGCAGATGAATGGGGATTGA
dio1 R	TGTCTAACACTACTGGGCAAGAAGGT
dio2 F	AGGCTGAGTGTGGACTTGCT
dio2 R	TGACCTGCTTGTAGGCATCCA
dio3 F	CACAAAAAGTGCGACCAAACG
dio3 R	GCCTTGTTGCAGTTTACT
thibz F	ACCTCCACAGAATCAGCAGC
thibz R	GCAGAGAACGAGCAAGGAGT
klf9 F	TGTGGCAAAGTTTATGGGAAGTCT
klf9 R	GGCGTTCACCTGTATGGACTCT
lat2 F	CAAGAGATGCACTAAAGCTGCC
lat2 R	CCTTGCTTCCAACACCCGAT
odc F	TGAAAACATGGGTGCCTACA
odc R	AAGTTCCATTCCGCTCTCCT
ef1a F	TGGATATGCCCCTGTGTTGGATT
ef1a R	TCCACGCACATTGGCTTTCCT

3 Methods

All the *X. laevis* handling, breeding, and dissection protocols can be found on Xenbase.org (Online resource for Xenopus by the scientific community). All the experiments are carried out room temperature unless stated otherwise.

3.1 Natural Breeding for NF45 Collection

1. Assemble the breeding tank (Glass aquarium, 25 L) with 21 °C dechlorinated aquarium water in a quiet corner. Ensure the tank is covered with black paper to mirror natural dark/night conditions. Place a mesh at the bottom of the tank to ensure adequate space for eggs (*see* **Note 1**).

2. Prepare 1000 Units/mL of hCG and prepare syringes for injections.
3. Inject the wild-type or transgenic *X. laevis* female (*see* **Note 2**) with 500–900 Units of hCG depending on its weight and inject the wild-type or transgenic *X. laevis* male (*see* **Note 2**) with 50–100 Units of hCG. For both, inject into the dorsal lymphatic sac as per the *X. laevis* care and injection guidelines.
4. Place one *X. laevis* couple (male and female) per breeding tank and cover with black paper lid to ensure dark conditions and leave overnight. Ensure the water and aquarium temperature is at 21 °C.
5. Next day, verify amplexus (Xenopus mating hold) and fertilized eggs laid. Remove the adult *X. laevis* couple and leave the fertilised eggs to develop in the tank until stage NF45 [13]. Renew the water in the breeding chamber if it is turbid and maintain a stable water temperature 21–23 °C.
6. Place 15 fertilised embryonic tadpoles per well in a 6-well plate at 23 °C (incubator) ready for chemical exposure.

3.2 In Vitro Fertilization (IVF) for Stage NF1 Collection

1. Prepare holding tanks (Glass aquarium, 25 L) with 21 °C dechlorinated aquarium water. Separate holding tanks for *X. laevis* females and males.
2. Prepare 1000 Units/mL of hCG and prepare syringes for injections.
3. Inject the wild-type *X. laevis* female (*see* **Note 2**) with 500–900 Units of hCG depending on its weight and inject the wild-type *X. laevis* male (*see* **Note 2**) with 50–100 Units of hCG. For both, inject into the dorsal lymphatic sac in accordance with *X. laevis* care and injection guidelines.
4. Place the wild-type *X. laevis* females and males in separate individual holding tanks (1 per tank).
5. Next day, verify *X. laevis* females are laying eggs. If they are, proceed with *X. laevis* male sacrifice in order to obtain testes.
6. To collect testes follow the *X. laevis* recommended dissection protocol. In summary, kill the frog by placing in 4 g/L MS-222 for at least 20 min. Decapitate the spinal cord and dissect the testes (small, white, bean-shaped organs attached to the fat bodies on either side of the midline).
7. Place the testes in 1× MMR on ice or store at 16 °C in an incubator until use.
8. Collect the eggs from the *X. laevis* tanks by squeezing the female. Place the eggs in order to form a monolayer in the dish.

9. Dilacerate the testes in small volume of 0.4× MMR and pass the homogenate over the eggs. Allow the eggs to sit for 10 min and then flood the petri dish with 0.1× MMR (*see* **Note 3**).
10. De-jelly the eggs by replacing the 0.1× MMR with de-jellying solution. Gently agitate the embryonic tadpoles to ensure even exposure. Thoroughly rinse 3+ times with 0.1× MMR.
11. Place 15 fertilized eggs per well in a 6-well plate at 23 °C (incubator) ready for chemical exposure (*see* **Note 4**).

3.3 Chemical Exposure

1. Ensure the chemicals of interest have been dissolved in a correct solvent at a 10,000× concentration (DMSO will be the standard solvent referred here onward) (*see* **Note 5**).
2. For co-exposure with T_3, prepare a final 5 nM T_3 solution in Evian water in a glass flask.
3. Use serial dilution from the 10,000× stock to desired concentration to ensure reproducibility and to avoid any pipetting errors. Use Evian water, with or without T_3, in 50 mL tubes. Prepare a final volume of 10 mL solution of desired concentration (*see* **Note 6**).
4. Control solution should be prepared in a similar fashion where the final 10 mL solution contains the same volume of solvent.
5. Remove the excess water from the 6-well plate containing either stage NF1 or NF45 embryonic tadpoles. Add the 8 mL of chemical solution prepared in the previous step. One well of 15 embryonic tadpoles serves as one condition replicate (Control or chemical concentration).
6. Place the 6-well plate in the 23 °C incubator for the next 72 h.
7. Renew the chemical solutions at precisely 24 h intervals during the 72 h exposure (prepare fresh chemical solutions from stock every day). Remove any dead tadpoles. If the mortality exceeds 20%, stop and repeat the experiment (*see* **Note 7**).

3.4 XETA Assay

Note: This is for stage NF45–48 exposed transgenic (TH/bZIP-GFP) *X. laevis* tadpoles.

1. After 72 h exposure, rinse the tadpoles with Evian water once and place them in 100 mg/L MS-222 (1 g of MS-222 and 1 g of sodium bicarbonate, pH 7.2–8).
2. Transfer tadpoles to black conic based 96-well plates, one tadpole per well.
3. Position the tadpole dorsally using a fine pipette (*see* **Note 8**) so the ventral surface is atop. Make sure that the head of the tadpole is positioned in the middle of the well and the tail is curled around the body. Remove the majority of liquid.

4. Using a fluorescence microscope equipped with GFP long-pass filters, image individual tadpoles at exact zoom levels and time of exposure. Most often the most appropriate is a 25× objective and 3 s exposure (*see* **Note 9**).
5. Once all the images have been taken, they can be analyzed using ImageJ software. Open all the images of condition (control or one concentration) in ImageJ.
6. Exclude non specific signals (for instance intestine and yolk signals) by splitting the image into three layers; red, blue, and green channels. Subtract the red and blue channels from the green one (*see* **Note 10**).
7. Calculate the integrated density of the clean green channel. This is expressed in relative units of fluorescence (RFU).
8. Use appropriate statistical program for data analysis.

3.5 Brain Dissection, RNA Extraction, and RT-qPCR

Note: This is for stage NF1–41 and NF45–48 exposed wild-type *X. laevis* tadpoles. In case of whole tadpoles or different organs, the dissection methods can be modified accordingly.

1. After 72 h exposure, rinse the tadpoles with Evian water once and place the tadpoles in 100 mg/L MS-222.
2. Rinse with RNAse-free water and transfer the tadpoles to petri dish for brain dissection using sterile dissection kits. The latter step can be skipped for whole embryo collection (*see* **Note 11**).
3. Collect two brains per 1.5 mL tube (or one whole tadpole per tube) and snap freeze in liquid nitrogen (*see* **Note 12**).
4. Store the samples at −80 °C until RNA extraction. Avoid storing non-extracted samples for more than 3 months.
5. After RNA extraction, verify the quantity and quality of extracted RNA using the RNA analyzer of your choice (*see* **Note 13**).
6. Extracted RNA can be stored at −80 °C until use.
7. cDNA synthesis should be carried out using supercript RT or using any kit following the manufacturer's instructions.
8. RT-qPCR can be assembled using your kit of choice (e.g., High-Capacity cDNA Reverse Transcription Kit).
9. List of TH signaling primers have been provided in Table 1.
10. Analyze results according to the recommended method by RT-qPCR manufacturer.

3.6 Behavior Analysis

Note: This is for stage NF45–48 exposed wild-type *X. laevis* tadpoles.

1. After 72 h exposure, rinse the tadpoles with Evian water once and place one tadpoles per well in 24-well plate containing 4 mL of Evian water.

2. Allow tadpoles to adapt in the 24-well plate for at least 10 min (*see* **Note 14**).
3. Use a program of alternating 30 s light and 30 s dark period as a stimulus for a total of 10 min.
4. Total distance traveled as well as distance traveled during each period is analyzed (*see* **Note 15**).

4 Notes

1. By placing a mesh at the bottom of the tank, a buffer zone is created where the eggs can accumulate. This also prevents the adults from accessing and eating the eggs.
2. Select *Xenopus laevis* females with a prominent cloaca and pear shaped body. Select males with dark nuptial pads on the forelimbs. Respect the recommended intervals between subsequent breeding, i.e., 3+ months for females and 2+ weeks for males.
3. A brief contraction of the dark pigment at the animal pole of the zygote, an outward sign of fertilization, should be visible within 10 min of adding the 0.1× MMR.
4. Ideally, the chemical preparation from Subheading 3.3 is done beforehand and the chemical solution added at NF1 stage right after the in vitro fertilization (IVF) at this step.
5. DMSO is a common solvent used. In case of other solvents such as ethanol, ensure same solvent control is used. Ideally, the stock solution of chemical of interest will be aliquoted into smaller tubes and frozen at −20 °C without any exposure to light. This ensures no repeated thawing and freezing, and thereby less chance of degradation into metabolites.
6. Rinse every 50 mL plastic tube with Evian water before. This ensures that left over chemicals from plastic tube manufacturing are washed away, if not, diluted.
7. If a mortality is due to a bad breeding or IVF, then stop the experiment. If mortality is only observed at certain concentrations, then the experiment can continue with the morality rate noted per day. Placing the tadpoles into 6-well plates 24 h before start of exposure and keeping them in a 23 °C incubator overnight ensures a smooth transition.
8. The current XETA assay relies on GFP fluorescence in the brain. This procedure can be applied to other fluorescent tagged lines. In such cases, the tadpoles are positioned to ensure a direct view of the organ.
9. This process can be automated using an automated reader such as Tecan Ultra Evolution Microplate Reader (MTX Lab systems, USA).

10. Yellow color of tadpoles may interfere with the image analysis and create false positive results. Yellow signal in head region indicates a dead tadpoles. Such tadpoles should be removed from analysis. Yellow signal at the level of the intestines/yolk is frequent in tadpoles younger stage than stage 46. You might want to crop away the gut area using the program Image J prior to further image analysis.
11. Ensure the dissection tools and surfaces are cleaned through with RNaseZAP. Sterile conditions will ensure high-quality RNA. Washing dissection tools with detergent, 0.1 M NaOH followed by overnight sterilization at 180 °C is a common sterilization technique.
12. According to the RNA extraction kit used, the lysis buffer can be placed in the 1.5 mL tube before snap freezing. Similarly, the number of samples per tube can be adjusted to ensure enough good quality RNA can be extracted. Two brains are providing enough material using Ambion RNAqueous kit.
13. Quality of the extracted RNA can be assessed using Agilent BioAnalyzer. Only samples with RIN higher than 7.5 should be used for further experiments.
14. Transferring and rinsing is highly stressful for the tadpole and could lead to impaired behavioral results.
15. Additional parameters can be investigated such as angle of swimming, motility and acceleration rate depending on the software used.

Acknowledgments

We thank Gérard Benisti, Philippe Durand and Jean-Paul Chaumeil for excellent animal care and thank Sébastien Le Mével for his input in the methods which are routinely used. This protocol has been refined thanks to work supported by grants from Centre National de la Recherche Scientifique (CNRS), Muséum National d'Histoire Naturelle (MNHN), and from European Union DevCom FP7-People-2013-ITN N°607142.

References

1. Delange F (1989) Iodine nutrition and congenital hypothyroidism. In: Research in congenital hypothyroidism. Springer US, Boston, MA, pp 173–185
2. Gaitan E, Lindsay RH, Reichert RD et al (1989) Antithyroid and goitrogenic effects of millet: role of C-glycosylflavones. J Clin Endocrinol Metab 68:707–714
3. Pop VJ, De Vries E, Van Baar AL et al (1995) Maternal thyroid peroxidase antibodies during pregnancy: a marker of impaired child development? J Clin Endocrinol Metab 80:3561–3566
4. Pop VJ, Brouwers EP, Vader HL et al (2003) Maternal hypothyroxinaemia during early pregnancy and subsequent child development:

a 3-year follow-up study. Clin Endocrinol 59:282–288

5. Kooistra L (2006) Neonatal effects of maternal Hypothyroxinemia during early pregnancy. Pediatrics 117:161–167
6. Henrichs J, Bongers-Schokking JJ, Schenk JJ et al (2010) Maternal thyroid function during early pregnancy and cognitive functioning in early childhood: the generation R study. J Clin Endocrinol Metabol 95:4227–4234
7. Costeira MJ, Oliveira P, Santos NC et al (2011) Psychomotor development of children from an iodine-deficient region. J Pediatr 159:447–453
8. Finken MJJ, Van Eijsden M, Loomans EM et al (2013) Maternal hypothyroxinemia in early pregnancy predicts reduced performance in reaction time tests in 5- to 6-year-old offspring. J Clin Endocrinol Metab 98:1417–1426
9. Julvez J, Alvarez-Pedrerol M, Rebagliato M et al (2013) Thyroxine levels during pregnancy in healthy women and early child neurodevelopment. Epidemiology 24:150–157
10. Ghassabian A, El Marroun H, Peeters RP et al (2014) Downstream effects of maternal hypothyroxinemia in early pregnancy: nonverbal IQ and brain morphology in school-age children. J Clin Endocrinol Metab 99:2383–2390
11. Korevaar TIM, Muetzel R, Medici M et al (2016) Association of maternal thyroid function during early pregnancy with offspring IQ and brain morphology in childhood: a population-based prospective cohort study. Lancet Diabetes Endocrinol 4:35–43
12. Fini JB, Le Mevel S, Turque N et al (2007) An in vivo multiwell-based fluorescent screen for monitoring vertebrate thyroid hormone disruption. Environ Sci Technol 41:5908–5914
13. Nieuwkoop PD, Faber J (1994) Normal table of *Xenopus laevis* (Daudin): a systematical & chronological survey of the development from the fertilized egg till the end of metamorphosis Garland Science, 1994 2. ed. Amsterdam : North-Holland Pub. Co., 1967. ISBN 10: 0815318960 ISBN 13: 9780815318965
14. Fini JB, Riu A, Debrauwer L et al (2012) Parallel biotransformation of tetrabromobisphenol A in *Xenopus laevis* and mammals: Xenopus as a model for endocrine perturbation studies. Toxicol Sci. 125(2):359–67
15. Fini J-B, Mughal BB, Le Mével S et al (2017) Human amniotic fluid contaminants alter thyroid hormone signaling and early brain development in Xenopus embryos. Sci Rep 7:43786

Chapter 16

Quantification of Thyromimetic Sobetirome Concentration in Biological Tissue Samples

Jordan Devereaux, Skylar J. Ferrara, and Thomas S. Scanlan

Abstract

Thyroid hormone is a principal regulator of essential processes in vertebrate physiology and homeostasis. Synthetic derivatives of thyroid hormone, known as thyromimetics, display desirable therapeutic properties. Thoroughly understanding how thyromimetics distribute throughout the body is crucial for their development and this requires appropriate bioanalytical techniques to quantify drug levels in different tissues. Here, we describe a detailed protocol for the quantification of the thyromimetic sobetirome using liquid chromatography tandem-mass spectrometry (LC-MS/MS).

Key words Sobetirome, Thyromimetic, Thyroid hormone, LC-MS/MS, Drug quantification, Pharmacokinetics, PK-ADME

1 Introduction

The modern era of thyroid hormone analogs was predicated on a small number of biological techniques developed during the 1980s and 1990s that combined to produce extremely potent agonists. Purified thyroid hormone receptors (TRs) expressed in heterologous systems for accurate measurements of selective binding affinity, cell-based transactivation assays for assessing activity and subtype selectivity, and x-ray crystal structures of TR subtypes bound to selective ligands to illuminate their binding modes and interactions.

The sequence of a thyroid hormone receptor was first reported simultaneously by Weinberger [1] and Sap [2] in 1986, who determined that it was homologous with the retroviral oncogene *c-erbA*. Researchers quickly determined that other variants of the thyroid hormone receptor could be found endogenously and that they had the same binding constants for T3 when expressed in vitro [3]. The surprising result was that while T3 and T4 had roughly equal affinities for the α and β subtypes, other analogs such as tiratricol had different binding affinities for each subtype [4]. Combined with the

Michelina Plateroti and Jacques Samarut (eds.), *Thyroid Hormone Nuclear Receptor: Methods and Protocols*, Methods in Molecular Biology, vol. 1801, https://doi.org/10.1007/978-1-4939-7902-8_16,
© Springer Science+Business Media, LLC, part of Springer Nature 2018

discovery that the subtypes had varying expression levels in each tissue [5, 6], this suggested that ligand subtype selective binding and activation could be a powerful route for producing selective effects. Reports in the early 1990s demonstrated that functional thyroid hormone receptors could be expressed heterologously in *E. coli*—the α-subtype ligand binding domain (LBD) [7] and the full-length β-receptor by itself [8] or as a fusion with GST [9]. This made it practical to perform binding assays on individual TRs.

The ability to express TRs in heterologous systems also made it possible to obtain them in the quantities necessary for growing single crystals. An x-ray crystal structure of the TRα LBD bound to L-T3 was reported by Wagner in 1995 [10]. The TRβ LBD bound to T3 was reported by Darimont in 1998, which allowed researchers to make structural comparisons between the two TR LBDs [11, 12]. More recently, crystal structures of a β-selective thyromimetic bound to the TRα and TRβ LBDs by Bleicher in 2008 gave significant insight into the conformational changes between the receptors that are believed to be the primary basis of the ligand's β-selectivity [13].

At roughly the same time reports were published demonstrating that thyroid hormone could stimulate the expression of specific protein products. A report by Evans in 1982 detailed how thyroid hormones regulate the expression of growth hormone genes in cultured rat pituitary tumor cells—addition of thyroid hormone increased growth hormone mRNA levels fourfold [14]. Reports by Larsen in 1986 and Wight and Glass in 1987 delineated some of the promoter sequences that are recognized and bound by the thyroid hormone receptor [15–17]. They also demonstrated that the sequence was modular and could be inserted upstream of a different gene to render it responsive to thyroid hormone activation. With these molecular tools in hand, Thompson reported in 1989 that it was possible to co-transfect mammalian cells with plasmids coding for a thyroid hormone receptor and a synthetic construct containing a reporter gene driven by a thyroid hormone response element [18]. Because each receptor subtype could be overexpressed separately, this created a powerful tool for assessing compound selectivity and potency. The method was patented by The Salk Institute in 1991, which largely prevented the use of this tool in industry for many years, though it was taken up more readily by academics [19].

The first compound to fully exploit these methods was **GC-1** (Fig. 1), reported by Chiellini in 1998 [20]. Building on the work of Yokoyama [21], it replaced the diaryl ether with a methylene bridge to produce a novel compound that demonstrated tenfold selectivity for TRβ compared to TRα in both in vitro receptor binding and cell-based transactivation assays. The selectivity was demonstrated unequivocally in vivo with a *Xenopus laevis* metamorphosis assay [22] and in mice, rats, and cynomolgus monkeys it was as or

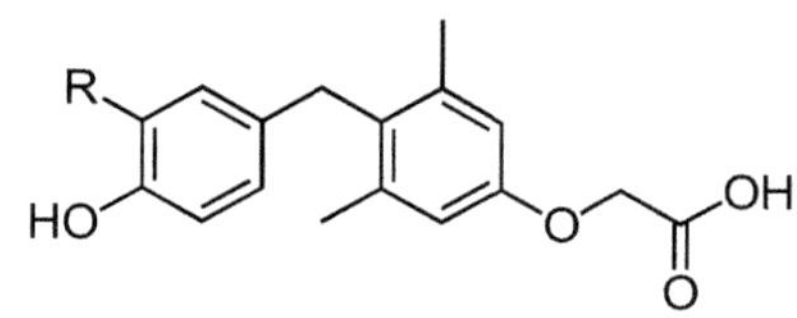

GC-1 R = iPr
GC-24 R = benzyl

Fig. 1 The GC series developed by the Scanlan group

more effective than L-T3 at producing decreases in serum cholesterol and triglycerides with almost no changes in hemodynamic parameters [23, 24]. Clinical development of GC-1 for cholesterol lowering was initiated by QuatRx Pharmaceuticals in 2005. Single and multiple dose Phase 1 studies demonstrated significant decreases in serum cholesterol without affecting TSH or circulating T4 levels. While development of GC-1 as a treatment for hypercholesterolemia was discontinued, a new effort is underway to study GC-1 in the orphan genetic disease x-linked adrenoleukodystrophy [25].

Despite the fact that most chemical features of GC-1 had been chosen primarily for synthetic tractability rather than through rational design [26], many of them turned out to be critical for its potency and selective effects. The diaryl ether and propionic acid analogs first reported by Yokoyama [21] were much less selective than GC-1 [27]. These results suggested that GC-1 was already significantly optimized, but one site for refinement was found in the synthesis of **GC-24** (Fig. 1) [28]. Replacing the 3′-isopropyl with a benzyl group significantly reduced the binding affinity of GC-24 for TRα while leaving the affinity for TRβ intact, increasing the TR subtype selectivity of the compound. The molecular basis of this selectivity was determined by mutational and crystallographic analysis of the receptors [29]. While these features made GC-24 an excellent tool for probing thyromimetic function, it has not supplanted GC-1 as a clinical candidate.

This period saw many of the major pharmaceutical firms pursue thyromimetics programs, largely in an effort to develop LDL cholesterol lowering agents. This led to the exploration of novel chemotypes as research groups worked around the patented chemical space, vastly expanding the range of known active motifs. While most investigated one or two novel chemical modifications, a few went much further.

Karo Bio reported its first series of analogs in 2003 [30] based on patents filed in 1999 and 2000 [31, 32] in partnership with Bristol-Myers Squibb. This series built on earlier iodine-based thyroalkanoic acid analogs by replacing the iodines with alkyl and halogen groups to improve metabolic stability. TRβ selectivity peaked at 14-fold in the 3,5-dichloro series with the acetic acid

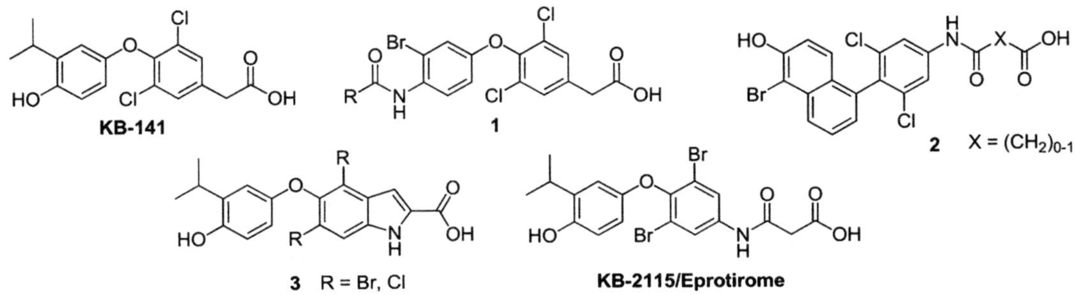

Fig. 2 Karo Bio thyromimetics

analog (Fig. 2—**KB-141**), which also demonstrated tenfold selectivity for cholesterol lowering compared to cardiotoxicity in rats and cynomolgus monkeys [33]. An extension of the series was reported by Hangeland in 2004 that retained the 3,5-dichloro or 3,5-dibromo acetic acid core and varied the 3′ position with substituted phenyl and heterocyclic groups [34]. Another attempt to modulate the selectivity of KB-141 was reported by Garg in 2007 that coupled a series of amino acids through the acetic acid group and added 5′-substituents [35]. A variation on the series (Fig. 2—**1**) was a set of compounds reported by Li in 2006 which replaced the 4′-hydroxyl group with a series of straight and branched chain amides built on a 4′-amino group [36]. No compound in this series improved on the selectivity or potency of KB-141. A new set of compounds (Fig. 2—**2**) reported by Karo Bio took the novel approach of replacing the standard biaryl core with a phenyl-naphthalene core, reported by Hangeland in 2005 [37]. Following this Karo Bio pursued a similar strategy to Bayer in synthesizing thyromimetics with fused ring heterocycles (Fig. 2—**3**). Collazo reported a series in 2006 where the inner ring was replaced by bicyclic nitrogen five- and six-membered rings with carboxylic acid appendages [38]. While a fair amount of chemical space around this basic structure was explored, the relatively simple KB-141 based on the structure of tiratricol remained the best out of these early efforts.

Karo Bio's best TRβ-selective compound KB2115 (Eprotirome) (Fig. 2) was patented in 2004 under Bristol-Myers Squibb [39]. An in vitro binding assay of KB-2115 demonstrated 22-fold TRβ-selectivity [40], but suggestions have also been made that preferential liver uptake played a role in its selective effects in vivo [41]. A Phase I 14-day, once-daily dosing clinical trial was reported in 2008, which showed that KB-2115 could lower plasma total and LDL cholesterol with no observed drug-induced effects on cardiac parameters in overweight and hypercholesterolemic subjects [42]. A Phase II clinical trial involving once-daily dosing over 12 weeks was reported in 2010 that investigated the effects of KB2115 in addition to simvastatin or atorvastatin [43]. KB-2115 plus a statin produced significantly larger decreases in serum cholesterol and LDL compared to a statin alone without cardiotoxicity or biomarkers of bone turnover. The positive results obtained in the first two trials led to a larger Phase III trial with heterozygous familial hypercholesterolemia patients currently receiving the standard of care [44]. The study was initiated in 2011, but terminated early due to induction of chondrodysplasia in dogs during a large animal toxicology study. Though it is unclear how the drug induced cartilage damage, or whether this finding was TR mediated, further development of the drug was halted.

Metabasis Therapeutics patented a series of thyromimetics in 2004 that took a novel approach to selectivity—prodrugs designed to be metabolized in the liver into an active ligand to reduce peripheral exposure and increase selectivity [45]. The compounds were a series of phosphonic acid derivatives coupled to a standard thyromimetic core. Due to the high charge density of phosphonic acid groups reducing their passive transport through plasma membranes, an array of phosphonate ester and phosphonamide derivatives of the series were synthesized to increase the membrane permeability of the prodrug while releasing the active phosphonic acid either through simple hydrolysis or active metabolism in the liver. The in vitro binding assays reported by Boyer in 2008 found that the active forms of the series had low binding affinity and selectivity in comparison to benchmarks such as L-T3, GC-1, KB-141, and KB-2115 [46]. In vivo results were more promising—a number of compounds in the series reduced serum cholesterol by 20–48% in cholesterol-fed rats at doses of 0.2 mg/kg. The cyclic 1-(3-chlorophenyl)-1,3-propanyl prodrug (Fig. 3—**MB07811**) was selected as the best of the series due to acceptable oral bioavailability (10%) and low serum esterase activity resulting in low extrahepatic concentrations of the active form due to specific CYP3A-mediated metabolism of the cyclic ester (Fig. 3). Full in vivo characterization was reported by Erion in 2007 [47]. MB07811 administered once daily to rats at 50 mg/kg did not evoke significant increases in heart rate, hemodynamic parameters, or heart weight. MB07811 entered into Phase 1a and 1b clinical trials in 2006 and reports in 2008 indicated that the drug was well tolerated with healthy patients in both rising single-dose and rising multiple-dose trials with significant reductions in serum cholesterol at doses ranging from 0.25 to 5 mg/kg. However, these positive

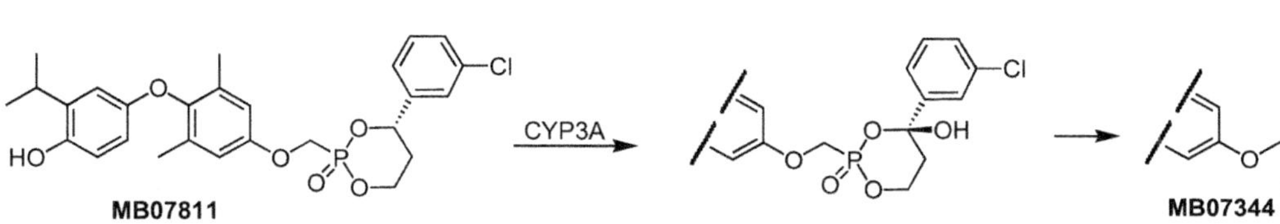

Fig. 3 Metabolic activation of thyromimetic prodrug MB07811 developed by Metabasis Therapeutics

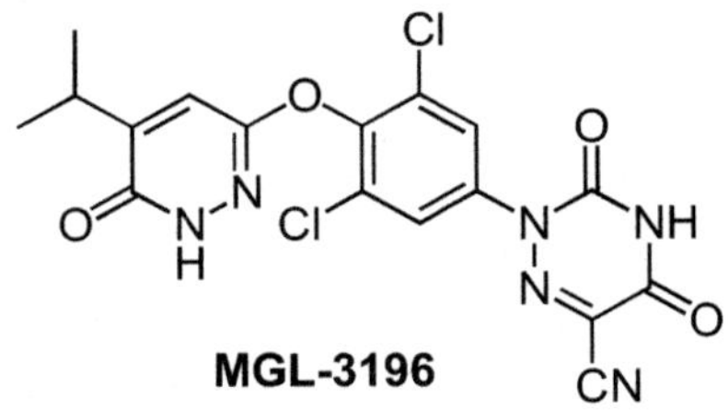

Fig. 4 Thyromimetic developed by Madrigal Pharmaceuticals

results were accompanied by dose-dependent reductions in endogenous thyroid hormone levels, suggesting that the liver-selective prodrug strategy was not effective at eliminating side effects. A phase 2a trial in hypercholesterolemic patients was withdrawn before initiation in 2009.

Madrigal Pharmaceuticals, previously known as VIA Pharmaceuticals, first reported their clinical candidate thyromimetic **MGL-3196** (Fig. 4) in an abstract in 2009 [48]. The compound had initially been discovered by Roche, who licensed it to Madrigal for further development. Synthetic, in vitro, and preclinical in vivo data were reported by Kelly in 2014 [49]. One compound was selected for in vivo testing in diet-induced obesity (DIO) mice and demonstrated significant reductions in serum cholesterol without reducing bone mineral density, increasing heart weight, or perturbing the HPT axis. Madrigal has carried out three separate Phase 1 clinical trials with MGL-3196 in 2011, 2012, and 2015. The results of the first two studies were reported by Taub in 2013 and indicated that MGL-3196 produced the desired outcomes—reductions in serum cholesterol and lipids without changes in heart rate or TSH [50]. Madrigal was granted a new patent for use of MGL-3196 to treat resistance to thyroid hormone in 2015, but no trial results have been reported [51].

A major lesson from these efforts to produce clinical thyromimetics is that target engagement is necessary but insufficient for success. While efforts were made to produce safe drugs through TRβ selectivity and targeting to the liver, their side effects suggest that their distribution was often broader than intended. Oral administration would ensure that some fraction of the drug reached the liver and activate TRs, but it was difficult to constrain these compounds from distributing to extrahepatic tissues. Accurately determining the pharmacokinetic properties of clinical candidates is critical for understanding their effects in vivo and deciding which should advance to trials. Modern bioanalytical techniques have increased the speed and accuracy of these measurements significantly.

Sobetirome (GC-1) is currently being investigated as a therapy for a range of neurological diseases [52]. Few thyromimetics partition to the central nervous system, so it is critical to understand its distribution, so a robust and reproducible bioanalytical technique is necessary to probe its pharmacokinetic properties. Optimized extraction techniques coupled with liquid chromatography tandem-mass spectrometry (LC-MS/MS) can provide quantitative concentrations of the drug in a range of tissues. In this chapter we describe a detailed protocol for analyzing the thyromimetic sobetirome by LC-MS/MS.

2 Materials

2.1 Tissue Sample

Mouse cohorts of three or more are administered a specific dose of sobetirome (typically in μmol/kg) either via tail vein injection (i.v.), orally (p.o.), or by injection (i.p.) and the mice are euthanized at some time post-injection dictated by the particular experiment. Blood and whole brain are the most common tissues collected for analysis and are removed promptly after the animal is euthanized (*see* **Note 1**). Blood is immediately placed on ice (0 °C) then processed into serum and brain is directly frozen on dry ice (−80 °C). After collection tissue samples are stored at −80 °C.

2.2 Internal Standard

Isotopically labeled d_6-sobetirome [53] was used as the internal standard.

2.3 Tissue Homogenization

1. Screw cap microcentrifuge tube (2 mL, Thermo Scientific™, Waltham, MA).
2. Three GoldSpec 1/8 chrome steel balls per tube (Applied Industrial Technologies, Cleveland, OH).
3. Beadbug Microtube Homogenizer (Benchmark Scientific, Edison, NJ) or Bead Ruptor 24 (Omni International, Kennesaw, GA).

2.4 Protein Crash and Sobetirome Extraction

1. Centrifugation. Eppendorf 5415R (Eppendorf, Hauppage, NY) and Beckman Coulter Avanti J-20 XPI with a JA-25.5 fixed angle rotor (Beckman Coulter, Fullerton, CA).
2. Thermo Scientific Savant SPD111V SpeedVac Concentrator attached to a Thermo Scientific Savant RVT4104 Refrigerated Vapor Trap connected to a Thermo Scientific VLP200 vacuum pump (Thermo Scientific, Waltham, MA).

2.5 Sample and Standard Curve Preparation for LC-MS/MS

1. Sobetirome was obtained according to the following literature procedure [54].
2. Screw thread amber vials (2 mL), glass inserts (0.2 mL), and polypropylene assembled screw threaded caps with PTFE/silicone septum (SUN-SRi, Rockwood, TN).

2.6 LC-MS/MS Analysis

QTRAP 4000 hybrid/triple quadrupole linear ion trap mass spectrometer (Applied Biosystems) with electrospray ionization (ESI) in negative mode. The mass spectrometer was interfaced to a Shimadzu (Columbia, MD) SIL-20AC XR auto-sampler followed by 2 LC-20AD XR LC pumps and analysis on an Applied Biosystems/SCIEX QTRAP 4000 instrument (Foster City, CA). A Hamilton PRP-C18 column: 5 μm particle size (50 mm × 2.1 mm stainless steel) was used and kept at 40 °C, and the autosampler was kept at 30 °C. Data were acquired using SCIEX Analyst 1.6.2 software (Framingham, MA, USA) and analyzed using Multiquant 3.0.2.

3 Methods

Blood and brain tissues are collected and stored until processing as described in the above Subheading 2.1. Blood was kept on ice (0 °C) for a minimum of 30 min and then spun down at 7500 × *g* for 15 min. Serum (100 μL) was collected and stored with tissues at −80 °C until samples were processed. Brain tissue was stored at −80 °C until processed.

3.1 Addition of Internal Standard and Tissue Homogenization

Serum samples were warmed to rt and 10 μL of 2.99 μM internal standard (d_6-sobetirome) were added per sample. Brain samples were warmed to rt and transferred to a homogenizer tube with three GoldSpec 1/8 chrome steel balls (Applied Industrial Technologies). The resulting tube was weighed, then 1 mL of H_2O was added, followed by 10 μL of 2.99 μM internal standard (d_6-sobetirome). The tube was homogenized with a Beadbug for 30 s (highest setting) or a Bead Ruptor 24 (*see* Subheading 4, *see* **Note 7** for settings).

3.2 Protein Crash and Sobetirome Extraction

500 μL of acetonitrile were added to each serum sample containing internal standard, then the mixtures were vortexed for 20 s. The samples were centrifuged at 10,000 × *g* for 15 min at 4 °C. 90% of the upper supernatant in each sample was transferred to a glass test tube (13 × 100 mm, Fisher brand) and concentrated using a speedvac for 1.5 h at 45 °C. Brain homogenates were each transferred to a 15 mL falcon tube containing 3 mL of acetonitrile. Additional acetonitrile (1 mL) was used to wash the homogenizer tube and the solution was transferred back to the falcon tube. The samples were then centrifuged at 10,000 × *g* for 15 min at 4 °C. Ninety percentage of the upper supernatant in each sample was transferred to a glass test tube (13 × 100 mm, Fisher brand) and concentrated using a speedvac for 4 h at 45 °C. All dried samples (brain and serum) were then dissolved in 400 μL of 50:50 acetonitrile/H_2O and vortexed for 20 s. The resulting mixtures were each transferred to Eppendorf tubes and centrifuged at 10,000 × *g* for 15 min. The supernatants were filtered with 0.22 μm centrifugal filters (*see* **Note 6**).

3.3 Sample and Standard Curve Preparation for LC-MS/MS

Each extracted sample (from Subheading 3.2) was added to a screw thread amber vial (2 mL) containing a glass insert (0.2 mL) and a polypropylene assembled screw threaded cap with PTFE/silicone septum (SUN-SRi, Rockwood, TN). The standard curve was made with 100 μL of serum from a mouse not injected with sobetirome (vehicle control, *see* **Note 2**). Processing was performed as above, but after filtering the sample was aliquoted into six vials and sobetirome was added to five out of the six vials to give final concentrations in matrix of 0.1, 1, 10, 100, and 1000 pg/μL (*see* **Notes 3**, **4**, and **5**).

3.4 LC-MS/MS Analysis

All samples were run on a QTRAP 4000 hybrid/triple quadrupole linear ion trap mass spectrometer (Applied Biosystems) with electrospray ionization (ESI) in negative mode with a Hamilton PRP-C18 column: 5 μm particle size (50 mm × 2.1 mm stainless steel, *see* **Note 8**). Data were acquired using SCIEX Analyst 1.6.2 software (Framingham, MA, USA) and analyzed using Multiquant 3.0.2.

4 Notes

Quantitative analysis of the thyromimetic sobetirome by LC-MS/MS allows for the determination of tissue distribution and other pharmacokinetic properties of the drug (*see* Figs. 5 and 6) and is relatively straightforward. Particular care must be taken when preparing the standard curve and the following notes address these issues:

1. Sobetirome levels can be measured in a variety of different tissues. Quantitation in serum and brain are described here as examples and can be extended methodologically to other

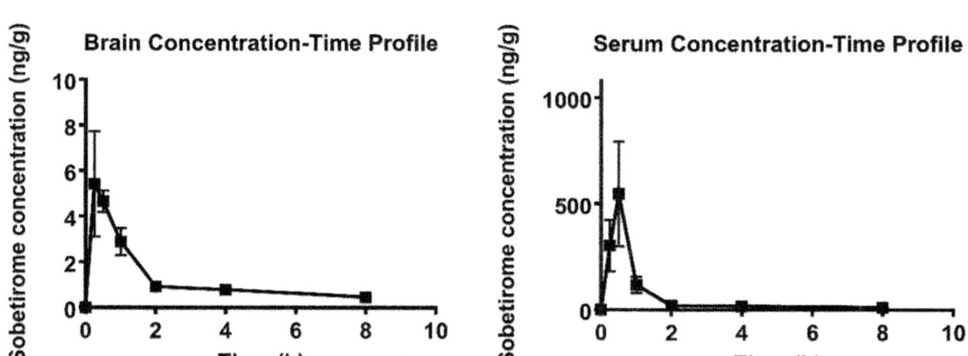

Fig. 5 Brain (left) and serum (right) concentration-time profile curves for sobetirome dosed at 3.05 μmol/kg by i.p. injection

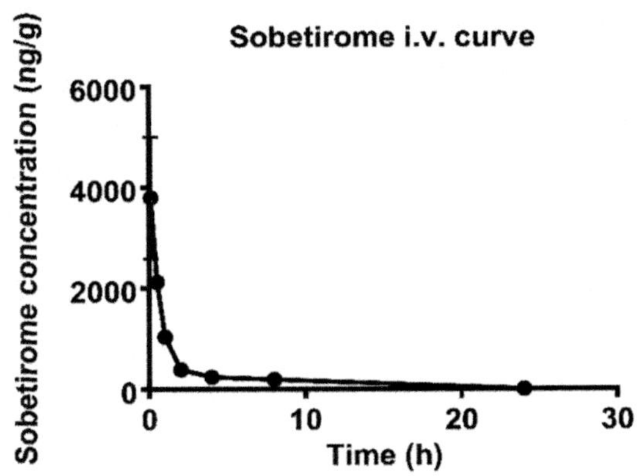

Fig. 6 Concentration-time profile from intravenously dosed sobetirome (3.05 μmol/kg)

tissues of interest. Concentration-time profiles provide pharmacokinetic parameters such as: area under the curve (AUC), half-life, clearance, and volume of distribution.

2. Careful handling and processing of the matrix sample (vehicle control, not administered sobetirome) should be taken as to not contaminate with sobetirome.
3. The standard curve concentrations: 0, 0.1, 1, 10, 100, and 1000 pg/μL are reproducibly useful for analysis of tissue sobetirome concentrations after dosing mice (approximate average weight = 26 g) at a 1 mg/kg or 3.05 μmol/kg dose. The standard curve must be adjusted accordingly if analyzing tissues from dosing much lower or higher than approximately 1 mg/kg.
4. Common complications arise from poor accuracy at the extreme low and the extreme high ends of the standard curve. If this problem occurs, it is helpful to add additional points to the standard curve around these low and high ends.
5. Resulting calculated concentrations in pg/μL from the standard curve are normalized by tissue weight and converted to ng/g or ng/mL.
6. Final extracted brain sample solutions typically contain rather large quantities of endogenous components carried through the extraction given their solubility in acetonitrile and water combinations. Therefore, it is best not to keep these sample solutions at low temperature for any extended period of time to avoid precipitation of these endogenous components. The best results are obtained by preparing the sample solutions within hours of running them on the mass spectrometer.

7. Bead Ruptor 24 (Omni International, Kennesaw, GA) settings for brain homogenization are as follows: Speed: $S = 4.00$, Cycle Time: $T = 1.00$, Number of Cycles: $C = 01$, and Dwell Interval: $D = 0.00$ at 20 °C.
8. Sobetirome and d_6-sobetirome internal standard were analyzed using a QTRAP 4000 hybrid/triple quadrupole linear ion trap mass spectrometer (Applied Biosystems) with electrospray ionization (ESI) in negative mode. The mass spectrometer was interfaced to a Shimadzu (Columbia, MD) SIL-20AC XR auto-sampler followed by 2 LC-20AD XR LC pumps and analysis on an Applied Biosystems/SCIEX QTRAP 4000 instrument (Foster City, CA). The instrument was operated with the following settings: source voltage −4500 kV, GS1 50, GS2 60, CUR 15, TEM 650, and CAD MEDIUM. The scheduled multiple-reaction-monitoring (MRM) transitions are based on the precursor ion m/z and their corresponding diagnostic product ions. Compounds were infused individually and instrument parameters optimized for each MRM transition (Table 1). The gradient mobile phase was delivered at a flow rate of 0.5 ml/min, and consisted of two solvents, A: 10 mM ammonium formate in water and B: 10 mM ammonium formate in 90% acetonitrile, 10% water. A Hamilton PRP-C18 column: 5 μm particle size (50 mm × 2.1 mm stainless steel) was used and kept at 40 °C, and the autosampler was kept at 30 °C. Gradient was as follows, initial concentration of B was 10%, held for 0.5 min, followed by an increase to 98% B over 4.5 min, held for 0.9 min, dropping back to 10% B over 0.1 min, and held at 10% B for 2 min for a total run time of 8 min. Data were acquired using SCIEX Analyst 1.6.2 software (Framingham, MA, USA) and analyzed using Multiquant 3.0.2.

Table 1
Multiple reaction monitoring (MRM) information for sobetirome (GC-1) and deuterated sobetirome (d_6-GC-1)

Compound	Retention time (min)	Q1 Mass	Q3 Mass	DP	EP	CE	CXP
GC-1	*2.88*	*327.3*	*269.3*	*−80*	*−10*	*−28*	*−21*
GC-1	2.88	327.3	269.0	−80	−10	−26	−26
GC-1	2.88	327.3	135.0	−80	−10	−48	−11
d_6 GC-1	*2.86*	*333.0*	*275.2*	*−80*	*−10*	*−34*	*−5*
d_6 GC-1	2.86	333.0	141.1	+90	−10	−48	−7

References

1. Weinberger C, Thompson CC, Ong ES, Lebo R, Gruol DJ, Evans RM (1986) The c-erb-A gene encodes a thyroid hormone receptor. Nature 324:641–646
2. Sap J, Munoz A, Damm K, Goldberg Y, Ghysdael J, Leutz A, Beug H, Vennström B (1986) The c-erb-A protein is a high-affinity receptor for thyroid hormone. Nature 324:635–640
3. Thompson CC, Weinberger C, Lebo R, Evans RM (1987) Identification of a novel thyroid hormone receptor expressed in the mammalian central nervous system. Science 237:1610–1614
4. Schueler PA, Schwartz HL, Strait KA, Mariash CN, Oppenheimer JH (1990) Binding of 3,5,3′-triiodothyronine (T3) and its analogs to the invitrotranslational products of c-erbA protooncogenes: differences in the affinity of the α- and β -forms for the acetic acid analog and failure of the human testis and kidney α-2 products to bind T3. Mol Endocrinol 4:227–234
5. Macchia E, Nakai A, Janiga A, Sakurai A (1990) Characterization of site-specific polyclonal antibodies to c-erbA peptides recognizing human thyroid hormone receptors {alpha}1, {alpha}2, and {beta} and native 3, 5, 3'-triiodothyronine receptor, and study of tissue distribution of the antigen. Endocrinology 126(6):3232–3239
6. Lazar MA (1993) Thyroid hormone receptors: multiple forms, multiple possibilities. Endocr Rev 14:184–193
7. Apriletti JW, Baxter JD, Lau K (1995) Expression of the rat α1 thyroid hormone receptor ligand binding domain in Escherichia coli and the use of a ligand-induced conformation change as a method for its purificiation to homogeneity. Protein Expr Purif
8. Lin K-H, Fukuda T, Cheng SY (1990) Hormone and DNA binding activity of a purified human thyroid hormone nuclear receptor expressed in Escherichia coli. J Biol Chem 265:5161–5165
9. Ball EH, Shephard LB, Gill GN (1995) Purification and properties of thyroid hormone receptor beta 1 expressed in Escherichia coli as a fusion protein. Protein Expr Purif 6:33–38
10. Wagner RL, Apriletti JW, McGrath ME, West BL, Baxter JD, Fletterick RJ (1995) A structural role for hormone in the thyroid hormone receptor. Nature 378:690–697
11. Darimont BD, Wagner RL, Apriletti JW, Stallcup MR, Kushner PJ, Baxter JD, Fletterick RJ, Yamamoto KR (1998) Structure and specificity of nuclear receptor-coactivator interactions. Genes Dev 12:3343–3356
12. Ribeiro RCJ, Apriletti JW, Wagner RL, Feng W, Kushner PJ, Nilsson S, Scanlan TS, West BL, Fletterick RJ, Baxter JD (1998) X-ray crystallographic and functional studies of thyroid hormone receptor. J Steroid Biochem Mol Biol 65:133–141
13. Bleicher L, Aparicio R, Nunes FM, Martínez L, Gomes Dias SM, Figueira ACM, Santos MAM, Venturelli WH, da Silva R, Donate PM, Neves FA, Simeoni LA, Baxter JD, Webb P, Skaf MS, Polikarpov I (2008) Structural basis of GC-1 selectivity for thyroid hormone receptor isoforms. BMC Struct Biol 8:8
14. Evans RM, Birnberg NC (1982) Glucocorticoid and thyroid hormones transcriptionally regulate growth hormone gene expression. Proc Natl Acad Sci U S A 79:7659–7663
15. Larsen PR, Harney JW, Moore DD (1986) Sequences required for cell-type specific thyroid hormone regulation of rat growth hormone promoter activity. J Biol Chem 261:14373–14376
16. Wight PA, Crew MD, Spindler SR (1987) Discrete positive and negative thyroid hormone-responsive transcription regulatory elements of the rat growth hormone gene. J Biol Chem 262:5659–5663
17. Glass CK, Franco R, Weinberger C, Albert VR, Evans RM, Rosenfeld MG (1987) A c-erb-A binding site in rat growth hormone gene mediates trans-activation by thyroid hormone. Nature 329:738–741
18. Thompson CC, Evans RM (1989) Trans-activation by thyroid hormone receptors: functional parallels with steroid hormone receptors. Proc Natl Acad Sci U S A 86:3494–3498
19. Evans RM, Weinberger CA (1991) Hollenberg, Hormone receptor-related bioassays. SM US Patent Office
20. Chiellini G, Apriletti JW, Yoshihara HA, Baxter JD, Ribeiro RCJ, Scanlan TS (1998) A high-affinity subtype-selective agonist ligand for the thyroid hormone receptor. Chem Biol 5:299–306
21. Yokoyama N, Walker GN, Main AJ (1995) Synthesis and structure-activity relationships of oxamic acid and acetic acid derivatives related to Lthyronine. J Med Chem 38(4):695–707
22. Furlow JD, Yang HY, Hsu M, Lim W, Ermio DJ, Chiellini G, Scanlan TS (2004) Induction of larval tissue resorption in Xenopus laevis tadpoles by the thyroid hormone receptor agonist GC-1. J Biol Chem 279:26555–26562
23. Trost SU, Swanson E, Gloss B, Wang-Iverson DB, Zhang H, Volodarsky T, Grover GJ,

Baxter JD, Chiellini G, Scanlan TS, Dillmann WH (2000) The thyroid hormone receptor-beta-selective agonist GC-1 differentially affects plasma lipids and cardiac activity. Endocrinology 141:3057–3064

24. Grover GJ, Egan DM, Sleph PG, Beehler BC, Chiellini G, Nguyen N-H, Baxter JD, Scanlan TS (2004) Effects of the thyroid hormone receptor agonist GC-1 on metabolic rate and cholesterol in rats and primates: selective actions relative to 3,5,3'-triiodo-L-thyronine. Endocrinology 145:1656–1661
25. Scanlan TS, Hartley M, Placzek AT, Righi M (2014) Use of sobetirome in the treatment of x-linked adrenoleukodystrophy. World Intell Prop Org 6, WO 2014/178892 A1
26. Scanlan TS (2010) Sobetirome: a case history of bench-to-clinic drug discovery and development. Heart Fail Rev 15(2):177–182
27. Yoshihara HAI, Apriletti JW, Baxter JD, Scanlan TS (2003) Structural determinants of selective thyromimetics. J Med Chem 46:3152–3161
28. Borngraeber S, Budny M-J, Chiellini G, Cunha Lima ST, Togashi M, Webb P, Baxter JD, Scanlan TS, Fletterick RJ (2003) Ligand selectivity by seeking hydrophobicity in thyroid hormone receptor. Proc Natl Acad Sci U S A 100:15358–15363
29. Wagner RL, Huber BR, Shiau AK, Kelly A, Cunha Lima ST, Scanlan TS, Apriletti JW, Baxter JD, West BL, Fletterick RJ (2001) Hormone selectivity in thyroid hormone receptors. Mol Endocrinol 15:398–410
30. Ye L, Li Y, Mellström K, Mellin C, Bladh L-G, Koehler K, Garg N, Collazo A, Litten C, Husman B (2003) Thyroid receptor ligands 1 Agonist ligands selective for the thyroid receptor β 1. J Med Chem 46:1580–1588
31. Karo Bio AB (1999) Thyroid receptor ligands and method II. US Patent Office December 23
32. Karo Bio AB (2000) Thyroid receptor ligands and method. US Patent Office March 20
33. Grover GJ, Mellström K, Ye L, Malm J, Li Y-L, Bladh L-G, Sleph PG, Smith MA, George R, Vennström B, Mookhtiar K, Horvath R, Speelman J, Egan D, Baxter JD (2003) Selective thyroid hormone receptor-beta activation: a strategy for reduction of weight, cholesterol, and lipoprotein (a) with reduced cardiovascular liability. Proc Natl Acad Sci U S A 100:10067–10072
34. Hangeland JJ, Doweyko AM, Dejneka T, Friends TJ, Devasthale P, Mellström K, Sandberg J, Grynfarb M, Sack JS, Einspahr H, Färnegårdh M, Husman B, Ljunggren J, Koehler K, Sheppard C, Malm J, Ryono DE (2004) Thyroid receptor ligands Part 2: Thyromimetics with improved selectivity for the thyroid hormone receptor beta. Bioorg Med Chem Lett 14:3549–3553
35. Garg N, Li Y-L, Garcia Collazo AM, Litten C, Ryono DE, Zhang M, Caringal Y, Brigance RP, Meng W, Washburn WN, Agback P, Mellström K, Rehnmark S, Rahimi-Ghadim M, Norin T, Grynfarb M, Sandberg J, Grover G, Malm J (2007) Thyroid receptor ligands Part 8: Thyromimetics derived from N-acylated-alpha-amino acid derivatives displaying modulated pharmacological selectivity compared with KB-141. Bioorg Med Chem Lett 17:4131–4134
36. Li Y, Litten C, Koehler K, Mellstrom K (2006) Thyroid receptor ligands Part 4: 4'-amido bioisosteric ligands selective for the thyroid hormone receptor beta. Bioorg Med Chem 16(4):884–886
37. Hangeland JJ, Friends TJ, Doweyko AM, Mellström K, Sandberg J, Grynfarb M, Ryono DE (2005) A new class of high affinity thyromimetics containing a phenyl-naphthylene core. Bioorg Med Chem Lett 15:4579–4584
38. Collazo A-MG, Koehler K, Garg N, Färnegårdh M, Husman B, Ye L, Ljunggren J, Mellström K, Sandberg J, Grynfarb M (2006) Thyroid receptor ligands Part 5: Novel bicyclic agonist ligands selective for the thyroid hormone receptor [beta]. Bioorg Med Chem Lett 16:1240–1244
39. Friends TJ, Ryono DE, Zhang M (2004) Aniline-derived ligands for the thyroid receptor. US Patent Office
40. Shiohara H, Nakamura T, Kikuchi N, Ozawa T, Nagano R, Matsuzawa A, Ohnota H, Miyamoto T, Ichikawa K, Hashizume K (2012) Discovery of novel indane derivatives as liver-selective thyroid hormone receptor β (TRβ) agonists for the treatment of dyslipidemia. Bioorg Med Chem 20:3622–3634
41. Angelin B, Kristensen JD, Eriksson M, Carlsson B, Klein I, Olsson AG, Chester Ridgway E, Ladenson PW (2015) Reductions in serum levels of LDL cholesterol, apolipoprotein B, triglycerides and lipoprotein(a) in hypercholesterolaemic patients treated with the liver-selective thyroid hormone receptor agonist eprotirome. J Intern Med 277:331–342
42. Berkenstam A, Kristensen J, Mellström K, Carlsson B, Malm J, Rehnmark S, Garg N, Andersson CM, Rudling M, Sjöberg F, Angelin B, Baxter JD (2008) The thyroid hormone mimetic compound KB2115 lowers plasma LDL cholesterol and stimulates bile acid synthesis without cardiac effects in humans. Proc Natl Acad Sci U S A 105:663–667

43. Ladenson PW, Kristensen JD, Ridgway EC, Olsson AG, Carlsson B, Klein I, Baxter JD, Angelin B (2010) Use of the thyroid hormone analogue eprotirome in statin-treated dyslipidemia. N Engl J Med 362:906–916
44. Sweetlove M (2012) Phase III trial of eprotirome. Pharm Med 26:185–187
45. Metabasis Therapeutics, Inc. (2004) Phosphorus-containing thyromimetics. US Patent Office. November 16, 2004
46. Boyer SH, Jiang H, Jacintho JD, Reddy MV, Li H, Li W, Godwin JL, Schulz WG, Cable EE, Hou J, Wu R, Fujitaki JM, Hecker SJ, Erion MD (2008) Synthesis and biological evaluation of a series of liver-selective phosphonic acid thyroid hormone receptor agonists and their prodrugs. J Med Chem 51:7075–7093
47. Erion M, Cable E, Ito B (2007) Targeting thyroid hormone receptor-β agonists to the liver reduces cholesterol and triglycerides and improves the therapeutic index. Proc Natl Acad Sci U S A 104(39):15490–15495
48. Taub R, Grimsby J, Larigan JD, Pietranico S, Dvorozniak M, Conde-Knape K (2009) VIA-3196, a liver-directed thyroid beta agonist for treating cardiometabolic disease. Circulation 120:S1095
49. Kelly MJ, Pietranico-Cole S, Larigan JD, Haynes N-E, Reynolds CH, Scott N, Vermeulen J, Dvorozniak M, Conde-Knape K, Huang K-S, So S-S, Thakkar K, Qian Y, Banner B, Mennona F, Danzi S, Klein I, Taub R, Tilley J (2014) Discovery of 2-[3,5-dichloro-4-(5-isopropyl-6-oxo-1,6-dihydropyridazin-3-yloxy) phenyl]-3,5-dioxo-2,3,4,5-tetrahydro[1,2,4] triazine-6-carbonitrile (MGL-3196), a highly selective thyroid hormone receptor β agonist in clinical trials for the treatment of dyslipidemia. J Med Chem 57:3912–3923
50. Taub R, Chiang E, Chabot-Blanchet M, Kelly MJ, Reeves RA, Guertin M-C, Tardif J-C (2013) Lipid lowering in healthy volunteers treated with multiple doses of MGL-3196, a liver-targeted thyroid hormone receptor-β agonist. Atherosclerosis 230:373–380
51. Hester DK II, Duguid RJ, Kelly MJ, Chasnoff A, Dong G, Crow EL, Taub R, Reynolds CH, Choi DS, Wang P (2015), Method of synthesizing thyroid hormone analogs and polymorphs thereof, US Patent Office July 23, 2015
52. Hartley MD, Kirkemo LL, Banerji T, Scanlan TS (2017) A thyroid hormone–based strategy for correcting the biochemical abnormality in XLinked adrenoleukodystrophy. Endocrinology 158:1328–1338
53. Placzek AT, Scanlan TS (2015) New synthetic routes to thyroid hormone analogs: d6-sobetirome, 3H-sobetirome, and the antagonist NH-3. Tetrahedron
54. Placzek AT, Ferrara SJ, Hartley MD, Sanford-Crane HS, Meinig JM, Scanlan TS (2016) Sobetirome prodrug esters with enhanced bloodbrain barrier permeability. Bioorg Med Chem 24:5842–5854

Chapter 17

Analysis of Thyroid Tumorigenesis in Xenograft Mouse Model

Xuguang Zhu and Sheue-Yann Cheng

Abstract

Analysis of thyroid tumorigenesis in xenograft mouse model is important to study human thyroid cancer. Recent studies have made big strides toward understanding the molecular mechanisms by which thyroid hormone nuclear receptors (TR) act to maintain normal cellular functions in growth, differentiation, and development. Despite growing interest, the role of TR in oncogenesis remains to be fully elucidated. Two TR genes give rise to three major TR isoforms: TRα1, TRβ1, and TRβ2. These TR subtypes express in a tissue- and development-dependent manner. Research has been directed at understanding the mechanisms by which TR could mediate aberrant cellular signaling that contributes to oncogenesis, at dissecting possible distinct roles of TR isoforms in oncogenesis, and at the differential susceptibility of target tissues to the oncogenic actions of TR. This chapter gives a brief overview of the current undersatanding of known molecular oncogenic actions of TR. Here, we describe analysis of thyroid tumorigenesis used in interrogating the in vivo oncogenic actions of TR.

Key words Thyroid hormone receptor, Oncogenesis, Thyroid tumorigenesis

1 Introduction

Early indications that TR could be involved in oncogenic actions came from association studies. A loss in the expression of the *THRB* gene because of the truncation/deletion of chromosome 3p where the *THRB* gene is located was reported in many malignancies including lung, melanoma, breast, head and neck, renal cell, uterine cervical, ovarian, and testicular tumors [1–6]. The *THRA* gene locus undergoes frequent loss of heterozygosity in sporadic breast cancer, and rearrangement of the *THRA* gene has also been reported in leukemia, breast, and stomach cancer [7–9]. Somatic mutations of TRs have been found in human hepatocellular carcinoma [10], breast cancer [11], and pituitary tumor [12, 13]. TR mutants identified in human hepatocellular carcinoma lose T3 binding activity and transcription capacity, and exhibit dominant negativity [10].

Michelina Plateroti and Jacques Samarut (eds.), *Thyroid Hormone Nuclear Receptor: Methods and Protocols*, Methods in Molecular Biology, vol. 1801, https://doi.org/10.1007/978-1-4939-7902-8_17,
© Springer Science+Business Media, LLC, part of Springer Nature 2018

The association of the dominant negative action with oncogenic actions of TR was supported by the findings that TRα1 is the cellular counterpart of the retroviral v-ERBA involved in the neoplastic transformation leading to acute erythroleukemia and sarcomas [14, 15]. v-ERBA is a highly mutated chicken TRα1 that does not bind T3 and loses the ability to activate gene transcription. v-ERBA competes with TR for binding to thyroid hormone response elements (TREs) and interferes with the normal transcriptional activity of liganded-TR on several promoters [16, 17]. The discovery that male transgenic mice over-expressing *v-ErbA* develop hepatocellular carcinomas provided the direct evidence to show the v-ERBA oncoprotein can promote neoplasia in mammals through its dominant negative activity [18]. These studies suggested that mutated TRs are involved in tumorigenesis. Studies aiming to secure unequivocal evidence to demonstrate the oncogenic of mutant TR by cell-based studies and murine mouse models are described below.

That dominant negative TRβ mutants could function as an oncogene was demonstrated by studies using genetically engineered mice. A TRβ mutant isolated from a patient, known as PV, suffering from resistance to thyroid hormone was targeted to the *Thrb* gene locus in mice ($Thrb^{PV}$ mice; [19]). $Thrb^{PV/PV}$ mice spontaneously develop follicular thyroid cancer as they age [20]. Using $Thrb^{PV/PV}$ mice, a series of studies elucidated how TRβPV could function as an oncogene [21–23]. TRβPV propels thyroid carcinogenesis by activation of tumor promoters such as cyclin D1 [24], β-catenin [25], phosphatidylinositol 3-kinase (PI3K)/AKT [26–28], pituitary tumor transforming gene (PTTG) [29, 30], and SRC/FAK [31]. TRβPV could also promote thyroid cancer progression by repression of tumor suppressors such as the peroxisome proliferator-activated receptor γ (PPARγ) [32]. These altered signaling pathways during thyroid carcinogenesis of $Thrb^{PV/PV}$ mice are consistent with the changes reported for the carcinogenesis in human thyroid. Molecular mechanisms by which TRβPV acts to aberrantly affect these critical cellular regulators have been elucidated. TRβPV acts at the transcriptional level via dominant negative action to suppress the tumor suppressor PPARγ [33]. TRβPV can also act via direct protein-protein interaction to affect key signaling molecules. TRβPV, by complexing with the p85 subunit of PI3K, constitutively stimulates PI3K's activity to activate its immediate downstream AKT signaling [26, 27]. When complexing with PTTG, TRβPV stabilizes PTTG, resulting in a high accumulation of PTTG in tumor cells. Accumulated PTTG aberrantly delays mitotic progression [29]. In addition, TRβPV, by physically complexing with β-catenin, prevents proteasome degradation of β-catenin, resulting in constitutive activation of β-catenin signaling [34]. Gelsolin, an actin binding protein, is involved in controlling cell morphology, motility, growth, and apoptosis. Studies have indicated that physical association of TRβPV with gelsolin reduces its binding to actin, leading to disarrayed cytoskeletal architectures and increased cell motility, thus contributing to metastasis [35]. These findings reveal diverse molecular mechanisms by which TRβPV acts as an oncogene to promote thyroid carcinogenesis.

In addition to thyroid tumors, *Thrb*$^{PV/PV}$ mice also spontaneously develop thyroid stimulating hormone-secreting tumors (TSH-omas) [36]. The TRβPV induces over-expression of cyclin D1 to propel cell cycle progression by the activation of the cyclin D1-cyclin-dependent kinase-retinoblastoma protein-E2F pathway. The T3-bound wild-type TRβ represses *cyclin D1* expression via tethering to the cyclin D1 promoter in binding to the c-AMP response element binding protein. However, this repression effect is lost in mutant TRβPV which cannot bind T3. The loss of binding thus results in constitutive activation of *cyclin D1* to propel growth of the pituitary in *Thrb*$^{PV/PV}$ mice [36]. These results indicate that the oncogenic action of TRβPV is acting through indirect interaction with another transcriptional factor on the chromatin for transcription activation. At present, it is not clear whether this mechanism by which TRβPV mediates its oncogenic actions is unique in the pituitary or also operates in other target sites.

While correlative data suggested that TRβ mutations could increase the risk of mammary tumor development, unequivocal evidence was still needed. To this end, *Thrb*$^{PV/PV}$ mice were crossed with *Pten*$^{+/-}$ mice to explore how TRβPV could affect the susceptibility of *Pten*$^{+/-}$ mice in the development of mammary tumors [37]. The presence of two *Thrb*^{PV} alleles markedly augments the risk of mammary hyperplasia in *Pten*$^{+/-}$ mice. TRβPV increases the activity of signal transducer and activator of transcription (STAT5) to increase cell proliferation and the expression of the STAT5 target gene encoding β-casein in the mammary gland. Cell-based studies showed that in cells expressing TRβ, T3 suppresses STAT5-mediated transcription activity and downstream target gene expression. This T3-mediated suppression is lost in cells expressing TRβPV, resulting in aberrant activation of STAT5 signaling to stimulate mammary growth [37]. Collectively, these studies showed that TRβPV can act as an oncogene in several target tissues and via diverse mechanisms to promote carcinogenesis depending on the cellular context.

Oncogenic actions of TRβ mutants have also been elucidated by using murine xenograft models. Studies described above clearly demonstrated that a TRβ mutant (TRβPV) functions as an oncogene in several target tissues of *Thrb*$^{PV/PV}$ mice. A fundamental question was whether the oncogenic activity of mutated TRβ is uniquely dependent on the PV mutated sequence (C-terminal 14 amino acid frameshift mutation; [38]). Studies were carried out using the murine xenograft model approach to address this question [39]. Murine xenograft models are used to investigate the factors involved in malignant transformation, invasion, and metastasis, as well as to examine response to therapy. In this type of model, tumor cells expressing the oncogene are transplanted into immunocompromised mice, either under the skin or into the organ type in which the tumor originated.

Using the xenograft mouse model approach, investigators assessed the region of TRβ mutants imparting oncogenic actions in four C-terminal frame-shift mutants—TRβPV, Mkar, Mdbs, and AM. The C-terminal helixes 11 and 12 are critically involved in the structural changes of the ligand binding domain upon binding of T3 [40]. The frame-shift mutated sequence of TRβPV is located in helix 12. The Mkar mutation has a T insertion at nucleotide 1590–1591 that leads to a frameshift mutation in the C-terminal 28 amino acids encompassing helix 11 and 12 [41]. The Mdbs mutation has a C insertion at nucleotide 1643–1644 that leads to a frameshift mutation in the C-terminal 10 amino acids located in helix 12 [41]. AM is a mutant that was constructed to combine the part of the mutation from Mkar (amino acids 436–446) and revert the distal amino acid sequence back to that of wild-type TRβ1 (amino acids 447–461, located in helix 11; [41]). Remarkably, these C-terminal mutants induce similar growth of tumors in mouse xenograft models [39]. These four mutants similarly interact with the p85α regulatory subunit of PI3K to aberrantly activate PI3K-AKT-mTOR and PI3K-ERK-MMP signaling to increase cell proliferation and invasiveness. Further, they also activate the PI3K-STAT3-BIM pathway to decrease apoptosis [39]. These results argue against the idea that the oncogenic activity of TRβPV is uniquely dependent on the TRβPV mutated sequence. Rather, these four mutants could favor a C-terminal conformation that interacts with the C-terminal SH2 domain of p85α to initiate activation of PI3K to relay downstream signaling to promote tumorigenesis. These studies identified three other TRβ1 mutants as oncogenes in addition to TRβPV. Further, they uncovered an "onco-domain" in the mutated C-terminal region of TRβ1.

If mutated TRβ functions as an oncogene, that would support the earlier observations in which the loss in the expression of the *THRB* gene because of the truncation/deletion or epigenetic silencing is associated with many malignancies [1–6]. Mouse xenograft models have provided a valuable tool to address whether re-activation of the silenced *THRB* gene could attenuate thyroid carcinogenesis. A positive correlation between the extent of promoter hypermethylation of the *THRB* gene and the progression of differentiated thyroid cancer was found in patient tissue specimens and in several human thyroid cancer cell lines [42]. When human thyroid cancer cell lines in which the *THRB* gene was silenced by hypermethylation were treated with demethylation agents such as 5 -aza-2 -deoxycytidine (5-aza-CdR) and zebularine, the expression of the *THRB* gene was reactivated concurrently with inhibition of cancer cell proliferation, migration, and tumor growth in a xenograft model. Re-expression of the *THRB* gene inhibited proliferation and migration of thyroid cancer cells through suppressing the activation of β-catenin signaling pathway [42]. These studies demonstrated that TRβ acts as a tumor suppressor in thyroid cancer cells.

With the use of mouse xenograft models, how functional expression of the *THRB* gene could act to inhibit tumorigenesis was delineated in human thyroid cancer cells (FTC-133 and FTC-236) stably expressing of TRβ [43]. Expression of the *THRB* gene in FTC-133 cells, versus control FTC cells without TRβ, reduced cancer cell proliferation and impeded migration of tumor cells through inhibition of the AKT-mTOR-p70 S6K pathway. TRβ expression in FTC-133 and FTC-236 led to less tumor growth in xenograft models. Importantly, new vessel formation was significantly suppressed in tumors induced by FTC cells expressing TRβ versus control FTC cells without TRβ. The decrease in vessel formation was mediated by the downregulation of vascular endothelial growth factor in FTC cells expressing TRβ. These findings indicated that TRβ acts as a tumor suppressor through downregulation of the AKT-mTOR-p70 S6K pathway and decreases vasular endothelial growth factor expression in FTC cells [43]. These studies raise the possibility that TRβ could be considered a potential therapeutic target for thyroid cancer.

Mouse xenograft models were also used to evaluate whether TRβ could act as a tumor suppressor in breast cancer MCF-7 cells, as it does in thyroid cancer cells [44]. Parental MCF-7 cells express the estrogen receptor (ER) but not TRβ. The tumor suppression activity of TRβ was evidenced by the decreased tumor growth in MCF-7 cells stably expressing TRβ (MCF-7-TRβ). Cell-based studies indicated that the estrogen (E2)-dependent growth of MCF-7 cells was inhibited by the expression of TRβ in the presence of T3. In a xenograft mouse model, large tumors rapidly developed after inoculation of MCF-7-Neo cells in athymic mice. In contrast, markedly smaller tumors (98% smaller) were found when MCF-7-TRβ cells were inoculated in athymic nude mice, indicating that TRβ inhibits the E2-dependent tumor growth of MCF-7 cells. Further detailed molecular analysis showed that TRβ acted to activate apoptosis and decrease proliferation of tumor cells, resulting in inhibition of tumor growth. The TRβ-mediated inhibition of tumor growth was elucidated via down-regulation of the JAK-STAT-cyclin D pathways. This in vivo evidence shows that TRβ could act as a tumor suppressor in breast tumorigenesis and provides new insights into the role of TR in breast cancer.

The studies described here illustrate the usefulness of mouse xenograft models to understand tumor suppressor functions of TRβ and to provide unequivocal evidence that TRβ mutants function as an oncogene. Subheadings 2–4 describe the detailed methodology for using the mouse xenograft.

2 Materials

2.1 Making Stable Cell Lines

1. FTC-133 and FTC-236 cell lines.
2. Complete medium for FTC-133 and FTC-236 cells: Dulbecco's modified Eagle's medium/Ham's F12 (1:1) medium supplemented with 10% fetal bovine serum, 10 μg/mL bovine insulin, 1 mIU/mL bovine thyrotropin, 100 units/mL penicillin, and 100 μg/mL streptomycin.
3. Plasmids: pFH-IRESneo and pFH-IRESneo-TRβ1 plasmids containing the G418 resistant gene (the *Neo* gene).
4. G418 200 mg/mL stock solution.
5. 6-well plates, T25 flasks, and T75 flasks.
6. Opti-MEM medium.
7. Lipofectamine 2000 DNA Transfection Reagent.
8. Incubator for cell culture.
9. Culture Hood for cell culture.

2.2 Xenograft Tumor Assay

1. 0.05% Trypsin-EDTA.
2. Complete cell culture medium.
3. Trypan blue solution.
4. Corning Matrigel basement membrane matrix. Leave Matrigel on ice overnight 1 day before recollection of the cells.
5. 1 mL syringes, Gauge 27, length ½ inch.
6. Digital caliper.
7. Sorvall RT6000B refrigerated centrifuge.

2.3 Western Blot Analysis

1. The antibodies against phosphorylated (p-)retinoblastoma (Rb), p-AKT, total AKT, p-p70 S6K, total p70 S6K, p-eIF4B, total eIF4B, p-GSK3β, total GSK3β, p-4EBP1, total 4EBP1, and glyceraldehyde 3-phosphate dehydrogenase (GAPDH).
2. Antibodies against cyclin B1, cyclin D1, cyclin E, total Rb, and VEGF.
3. Tissue lysis buffer: 50 mM Tris–HCl pH 8.0, 150 mM NaCl, 1 mM EDTA, 1.0% NP-40; 0.1% Triton X-100, protease inhibitors.
4. Laemmli 2× buffer/loading buffer: 4% SDS, 10% 2-mercaptoethanol, 20% glycerol, 0.004% bromophenol blue, 0.125 M Tris–HCl, pH 6.8.
5. Running buffer: 25 mM Tris base, 192 mM glycine, 0.1% SDS.
6. Transfer buffer: 31 mM Tris pH 8.3, 240 mM glycine, 20% methanol, 0.04% SDS.
7. Tris-buffered saline with Tween-20 (TBST): 25 mM Tris, pH 7.6, 137 mM NaCl, 3 mM KCl, 0.1% Tween-20.

8. Blocking buffer: Add 5% milk to TBST buffer.
9. Misonix XL-2000 Sonicator.
10. Beckman Coulter DU 640 spectrophotometer.
11. Immobilon-P Membrane, PVDF, 0.45 μm.
12. Western Lightning® Plus-ECL Kit.
13. Electrophoresis Power Supply.
14. Novex XCell SureLock Mini-Cell and XCell II Blot Module, Invitrogen.
15. Sorvall Legend Micro 17 microcentrifuge.

2.4 Immunohistochemistry Analysis

1. Xylene, Sigma.
2. Ethanol (100%, 95%, 75%, and 50%).
3. Hematoxylin.
4. Wash buffer: 1× PBS.
5. Primary antibody against CD31 (as an example, CD31 is an endothelial marker for detecting angiogenesis).
6. 10× phosphate buffered saline (PBS): 137 mM NaCl; 2.7 mM KCl; 4.3 mM Na2 HPO4; 1.47 mM KH2PO4, pH 7.4.
7. 0.05% citraconic anhydride, pH 7.4.
8. 0.3% hydrogen peroxide: To prepare, add 2 mL 30% hydrogen peroxide to 198 mL deionized water.
9. Blocking solution: 10% normal goat serum in 1×PBS.
10. DAB reagent: Prepare according to the manufacturer's recommendations (Peroxidase Substrate Kit, Vector Laboratories, Inc.).
11. Glass slide staining jars.
12. Coverslip: Fisherbrand cover glasses: squares, size: 25 mm.

3 Methods

3.1 Making Stable Cell Lines

Many different approaches are possible for establishing stable cell lines, depending on the type of cells of interest and the constructs to be inserted into the cell genome. The protocol described below is specific for establishing cell lines via the lipofectamine transfection approach. The procedure is used to establish a cell line expressing the wild-type thyroid hormone receptor.

1. For a stable cell line to be created, plate cells in a 6-well tissue culture plate 24 h before transfection such that the cells reach 80% confluence at the time of transfection.
2. Transfect the cells in two wells within a 6-well plate with 2 μg of total plasmid DNA per well. Dilute 5 μL lipofectamine in 250 μL Opt-MEM and 2 μg of plasmid in 250 μL Opt-MEM. Add diluted plasmid to diluted lipofectamine reagent

and incubate for 15 min at room temperature. Add the DNA-lipid complex to the cells and incubate them for 5 h.

3. Add complete medium and culture the cells for 24 h in the incubator.
4. Add 3 mL of media containing 2 mg/mL G418. Change media every 2–3 days. The cells with plasmid integration will survive and form colonies by 7–14 days after transfection.
5. After single colonies form, transfer them first to a 24-well plate and then to a T25 flask or a T75 flask for further expansion of cells.

3.2 Xenograft Tumor Assay

Xenograft mouse models have been widely used to determine the in vivo activity of anti-cancer therapeutics before clinical development. Immunodeficient mice, such as athymic (nude) or severe combined immune deficient mice, are commonly used since these mouse strains exhibit very high growth rates for xenografts. The most common models are xenografts of cell lines grown subcutaneously. These xenograft tumors in nude mice can be easily monitored and measured for tumor progression. The following section describes the procedures for establishing a subcutaneous xenograft mouse model in nude mice.

3.2.1 Preparation of Tumor Cells

1. Use the 5×10^6 cells for injection.
2. Grow cells in complete medium and exclude any contamination.
3. Remove medium and wash cells with 1×PBS. In a T-75 flask, add 2 mL of 0.05% trypsin-EDTA and incubate at 37 °C for 1–5 min. Disperse cells and add complete medium. Collect cells in a 50 mL conical tube.
4. Count cells using a hemocytometer. Mix cells 1:1 with trypan blue solution.
5. Calculate the viable cell number by using the following formula: Average counts × 10,000 × dilution factor = cell numbers/mL medium.
6. Centrifuge at 800 rpm for 3 min. Remove the supernatant and suspend cells to achieve a concentration of 5×10^6 cells/100 μL medium. Add an equal volume of Matrigel and mix well; place on ice for injection.

3.2.2 Preparation of Mice

Nude mice should be 4–6 weeks old.

3.2.3 Preparation of the Injection

1. Sterilize the inoculation area of the mice with an ethanol prep pad. Use a 1 mL syringe with a needle (gauge 27, length ½ in.) for inoculation.
2. Mix cells and then draw into a syringe without a needle. Attach a needle and remove trapped bubbles.

3. Inject 200 μL cells (5×10^6) subcutaneously into the lower right flank of the mice.
4. Start treatment when the tumors have reached an average volume of 100 mm^3.
5. Tumor diameters are measured with a digital caliper.
6. Tumor volume in mm^3 is calculated by the formula: Volume = $(\pi/6) \times (\text{length}) \times (\text{width}) \times (\text{height})$.

3.3 Western Blot Analysis

Western blotting is used to visualize proteins that have been separated by gel electrophoresis based on protein molecular weights. The proteins, as well as the molecular protein markers indicating protein sizes, are separated in a gel and transferred from the gel to a PVDF (polyvinylidene fluoride) membrane. The membrane with transferred proteins can then be probed by primary antibodies specific against the target of interest and visualized using secondary antibodies and detection reagents.

3.3.1 Preparation of Lysate from Tissues

1. Dissect the xenograft tumors from nude mice as quickly as possible to prevent degradation by proteases.
2. Place the tumor tissues in Eppendorf tubes and immerse in liquid nitrogen to snap freeze. Keep tissue samples in a freezer at −80 °C for later use.
3. For tissue homogenization, cut an approximately 10 mg piece of tissue on dry ice. Place the tumor tissues in Eppendorf tubes and add 100 μL of ice-cold lysis buffer rapidly to the tube, homogenize with a homogenizer, and sonicate for two rounds of 10 s on ice with a 30 s interval.
4. Centrifuge for 5 min at 13,800 × *g* at 4 °C in a microcentrifuge. Remove the tubes from the centrifuge and place on ice. Transfer the supernatant to a fresh tube kept on ice and discard the tubes with pellet.

3.3.2 Sample Preparation

1. Use 1 to 2 μL tissue lysate to perform a protein quantification assay. Determine the protein concentration for each cell lysate.
2. Use 30 μg of total protein for the analysis. Calculate how much lysate to load and add an equal volume 2× Laemmli sample buffer.
3. Boil each cell lysate in sample buffer at 95 °C for 5 min to reduce and denature samples.

3.3.3 Loading and Running the Gel

1. Load denatured samples into the wells of the SDS-PAGE gel, along with molecular weight marker.
2. Run the gel for 1–2 h at 120 V until bromophenol blue is close to the bottom of the gel.

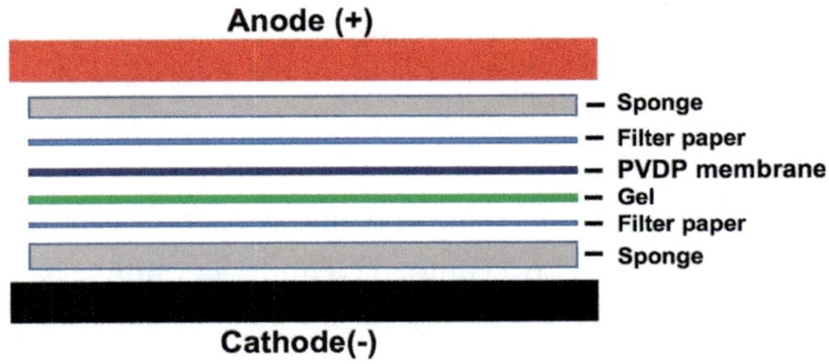

Fig. 1 Assembly for the transfer of proteins from SDS-Polyacrylamide gel to PVDP membrane. PVDP membrane and polyacrylamide gel are sandwiched between two filter paper, Sponges and two plate electrodes

3.3.4 Transferring the Protein from the Gel to the Membrane

Cut the PVDF membrane into a size matched to the size of the gel and activate PVDF with methanol for 5 s and rinse with transfer buffer. Prepare a "filter paper-gel-PVDF membernce-filter paper" stack" (Fig. 1). Set the current at 400 mA and run the transfer for 2–3 h.

3.3.5 Antibody Staining

1. Block the membrane for 1 h in the blocking buffer at room temperature.
2. Incubate the membrane with diluted primary antibody recommended by the manufacturer in blocking buffer for overnight incubation at 4 °C.
3. Wash the membrane in three washes of TBST for 5 min each.
4. Incubate the membrane with the recommended dilution of conjugated secondary antibody in blocking buffer at room temperature for 1 h.
5. Wash the membrane in three washes of TBST for 5 min each.
6. For signal development, follow the kit manufacturer's recommendations. Remove excess reagent and cover the membrane in transparent plastic wrap.
7. Acquire image using darkroom development techniques for chemiluminescence.

3.4 Immunohistochemistry Analysis

Immunohistochemistry analysis provides the most direct approach for identifying both the cellular and subcellular distribution of protein in situ.

3.4.1 Deparaffinization/ Rehydration

1. Bake paraffin slides at 60 °C oven for 3 h.
2. Incubate slides in three washes of xylene for 5 min each.
3. Incubate slides in two washes of 100% ethanol for 5 min each.
4. Incubate slides in two washes of 95% ethanol for 5 min each.
5. Incubate slides in one wash of 75% ethanol for 5 min.
6. Incubate slides in one wash of 50% ethanol for 5 min.
7. Wash sections twice in 1×PBS for 5 min each.

3.4.2 Antigen Unmasking

Incubate the slides in 0.05% citraconic anhydride at 98 °C for 60 min. Cool down on bench top for 30 min.

3.4.3 Staining

1. Wash slides in 1×PBS two times for 5 min each.
2. Incubate slides in 10% normal goat serum in 1xPBS at room temperature for 1 h.
3. Wash slides in 1×PBS twice for 5 min each.
4. Incubate the slides with 100–400 μL primary antibody diluted in 10% normal goat serum in 1×PBS. Incubate overnight at 4 °C.
5. Remove antibody solution and wash slides in 1×PBS six times for 5 min each.
6. Add 100–400 μL horseradish peroxidase conjugated secondary antibody diluted in 10% normal goat serum. Incubate for 1 h at room temperature.
7. Remove secondary antibody solution and wash sections six times with 1×PBS for 5 min each.
8. Add 100–400 μL DAB to each slide and monitor staining closely. As soon as the slides are stained, immerse them in 1×PBS two times for 5 min each.
9. Counterstain slides in hematoxylin for 1 min.
10. Wash slides in 1×PBS two times for 5 min each.

3.4.4 Dehydrate Sections

1. Incubate slides in 95% ethanol two times for 1 min each.
2. Incubate in 100% ethanol two times for 1 min each.
3. Incubate in xylene two times for 2 min each.

3.4.5 Mount Coverslips

To protect the stained tissues, pipet 12 μL of mounting medium permount (Fisher Scientific, Catalog#SP15-100) to the center of the stained tissues. Carefully place the coverslip on the mounting media. This procedure is carefully carried out so that no air bubbles are trapped under the coverslip. After the mount medium is dry, the slides may be used for subsequent observations and images recording.

4 Notes

1. Making stable cell lines
   (a) Only use healthy cells
   Passage cells 2–3 times after thawing before using them in transfection experiments. Do not allow the cells to become overgrown. Only use cells with more than 90% viability.
   (b) Prepare high-quality DNA for transfection.

Prepare DNA using an endotoxin-free protocol. Determine the DNA quality by measuring the OD 260/280 ratio, which should be between 1.7 and 1.9.

(c) Plate cells 1 day earlier.
Seed your cells at a density such that they will be 70–90% confluent at the time of transfection.

(d) Prepare lipid DNA complexes.
For maximum performance, prepare complexing the lipid and DNA in Opti-MEM medium.

(e) Use a positive control such as GFP or LacZ reporter plasmid to assess transfection efficiency.
The percent of cells transfected can be determined using fluorescence microscopy. Alternatively, LacZ expression can be used to determine transfection efficiency.

(f) Choice a selective marker to establish cell lines.
Several selection markers are available for the establishment of cell lines. If feasible, choice puromycin for fast selection.

2. Xenograft tumor assay

(a) Only use healthy cells for subcutaneous inoculation. Do not allow the cells to become overgrown. Use 1–5 × 10^6 cells with more than 90% viability for inoculation.

(b) The matrigel should be thawed overnight on ice in a 2 °C–6 °C refrigerator.

(c) The cells mixed with matrigel should be kept on ice before subcutaneous inoculation.

3. Western blot analysis

(a) Select best primary antibody.

Primary antibody is one of the keys for good experimental results. Search and find the best primary antibody for Western blot analysis.

(b) Keep up the protein transfer efficiency.
Use protein marker to monitor the transfer efficiency.

(c) Make sure to equilibrate membranes and gels on transfer solution.
For PVDF membranes, first immerse PVDF membrane in 100% methanol for a few seconds, then equilibrate with transfer buffer. During incubation with antibody, the membrane should not be allowed to be dry.

(d) Blocking solution
Appropriate blocking agent is critical for Western blot. Blocking solutions work better when supplemented with a mild detergent like Tween-20. Nonfat dry milk or BSA

fraction V at working concentrations ranging between 0.5% and 5% are often used for blocking agent.

(e) Optimize your incubation time
In many cases, a few hour incubation should be enough to visualize the protein of interest, however, overnight incubation at 4 °C will allow more sufficient time for the antigen-antibody reaction.

4. Immunohistochemistry analysis

(a) Sufficient deparaffinization
Due to its hydrophobicity, paraffin is insoluble in antibody solution. An insufficient deparaffinization will lead to uneven immunohistochemistry staining or enhancing background.

(b) Perform heat-induced or protease-induced epitope retrieval
Antigens can be masked as a result of the fixation process, which makes antibody impossible to access antigen. Antigen retrieval is needed to unmask antigens for the binding of antibody.

(c) Block endogenous peroxidases
To avoid staining artifacts, it is important to block endogenous peroxidases prior to using horseradish peroxidase (HRP) antibody conjugates.

(d) Block nonspecific binding sites
Blocking should be performed prior to incubation with the primary antibody to prevent nonspecific antibody binding.

(e) Incubate with primary antibody
Check on the manufacturer's datasheet that the antibody has been tested in the specific immunohistochemical method. During incubation, make sure that no bubbles are trapped inside hybridization solution. It is important to put the slides inside a moisturized chamber to prevent the dryup of solution during overnight incubation.

(f) Incubate with DAB or other substrate solution
In an immunoenzymatic staining, a colored precipitate is formed due to the reaction of an enzyme with its substrate. Make sure that all of slides are stained with same batch of the reagents and same length of time so that it is feasible to compare the intensity of staining among slides.

(g) Counterstain
Counterstaining is crucial for a IHC experiment as the counterstain provides background contrast.

(h) Mount coverslip

Mounting protects the specimen from damage while adding contrast during microscopy. It is important that no air bubbles are trapped under coverslip.

5 Perspectives

The understanding of oncogenic actions of TRβ mutants has been facilitated by the creation of mouse models in which thyroid cancer, and TSHomas spontaneously develop. Extensive elucidation of altered signaling pathways in thyroid carcinogensis in the mouse models has indicated that TRβPV acts as an oncogene via multiple molecular mechanisms. TRβPV can function by interfering with the transcription activity of TRβ by abnormal repression in the expression of tumor promoters (e.g., PPARγ). TRβPV can also act at the transcription level independent of TR, via "gain-of-function." Importantly, TRβPV can also act via extra-nuclear sites, for example by initiating the actions via direct protein-protein interaction with key cellular regulators such as PI3K. Identification of key signaling pathways and regulators propelling thyroid carcinogenesis provides new opportunities for potential molecular targets for diagnosis and treatments.

Moreover, by means of mouse xenograft models, oncogenic action of TRβPV can be studied in cultured cell lines. Using this approach, in addition to TRβPV, the oncogenic actions of several other C-terminal TRβ mutants were also demonstrated, indicating that the oncogenic actions of C-terminal TRβ mutants are not TRβPV-sequence dependent [39]. The mouse xenograft models were also used to elucidate the underlying mechanism by which wild-type TRβ functions as a tumor suppressor in cell lines derived from different tissues [43, 44]. In view of the ease and versatility of mouse xenograft models, one question that would be of interest to ascertain is whether the dominant negative TRα1 mutants could function as an oncogene. Earlier findings demonstrated that V-erbA is a mutated TRα1, which acts in neoplasia by blocking erythroid differentiation and by altering the growth properties of fibroblasts [45]. While the *v-erb*B locus alone is sufficient to induce erythroleukemia and sarcoma independent of the *v-erb*A gene, the *v-erb*A by itself is not capable of independently causing transformation in either erythroid cells or fibroblasts [45, 46]. It is likely that another oncogene such as v-erbB (a mutated version of epidermal growth factor receptor; EGFR) would be needed, as in AEV-induced erythroleukemia and sarcoma, to collaborate with dominant negative TRα1 mutants to bring out the transformed phenotypes. Addressing these challenges and others that may emerge subsequently will certainly lead to recognition and appreciation of the important roles of TR in cancer biology.

Acknowledgments

We regret any reference omissions due to length limitation. We wish to thank all colleagues and collaborators who have contributed to the work described in this review. The research described in this review by the authors and their colleagues at National Cancer Institute was supported by the Intramural Research Program of the Center for Cancer Research, National Cancer Institute, National Institutes of Health.

Reference

1. Ali IU, Lidereau R, Callahan R (1989) Presence of two members of c-erbA receptor gene family (c-erbA beta and c-erbA2) in smallest region of somatic homozygosity on chromosome 3p21-p25 in human breast carcinoma. J Natl Cancer Inst 81(23):1815–1820
2. Chen LC, Matsumura K, Deng G, Kurisu W, Ljung BM, Lerman MI, Waldman FM, Smith HS (1994) Deletion of two separate regions on chromosome 3p in breast cancers. Cancer Res 54(11):3021–3024
3. Gonzalez-Sancho JM, Garcia V, Bonilla F, Munoz A (2003) Thyroid hormone receptors/THR genes in human cancer. Cancer Lett 192(2):121–132
4. Huber-Gieseke T, Pernin A, Huber O, Burger AG, Meier CA (1997) Lack of loss of heterozygosity at the c-erbA beta locus in gastrointestinal tumors. Oncology 54(3):214–219
5. Leduc F, Brauch H, Hajj C, Dobrovic A, Kaye F, Gazdar A, Harbour JW, Pettengill OS, Sorenson GD, van den Berg A et al (1989) Loss of heterozygosity in a gene coding for a thyroid hormone receptor in lung cancers. Am J Hum Genet 44(2):282–287
6. Sisley K, Curtis D, Rennie IG, Rees RC (1993) Loss of heterozygosity of the thyroid hormone receptor B in posterior uveal melanoma. Melanoma Res 3(6):457–461
7. Futreal PA, Soderkvist P, Marks JR, Iglehart JD, Cochran C, Barrett JC, Wiseman RW (1992) Detection of frequent allelic loss on proximal chromosome 17q in sporadic breast carcinoma using microsatellite length polymorphisms. Cancer Res 52(9):2624–2627
8. Yokota J, Yamamoto T, Miyajima N, Toyoshima K, Nomura N, Sakamoto H, Yoshida T, Terada M, Sugimura T (1988) Genetic alterations of the c-erbB-2 oncogene occur frequently in tubular adenocarcinoma of the stomach and are often accompanied by amplification of the v-erbA homologue. Oncogene 2(3):283–287
9. Dayton AI, Selden JR, Laws G, Dorney DJ, Finan J, Tripputi P, Emanuel BS, Rovera G, Nowell PC, Croce CM (1984) A human c-erbA oncogene homologue is closely proximal to the chromosome 17 breakpoint in acute promyelocytic leukemia. Proc Natl Acad Sci U S A 81(14):4495–4499
10. Lin KH, Shieh HY, Chen SL, Hsu HC (1999) Expression of mutant thyroid hormone nuclear receptors in human hepatocellular carcinoma cells. Mol Carcinog 26(1):53–61
11. Silva JM, Dominguez G, Gonzalez-Sancho JM, Garcia JM, Silva J, Garcia-Andrade C, Navarro A, Munoz A, Bonilla F (2002) Expression of thyroid hormone receptor/erbA genes is altered in human breast cancer. Oncogene 21(27):4307–4316. https://doi.org/10.1038/sj.onc.1205534
12. Safer JD, Colan SD, Fraser LM, Wondisford FE (2001) A pituitary tumor in a patient with thyroid hormone resistance: a diagnostic dilemma. Thyroid 11(3):281–291. https://doi.org/10.1089/105072501750159750
13. Ando S, Sarlis NJ, Oldfield EH, Yen PM (2001) Somatic mutation of TRbeta can cause a defect in negative regulation of TSH in a TSH-secreting pituitary tumor. J Clin Endocrinol Metab 86(11):5572–5576. https://doi.org/10.1210/jcem.86.11.7984
14. Sap J, Munoz A, Damm K, Goldberg Y, Ghysdael J, Leutz A, Beug H, Vennstrom B (1986) The c-erb-a protein is a high-affinity receptor for thyroid hormone. Nature 324(6098):635–640. https://doi.org/10.1038/324635a0
15. Thormeyer D, Baniahmad A (1999) The v-erbA oncogene (review). Int J Mol Med 4(4):351–358
16. Yen PM, Ikeda M, Wilcox EC, Brubaker JH, Spanjaard RA, Sugawara A, Chin WW (1994) Half-site arrangement of hybrid glucocorticoid and thyroid hormone response elements

specifies thyroid hormone receptor complex binding to DNA and transcriptional activity. J Biol Chem 269(17):12704–12709

17. Chen HW, Privalsky ML (1993) The erbA oncogene represses the actions of both retinoid X and retinoid a receptors but does so by distinct mechanisms. Mol Cell Biol 13(10):5970–5980
18. Barlow C, Meister B, Lardelli M, Lendahl U, Vennstrom B (1994) Thyroid abnormalities and hepatocellular carcinoma in mice transgenic for v-erbA. EMBO J 13(18):4241–4250
19. Kaneshige M, Kaneshige K, Zhu X, Dace A, Garrett L, Carter TA, Kazlauskaite R, Pankratz DG, Wynshaw-Boris A, Refetoff S, Weintraub B, Willingham MC, Barlow C, Cheng S (2000) Mice with a targeted mutation in the thyroid hormone beta receptor gene exhibit impaired growth and resistance to thyroid hormone. Proc Natl Acad Sci U S A 97(24):13209–13214. https://doi.org/10.1073/pnas.230285997
20. Suzuki H, Willingham MC, Cheng SY (2002) Mice with a mutation in the thyroid hormone receptor beta gene spontaneously develop thyroid carcinoma: a mouse model of thyroid carcinogenesis. Thyroid 12(11):963–969. https://doi.org/10.1089/105072502320908295
21. Guigon CJ, Cheng SY (2009) Novel non-genomic signaling of thyroid hormone receptors in thyroid carcinogenesis. Mol Cell Endocrinol 308(1–2):63–69. https://doi.org/10.1016/j.mce.2009.01.007
22. Guigon CJ, Cheng SY (2009) Novel oncogenic actions of TRbeta mutants in tumorigenesis. IUBMB Life 61(5):528–536. https://doi.org/10.1002/iub.180
23. Kim WG, Cheng SY (2013) Thyroid hormone receptors and cancer. Biochim Biophys Acta 1830(7):3928–3936. https://doi.org/10.1016/j.bbagen.2012.04.002
24. Ying H, Suzuki H, Furumoto H, Walker R, Meltzer P, Willingham MC, Cheng SY (2003) Alterations in genomic profiles during tumor progression in a mouse model of follicular thyroid carcinoma. Carcinogenesis 24(9):1467–1479. https://doi.org/10.1093/carcin/bgg111
25. Guigon CJ, Kim DW, Zhu X, Zhao L, Cheng SY (2010) Tumor suppressor action of liganded thyroid hormone receptor beta by direct repression of beta-catenin gene expression. Endocrinology 151(11):5528–5536. https://doi.org/10.1210/en.2010-0475
26. Furuya F, Hanover JA, Cheng SY (2006) Activation of phosphatidylinositol 3-kinase signaling by a mutant thyroid hormone beta receptor. Proc Natl Acad Sci U S A 103(6):1780–1785. https://doi.org/10.1073/pnas.0510849103
27. Furuya F, Lu C, Willingham MC, Cheng SY (2007) Inhibition of phosphatidylinositol 3-kinase delays tumor progression and blocks metastatic spread in a mouse model of thyroid cancer. Carcinogenesis 28(12):2451–2458. https://doi.org/10.1093/carcin/bgm174
28. Kim CS, Vasko VV, Kato Y, Kruhlak M, Saji M, Cheng SY, Ringel MD (2005) AKT activation promotes metastasis in a mouse model of follicular thyroid carcinoma. Endocrinology 146(10):4456–4463. https://doi.org/10.1210/en.2005-0172
29. Ying H, Furuya F, Zhao L, Araki O, West BL, Hanover JA, Willingham MC, Cheng SY (2006) Aberrant accumulation of PTTG1 induced by a mutated thyroid hormone beta receptor inhibits mitotic progression. J Clin Invest 116(11):2972–2984. https://doi.org/10.1172/JCI28598
30. Kim CS, Ying H, Willingham MC, Cheng SY (2007) The pituitary tumor-transforming gene promotes angiogenesis in a mouse model of follicular thyroid cancer. Carcinogenesis 28(5):932–939. https://doi.org/10.1093/carcin/bgl231
31. Kim WG, Guigon CJ, Fozzatti L, Park JW, Lu C, Willingham MC, Cheng SY (2012) SKI-606, an Src inhibitor, reduces tumor growth, invasion, and distant metastasis in a mouse model of thyroid cancer. Clin Cancer Res 18(5):1281–1290. https://doi.org/10.1158/1078-0432.CCR-11-2892
32. Kato Y, Ying H, Zhao L, Furuya F, Araki O, Willingham MC, Cheng SY (2006) PPARgamma insufficiency promotes follicular thyroid carcinogenesis via activation of the nuclear factor-kappaB signaling pathway. Oncogene 25(19):2736–2747. https://doi.org/10.1038/sj.onc.1209299
33. Araki O, Ying H, Furuya F, Zhu X, Cheng SY (2005) Thyroid hormone receptor beta mutants: dominant negative regulators of peroxisome proliferator-activated receptor gamma action. Proc Natl Acad Sci U S A 102(45):16251–16256. https://doi.org/10.1073/pnas.0508556102
34. Guigon CJ, Zhao L, Lu C, Willingham MC, Cheng SY (2008) Regulation of beta-catenin by a novel nongenomic action of thyroid hormone beta receptor. Mol Cell Biol 28(14):4598–4608. https://doi.org/10.1128/MCB.02192-07
35. Kim CS, Furuya F, Ying H, Kato Y, Hanover JA, Cheng SY (2007) Gelsolin: a

novel thyroid hormone receptor-beta interacting protein that modulates tumor progression in a mouse model of follicular thyroid cancer. Endocrinology 148(3):1306–1312. https://doi.org/10.1210/en.2006-0923

36. Furumoto H, Ying H, Chandramouli GV, Zhao L, Walker RL, Meltzer PS, Willingham MC, Cheng SY (2005) An unliganded thyroid hormone beta receptor activates the cyclin D1/cyclin-dependent kinase/retinoblastoma/E2F pathway and induces pituitary tumorigenesis. Mol Cell Biol 25(1):124–135. https://doi.org/10.1128/MCB.25.1.124-135.2005
37. Guigon CJ, Kim DW, Willingham MC, Cheng SY (2011) Mutation of thyroid hormone receptor-beta in mice predisposes to the development of mammary tumors. Oncogene 30(30):3381–3390. https://doi.org/10.1038/onc.2011.50
38. Parrilla R, Mixson AJ, McPherson JA, McClaskey JH, Weintraub BD (1991) Characterization of seven novel mutations of the c-erbA beta gene in unrelated kindreds with generalized thyroid hormone resistance. Evidence for two "hot spot" regions of the ligand binding domain. J Clin Invest 88(6):2123–2130. https://doi.org/10.1172/JCI115542
39. Park JW, Zhao L, Willingham M, Cheng SY (2015) Oncogenic mutations of thyroid hormone receptor beta. Oncotarget 6(10):8115–8131. https://doi.org/10.18632/oncotarget.3466
40. Figueira AC, Saidemberg DM, Souza PC, Martinez L, Scanlan TS, Baxter JD, Skaf MS, Palma MS, Webb P, Polikarpov I (2011) Analysis of agonist and antagonist effects on thyroid hormone receptor conformation by hydrogen/deuterium exchange. Mol Endocrinol 25(1):15–31. https://doi.org/10.1210/me.2010-0202
41. Wu SY, Cohen RN, Simsek E, Senses DA, Yar NE, Grasberger H, Noel J, Refetoff S, Weiss RE (2006) A novel thyroid hormone receptor-beta mutation that fails to bind nuclear receptor corepressor in a patient as an apparent cause of severe, predominantly pituitary resistance to thyroid hormone. J Clin Endocrinol Metab 91(5):1887–1895. https://doi.org/10.1210/jc.2005-2428
42. Kim WG, Zhu X, Kim DW, Zhang L, Kebebew E, Cheng SY (2013) Reactivation of the silenced thyroid hormone receptor β gene expression delays thyroid tumor progression. Endocrinology 154(1):25–35. https://doi.org/10.1210/en.2012-1728
43. Kim WG, Zhao L, Kim DW, Willingham MC, Cheng SY (2014) Inhibition of tumorigenesis by the thyroid hormone receptor beta in xenograft models. Thyroid 24(2):260–269. https://doi.org/10.1089/thy.2013.0054
44. Park JW, Zhao L, Cheng SY (2013) Inhibition of estrogen-dependent tumorigenesis by the thyroid hormone receptor beta in xenograft models. Am J Cancer Res 3(3):302–311
45. Graf T, Beug H (1983) Role of the v-erbA and v-erbB oncogenes of avian erythroblastosis virus in erythroid cell transformation. Cell 34(1):7–9
46. Frykberg L, Palmieri S, Beug H, Graf T, Hayman MJ, Vennstrom B (1983) Transforming capacities of avian erythroblastosis virus mutants deleted in the erbA or erbB oncogenes. Cell 32(1):227–238

Chapter 18

Human Genetics of Thyroid Hormone Receptor Beta: Resistance to Thyroid Hormone Beta (RTHβ)

Theodora Pappa and Samuel Refetoff

Abstract

Resistance to thyroid hormone beta (RTHβ) is a syndrome characterized by reduced responsiveness of peripheral tissues to thyroid hormone (TH). Affected individuals have consistently high TH levels and non-suppressed thyrotropin in the absence of acute illness, drugs, or alterations in TH binding proteins. Depending on the tissue, features of TH excess and deficiency may coexist, although most individuals have a euthyroid, normal metabolic state at the expense of high TH levels. In most cases the disorder is associated with germline mutations in the *THRB* gene. In the last decades, advances in genetics have expanded our knowledge on the etiology and pathophysiology of the syndrome and have shed more light on the molecular mechanisms of TH action. This review provides an update on the genetics of RTHβ, summarizes the clinical and biochemical presentation of the syndrome, and describes the methodology used to diagnose and manage individuals with RTHβ.

Key words Resistance to thyroid hormone, Thyroid hormone receptor beta, Mutation, Thyroid hormone action

1 Introduction

Impaired sensitivity to thyroid hormone (TH) refers to a group of disorders that impair the effectiveness of TH and includes defects in TH action, cell transport, and metabolism. The most common category is resistance to thyroid hormone beta (RTHβ), a syndrome characterized by reduced action of T_3 at the cellular and tissue level.

The first observation of cases with apparent RTH was made in 1967. In that study, two out of six children of a consanguineous family presented with a combination of goiter, deaf-mutism, stippled epiphyses, high protein-bound iodine (an indirect method for measurement of the serum thyroid hormone concentration available at that time) and normal thyroxine binding capacity introducing the hypothesis of end-organ resistance to the action of thyroid hormones (TH) [1]. Yet, it was not until 1989 that the

Michelina Plateroti and Jacques Samarut (eds.), *Thyroid Hormone Nuclear Receptor: Methods and Protocols*, Methods in Molecular Biology, vol. 1801, https://doi.org/10.1007/978-1-4939-7902-8_18,
© Springer Science+Business Media, LLC, part of Springer Nature 2018

first mutation in the ligand binding domain of the *thyroid hormone receptor beta* (*THRB*) gene was identified in an individual with generalized RTH [2].

Fifty years after the report of the first case, resistance to thyroid hormone beta (RTHβ) is a well-characterized genetic syndrome and constitutes the most common cause of impaired sensitivity to TH. This chapter presents an overview of the genetics and pathophysiology of the disease. The clinical presentation, diagnostic methodology, and clinical management are also discussed in detail.

1.1 Thyroid Hormone Synthesis and Activation

The production and secretion of TH are regulated by the hypothalamus-pituitary-thyroid (HPT) axis. The hypothalamus secretes thyrotropin-releasing hormone (TRH) in the portal system, which stimulates the synthesis and secretion of thyrotropin (TSH) from the anterior pituitary gland. TSH in turn stimulates the synthesis and release of TH from the thyroid gland. Circulating TH levels complete the negative feedback loop regulating the synthesis and secretion of both TRH and TSH (Fig. 1A).

3,3′,5,5′-tetraiodothyronone (thyroxine, T_4) is the major secretory product of the thyroid gland with 3,3′5-triiodothyronine (T_3) being released in much smaller amounts. T_3 is the metabolically active TH and 80% is derived by T_4 conversion to T_3 outside the thyroid gland. At the tissue level, TH transport into cells is regulated by active transmembrane transporters and the local supply of biologically active TH is modulated by deiodinases, which are selenocysteine containing enzymes. T_4 is converted to T_3 by type 1 and type 2 iodothyronine deiodinases (D1 and D2), whereas the role of type 3 deiodinase (D3) is mainly the conversion of T_4 to the metabolically inactive reverse T_3 (rT_3) and to inactivate T_3 by converting it to diiodothyronine (T_2) (Fig. 1B). The reader is referred to the relevant book chapters for a comprehensive review on the HPT axis and thyroid physiology.

1.2 Thyroid Hormone Action

The key steps to ensure appropriate thyroid hormone action include: production of an adequate amount of intact TH, its transport across cell membrane, activation through intracellular metabolism, cytosolic processing and nuclear translocation, binding to TH receptors and interaction with cofactors and other post-receptor effects mediating TH action (Fig. 1B&C).

The Type I TR-dependent TH action is mediated through direct binding of the T_3 associated thyroid hormone receptors (TRs) to DNA; these are TRα1, TRβ1, and TRβ2. The roles of other forms of TH-mediated actions in humans are still under investigation [3]. Following translocation of T_3 to the nucleus, it binds to the TR, which forms a heterodimer with the retinoid X receptor (RXR). This results in conformational changes and dissociation of co-repressors, bound to TR in the absence of T_3. Co-activators are also recruited and specific sequences in the

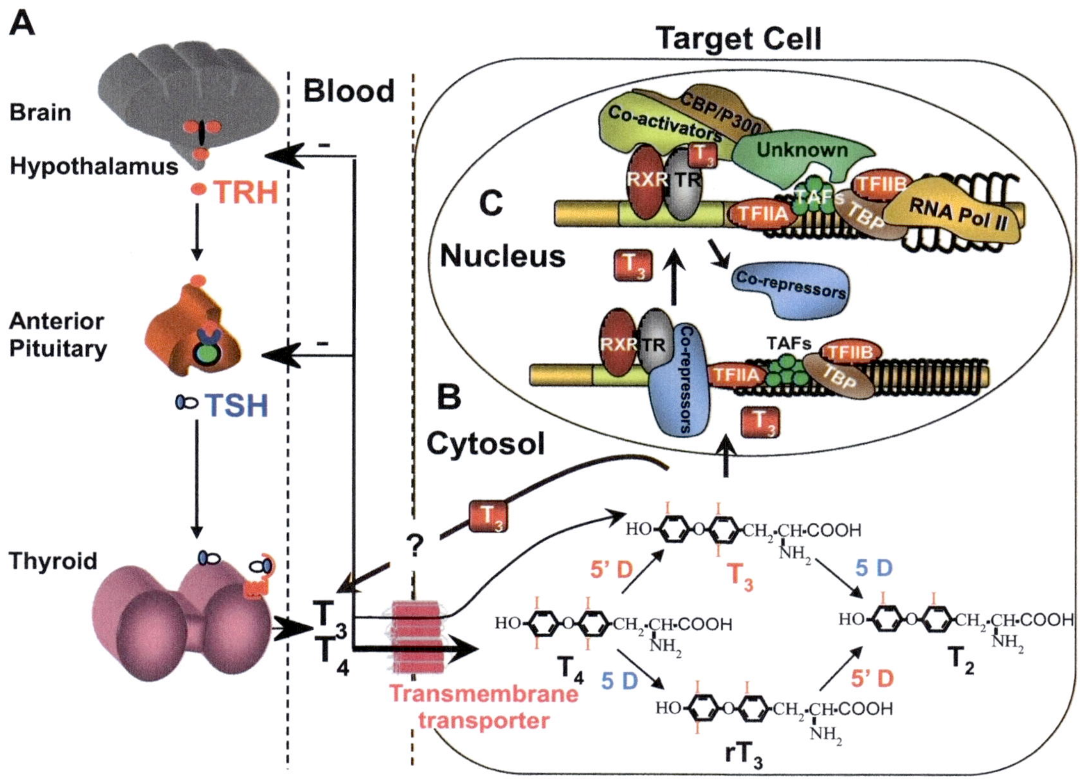

Fig. 1 Regulation of thyroid hormone (TH) supply, metabolism and type 1 TH-dependent action. (**a**) Central negative feedback loop regulating TH level in blood. (**b**) Intracellular TH metabolism through deiodination regulating TH bioactivity. (**c**) Nuclear type 1 TH action. *TRH* TSH releasing hormone, *TSH* thyroid stimulating hormone (thyrotropin), T_4 3,3′,5,5′-tetraiodothyronine (thyroxine, T_4); T_3 3,3′,5-triiodothyronine (T_3), rT_3 3,3′,5′-triiodothyronine (reverse T_3, rT_3), T_2 3,3′-dioidothyronine, 5′D deiodination by removal of an iodine from the 5′ position of iodothyronine, 5D 5-deiodination, T_2 3,3′-dioidothyronine, *TR* TH receptor, *RXR* retinod X receptor, CBP/P300 cAMPbinding protein/general transcription adaptor, *TFIIA and TFIIB* transcription intermediary factor II, A and B, *TBP* TATA-binding protein, *TAF* TBP-associated factor. Modified from: Refetoff S, Dumitrescu. A. Syndromes of reduced sensitivity to thyroid hormone: genetic defects in hormone receptors, cell transporters and deiodination. Best Pract Res Clin Endocrinol Metab 2007; 21(2):277

vicinity of the promoter region (thyroid hormone response elements or TREs) allow binding of TR to DNA driving gene expression and, eventually, protein synthesis. In addition, TRE positions that were occupied with TR homodimers in the absence of T_3, now become available as the TR homodimers are removed in order for active TR-RXR heterodimers to bind and, thus, activate transcription of target genes positively regulated by TH (Fig. 1C) [4].

TRβ is encoded by a gene on chromosome 3. Alternative transcription initiation results in two major isoforms, TRβ1 and TRβ2. TRβ1 is mainly expressed in liver, heart, and brain and TRβ2 is specific for hypothalamus, pituitary, and retina [5].

2 Resistance to Thyroid Hormone Beta (RTHβ)

2.1 Incidence and Inheritance

There is not much data on the incidence of RTHβ, since the diagnosis can easily be missed by neonatal screening based on TSH measurement, as values are normal or only slightly elevated. However, neonatal screening programs based on measurement of T_4 could detect a high value typical for RTHβ. In a study of infants screened by T_4 measurement in the state of Oregon, the prevalence of RTHβ was reported to be 1:40,000 live births [6]. There is no geographical or ethnic predilection and no difference in gender prevalence. With the exception of a few reported cases with recessive inheritance [7], the syndrome is inherited in an autosomal dominant mode.

2.2 THRB Gene Mutations

RTHβ is associated with *THRB* gene mutations located in the regions encoding the ligand (T_3) binding domain and the adjacent hinge region of the receptor molecule. The majority of the mutations are located in three clusters enriched with CpG dinucleotide hot spots in the carboxy terminus of TRβ (Fig. 2) [8, 9], and result in mutant proteins with reduced affinity for T_3, inability to dissociate from co-repressors or associate with co-activators in the presence of TH [9–11]. Most mutations are single nucleotide substitutions leading to an amino acid change or, less frequently, to a truncated protein. Less frequently nucleotide insertions, deletions, and duplications have been described, resulting in frameshifts and nonsense molecules. Mutations in the same codon producing different amino acids have also been reported, such as in codon 453, changing proline to either threonine, serine, alanine, asparagine, tyrosine, histidine, or isoleucine. Somatic *THRB* mutations have been reported in two TSH producing pituitary adenomas [12]. So far, there is no report of a *THRB* mutation in a noncoding region or a splice site, nor in the DNA binding domain. Theoretically, the latter would produce a phenotype only in the homozygous state, as loss of DNA binding will abrogate a dominant negative effect.

In most cases, RTHβ is caused by a germline mutation in *THRB* gene. However, 15% of families with the RTHβ phenotype have no *THRB* mutations, a subgroup referred to as nonTR-RTH. These cases may represent mosaicism in a de novo mutation or are the result of not yet identified causes.

Patients with RTHβ are typically heterozygotes for a mutation in the *THRB* gene, i.e., they have one normal and one mutant *THRB* allele with impaired T_3-binding. The presence of one mutant allele is sufficient to produce a clinical phenotype, because the mutant TRβ interferes with the function of the normal TRβ molecule, a phenomenon named dominant negative effect (DNE) [13]. The mutant TRβ retains the DNA binding capacity and can form dimers with normal TRs or RXRs. Therefore, in patients with

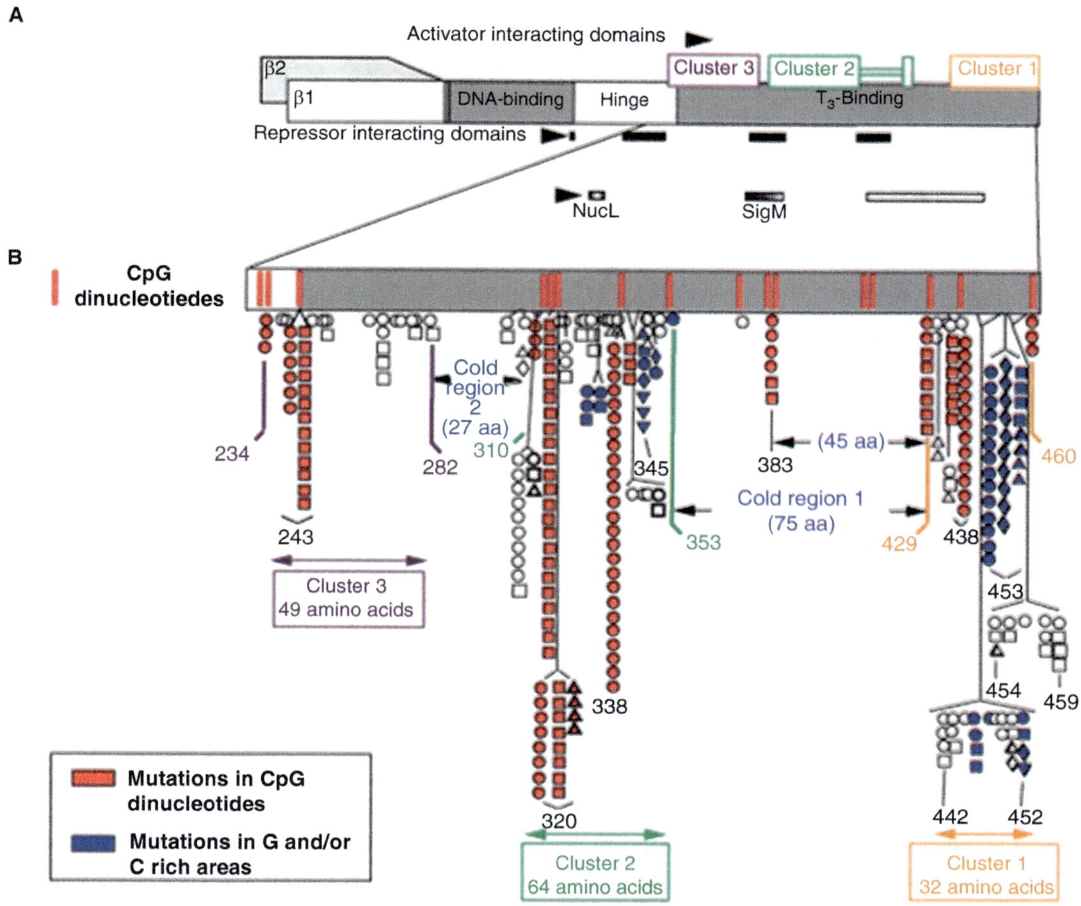

Fig. 2 (**a**) Schematic representation of thyroid hormone receptor beta (TRβ), its functional domains and their relationship to the three clusters of natural mutations. (**b**) The three mutation clusters in the T_3 binding domain and the hinge region are expanded depicting the CpG mutational "hot spots." The location of 124 different mutations (published and unpublished) is each indicated by a symbol. Identical mutations in members of unrelated families are indicated vertically by the same color and symbol. "Cold regions" are devoid of mutations associated with RTHβ. *RTH* resistance to thyroid hormone beta, *AF2* hormone-dependent activation function (12th amphipatic helix), *RBE* corepressor-binding enhancer, *RBI* corepressor-binding inhibitor, *SSD* silencing subdomain, *NucL* nuclear localization, *SigM* signature motif, *TRE* thyroid hormone response element. Modified from Refetoff S, Weiss RE Resistance to thyroid hormone (Chapter 6). IN: Molecular Genetics of Endocrine Disorders, T.V. Thakker (ed.), Chapman & Hill, London, pp. 85–122, 1997

RTHβ, TREs are occupied by inactive hetero- or homodimers containing mutant TRβ unable to bind to T_3 (Fig. 3). This explains why no mutations have been identified in CpGs encoding THRB regions involved in homodimerization (Fig. 2). Another example of how the dominant negative effect can be mediated involves impaired interaction of the mutant TRβ with a cofactor, i.e., increased affinity for a co-repressor [10] or reduced association with a co-activator [11].

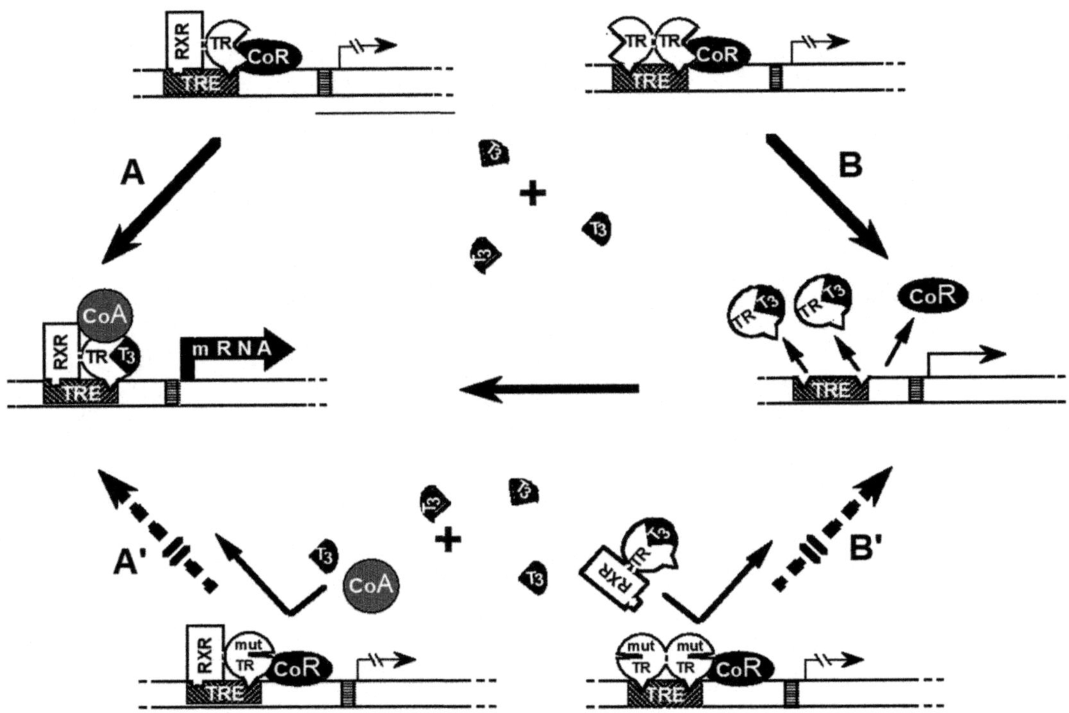

Fig. 3 Schematic overview of the mechanism of dominant negative effect. In the absence of T_3, TRE are occupied by TR hetero- or homodimers (TR-RXR and TR-TR respectively) and associated with a corepressor (CoR) leading to suppressed transactivation. (**a**) In the presence of T_3, corepressors are being released and coactivators (CoA) associate with TR-RXR heterodimers and (**b**) TR dimers are released from TRE allowing binding of active TR-RXR heterodimers. In the presence of a mutant TR (mTR) unable to bind with T_3, mTR-containing-dimers and heterodimers occupy TRE and, thus, T_3 is unable to activate the mTR-RXR heterodimer (A′) or release TREs from the inactive mTR homodimers (B′). Modified from Refetoff S, Weiss RE, Usala SJ. The syndromes of resistance to thyroid hormone. Endocr Rev. 1993;14:348–399

In five out of more than 1000 families RTHβ affected individuals were homozygous, either devoid of a *THRB* gene or expressing only a mutant *THRB*. In the first reported family with RTHβ, affected individuals harbored a deletion of the entire coding region of *THRB* gene in both alleles [7]. They presented with dysmorphic features, stippled epiphyses, deaf-mutism, and retinal changes suggesting color blindness. In a later report, the proband was homozygous for the T337del mutation and manifested with growth and mental retardation, heart abnormalities, and hearing impairment [14]. The third reported case with impaired intellectual development, hyperkinetic behavior and visual and hearing loss harbored only the mutant I280V TRβ. The mutant allele was inherited from the mother, whereas it was unclear if the paternal allele had the same mutation or was deleted [15]. Additionally, three individuals from two different families were found to be homozygous for two *THRB* gene mutations, namely G347E and R316C, and they similarly manifested with developmental delay, growth and mental retardation, hearing impairment and dysmorphic features [16]. All these characteristics are not commonly observed in heterozygous individuals with RTHβ, but were present in all individuals devoid of a normal (wild-type) TRβ molecule. The total absence of TRβ-mediated TH action, necessary for cochlear and photoreceptor development, explains the visual and hearing disorders in RTH individuals devoid of normal TRβ.

Patients with *THRA* gene mutations have a different phenotype than RTH (growth and developmental retardation, skeletal dysplasia, constipation) with distinct thyroid function tests (TFTs) (low T_4, high T_3, very low T_3) [17] (see Briet et al. in this book).

2.3 Clinical Presentation

RTHβ is characterized by a unique pattern of TFTs consisting of high free TH levels (fT_4 and usually also fT_3), but no TSH suppression. TSH is typically normal or slightly high. Nowadays, a more common reason triggering further investigation and testing for RTHβ is the incidental finding of abnormal TFTs.

Prior to the wide availability of thyroid testing, goiter used to be the most common initial finding (usually symmetric and diffuse), present in 65–85% of affected individuals, followed by hyperactivity (33–68%) and tachycardia (33–75%) (Fig. 4). The combination of goiter and tachycardia may lead to the erroneous diagnosis of autoimmune hyperthyroidism. Patients with RTHβ typically have no characteristic clinical signs or symptoms of thyrotoxicosis or hypothyroidism, or, when present, these may be subtle and not constant. Thyrotoxic symptoms include tachycardia, hyperactivity, and high basal metabolic rate. On the other hand, features of hypothyroidism may include delayed bone maturation

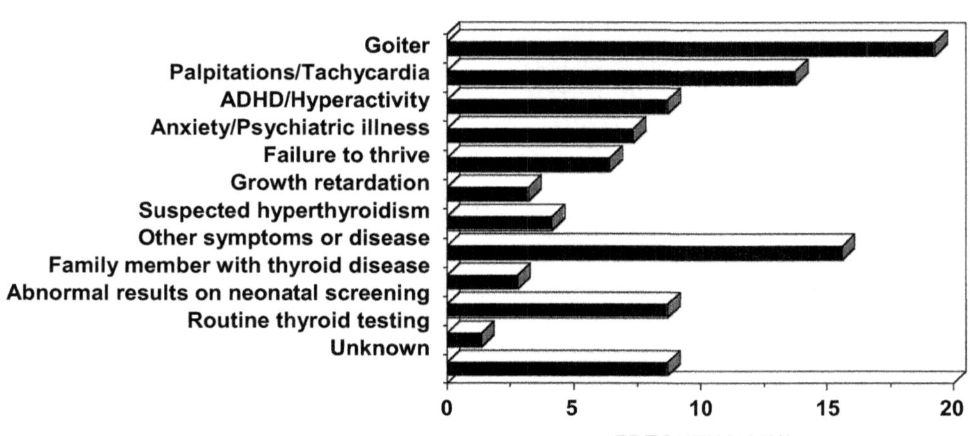

Fig. 4 Reasons for thyroid testing and investigation in patients with RTHβ. Modified from Refetoff S, Weiss RE, Usala SJ. The syndromes of resistance to thyroid hormone. Endocr Rev. 1993;14:348–399

and growth retardation, learning difficulty and developmental delay in children, hearing impairment and nystagmus. Hypothyroid findings are typically present in patients who have erroneously received ablative therapy with radioactive iodine or undergone thyroidectomy to normalize TH levels.

The prevalence of attention deficit hyperactivity disorder (ADHD), learning disabilities, hearing loss, and recurrent ear infections is increased in RTHβ [18]. Reports of co-occurrence of other thyroid disorders in individuals with RTHβ, such as Pendred's syndrome and lingual thyroid, are likely coincidental. It was recently shown that autoimmune thyroid disease is more prevalent in RTHβ. Individuals with RTHβ, especially males, have a higher likelihood to test positive for thyroglobulin (TG) and thyroperoxidase antibodies compared to their unaffected first-degree relatives [19]. A defect in the function of retinal photoreceptors has been recently identified [20].

The vast majority of individuals with RTHβ have a normal metabolic status at the expense of higher TH levels. They have normal growth and development and present only with a small goiter and high TH levels, whereas others have variable degrees of mental and growth retardation. In general, hyperactivity symptoms tend to improve with increasing age. Depending on the TRβ expression level in different tissues [4] there is coexistence of TH excess and deficiency. Hypothalamus and pituitary depend on TRβ expression and manifest relative hormone deficiency, whereas the heart depends mainly on TRα and exhibits signs of TH excess (e.g., sinus tachycardia).

There is inter- and intra-familial variability in the phenotype of RTHβ [21]. Possible explanations include variability in the severity of hormonal resistance, expression of mutant TRβ, effectiveness of the compensatory mechanisms, efficacy of prior therapy, and differences in the genetic background of modulating cofactors affecting TH action [8, 22].

2.4 Laboratory Findings

The cardinal TFT alterations include high serum fT_4 and non-suppressed TSH. Although this finding has been frequently referred to as "inappropriate TSH secretion" the term is rather incorrect, since TSH levels are appropriate for the reduced TH action mediated through the mutant TRβ. The high serum TH concentration counteracts for the defect. Of note, there is increased sialylation and biological potency in the secreted TSH of patients with RTHβ resulting in goiter even with normal serum TSH concentration [23]. The inverse log-linear correlation between serum TSH and fT_4 is preserved at a lower slope, and the latter seems to be dependent on the specific *THRB* mutation [24]. Usually T_3 and rT_3 are elevated. Compatible with TSH stimulation and thyroid gland enlargement, serum TG levels tend to be elevated and thyroidal radioiodide uptake high with normal perchlorate discharge test.

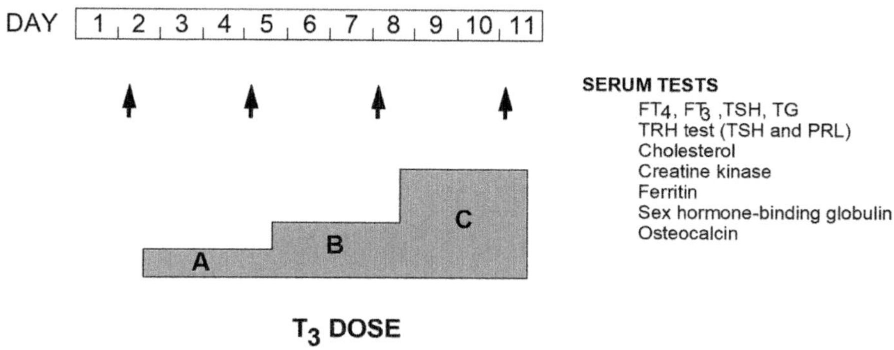

Fig. 5 Diagram of the T_3-suppression test to evaluate the sensitivity to TH using incremental doses of T_3. A, B and C correspond to the three incremental doses of L- T_3 as stated in the text. Modified from Refetoff S, Weiss RE, Usala SJ. The syndromes of resistance to thyroid hormone. Endocr Rev. 1993;14:348–399

A unique lab characteristic and a hint for differentiating RTHβ from autoimmune thyrotoxicosis in the near normal T_3:T_4 ratio [4].

Subjects with RTHβ have reduced TSH and peripheral responses to the administration of incremental doses of T_3 (Figs. 5 and 6). This contrasts with patients harboring a TSH producing pituitary adenoma whose TSH though not suppressible retains intact peripheral responses to T_3. A standardized protocol using short-term administration of liothyronine (L-T_3) to determine the sensitivity of central and peripheral tissues to TH has been devised [4] and adopted worldwide. It involves the administration of three incremental doses of L-T_3, each for three consecutive days; a replacement dose and two supraphysiologic doses. For adults these are 50, 100 and 200 μg/day, given in split daily doses. Laboratory measurements include thyroid function tests and the responses of TSH and prolactin to TRH stimulation test, as well as peripheral markers of TH action: cholesterol, creatine phosphokinase (CK), ferritin, and sex hormone binding globulin (SHBG) concentration. An attenuated stimulation of SHBG and ferritin and a reduced response of TSH to TRH and suppression of cholesterol and CK are compatible with RTHβ (Fig. 6). In pediatric patients the daily dose of T_3 equivalent to the adult 100 μg dose is: 25 μg for ages 1–3 years [8–15 kg body weight (BW)], 50 μg for ages 4–9 years (16–25 kg BW), and 75 μg for ages 10–14 years (26–45 kg BW) with the other doses altered accordingly.

Additionally, in RTHβ the TSH response to TRH is normal or exaggerated, contrary to TSH secreting adenomas, in which TSH classically does not respond to TRH stimulation [4]. The exaggerated TSH response to TRH stimulation can be observed even following administration of incremental doses of T_3 (Fig. 6).

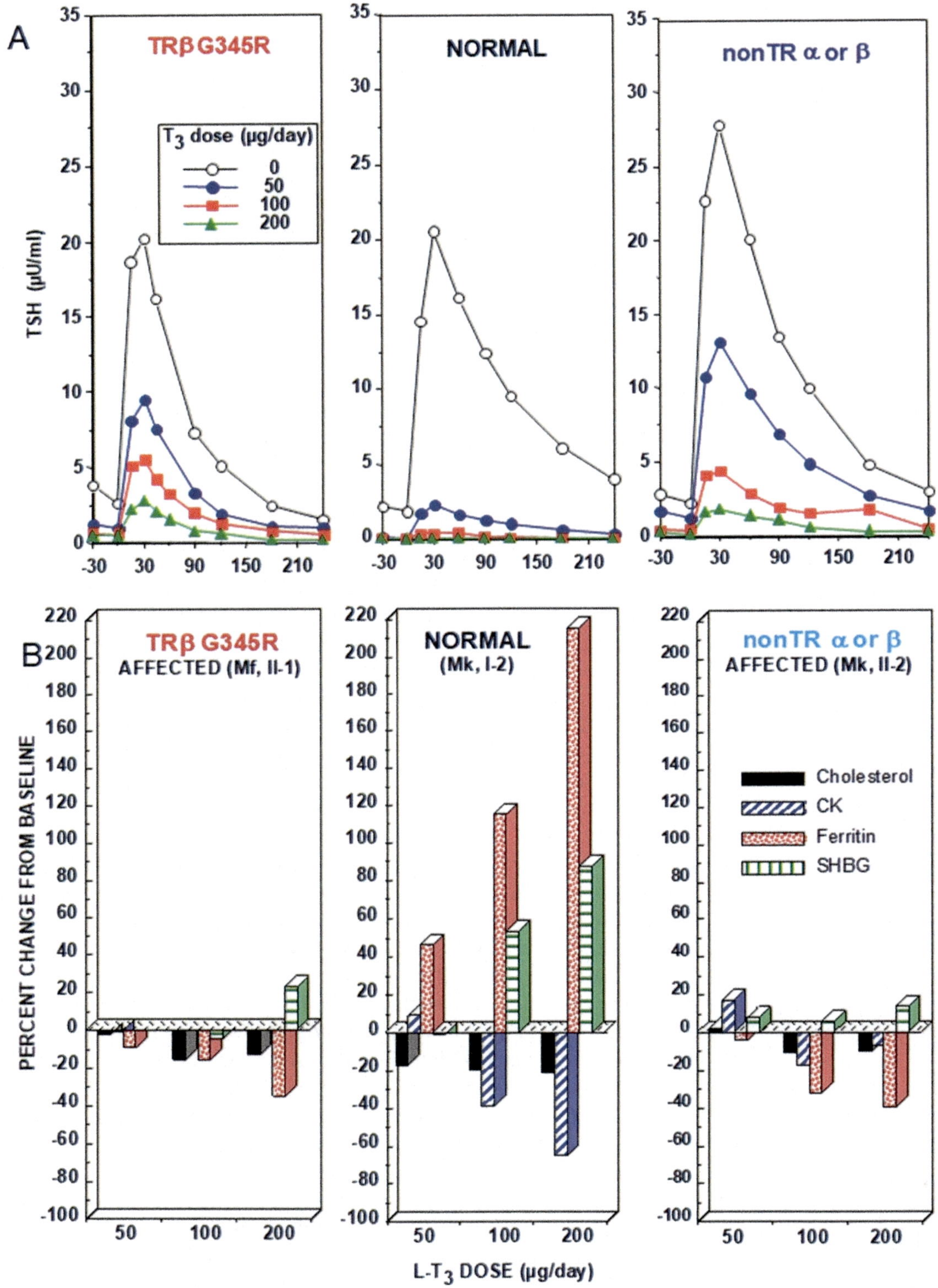

Fig. 6 Responses to the T_3-suppression test in three representative subjects: with RTHβ (due to *THRB* gene mutation) (left), RTH without a mutation in the *THRB* gene (nonTR-RTH, right), and a normal individual (center). T_3 was given in three incremental doses, each for 3 days followed by TRH stimulation test. (**a**) TSH responses to TSH-releasing hormone (TRH) stimulation, require higher doses of exogenous T_3 (L-T_3) for suppression in RTHβ. (**b**) Peripheral tissues responses, blunted (ferritin and SHBG), or paradoxical (cholesterol and creatine kinase) in RTHβ compared to

2.5 Diagnosis

The diagnosis of RTHβ requires a high level of suspicion. Elevation of serum TH levels could represent a laboratory error, a transient phenomenon due to non-thyroid related conditions or interference from medications or other compounds. Therefore, it is prudent to confirm the abnormal TFTs with repeat testing few weeks or months later. It is also important to measure fT_4 levels by equilibrium dialysis to exclude disorders in TH binding proteins, e.g., thyroxine binding globulin, albumin, and transthyretin. Of note, in familial dysalbuminemic hyperthyroxinemia fT_4 levels could be falsely elevated when measured using the direct method. In the presence of positive TG and/ or thyroperoxidase antibodies, there is the possibility of iodothyronine antibodies resulting in high serum TH values.

A cost effective measure is to perform TFTs in the parents and siblings of the individual suspected of having RTHβ, which is typically dominantly inherited. The presence of similar TFT abnormalities supports the diagnosis of RTHβ.

The diagnosis of RTHβ should be confirmed by genetic testing involving direct sequencing of DNA from the individual suspected of having RTHβ based on clinical and laboratory features. Identification of a *THRB* gene mutation provides a solid diagnosis in a patient with abnormal TFTs, allows prenatal diagnosis and counseling and prevents unnecessary treatment. However, absence of a mutation does not rule out the defect, such as in cases with mosaicism and nonTR-RTH. In these cases the L-T_3 suppression test could be useful to confirm the diagnosis of RTHβ. A proposed diagnostic algorithm in patients with suspected RTHβ is provided in Table 1.

Table 1
Diagnostic algorithm in individuals with clinical suspicion of RTHβ

Common presentation: High fT4+ non-suppressed TSH
Confirm elevated fT4; exclude TH transport defects; Measure fT4 with equilibrium dialysis
Thyroid testing in first-degree relatives; Parents, siblings and children
Sequencing of the *THRB* gene-> Positive result=RTHβ diagnosis
Negative result+ high clinical suspicion-> Measure serum α-subunit to exclude TSH secreting adenoma
Modified T3 suppression test: Blunted TSH suppression and peripheral responses = RTH

Fig. 6 (continued) the normal individual. *RTHβ* resistance to thyroid hormone beta, *TR* thyroid hormone receptor, *TRβ* thyroid hormone receptor beta molecule, *TRα* thyroid hormone receptor alpha molecule, *TSH* thyroid stimulating hormone, *TRH* thyrotropin-releasing hormone. Modified from Sadow PM, Reutrakul S, Weiss RE, Refetoff S. Resistance to thyroid hormone in the absence of mutations in the thyroid hormone receptor genes Current Op. Endocrinol. Diabetes, 7:253–259, 2000

The clinical manifestations of RTHβ vary and are nonspecific and the syndrome has to be differentiated from others with reduced sensitivity to TH, also characterized by high TH levels in association with non-suppressed TSH, normal or slightly elevated, such as TH metabolism defects due to *SBP2* gene mutations (Table 2). Furthermore, when an adult individual presents with TFTs suspicious of RTHβ, one should exclude the possibility of a TSH-producing pituitary adenoma, especially when both parents have normal TFTs, and measure the alpha-subunit of pituitary glycoproteins prior to pituitary imaging.

2.6 Management

It is important to remember that the treatment goal should not be to normalize the elevated TH levels. The clinician should factor in the age and symptoms of the patient and response to previous therapy.

Tachycardia is a very common clinical finding and can be managed with β-blockers, such as atenolol. The latter is preferred, as it does not interfere with the TH metabolism. It is also possible that the ADHD may improve with β-blocker treatment, although we recommend standard therapy for ADHD. Regarding goiter, supraphysiologic doses of T_3 given every other day aiming to suppress TSH can reduce the size of the gland without side effects. The effective level of TSH suppression is assessed by the suppression of serum TG [25]. This is particularly relevant in individuals who

Table 2
Thyroid function tests and features in clinical conditions to be differentiated from RTHβ

Defect	T4	T3	rT3	TSH	fT4 direct	fT4 dialysis	Prevalence[a]	Other manifestations
TBG excess	↑	↑	↑	N	N	N	Common	
Increased TTR[b]	↑	N	↑	N	N	N	Rare	
FDH	↑	↑ or N	↑	N	↑	N	Rare/common[c]	
RTH (*THRB* mutations or nonTR-RTH)	↑	↑ or N	↑	Sl ↑ or N	↑	↑	Uncommon	ADHD, tachycardia, goiter
TSHoma	↑	↑	↑	Sl ↑ or N	↑	↑	Rare	
SBP2/SECISBP2 mutation	↑	↓	↑	Sl ↑ or N	↑	↑	Unknown	Growth delay azoospermia
Acute NTI due to invection	↑	↓↓	↑	N	↑ or N	N	Common	

TBG thyroxine binding globulin, *TTR* transthyretin, *FDH* familial dysalbuminemic hyperthyroxinemia, *RTH* resistance to thyroid hormone, *TSHoma* TSH secreting pituitary adenoma, *SBP2* selenocysteine insertion binding sequence protein, *NTI* non-thyroidal illness, *sl* slight, *N* normal, ↑ increased, ↓ decreased, *ADHD* attention deficit hyperactivity disorder.
[a]"Common" indicates >1:300; "Uncommon" indicates >1:50,000; "Rare" indicates <1:50,000
[b]Transthyretin with high affinity for T4 and rT3
[c]Depending on ethnicity

have undergone thyroidectomies, because postoperative recurrence of the goiter is the rule.

A rare occasion of an infant with a severe form of RTHβ has been documented; the affected infant harbored a frameshift mutation producing a mutant TRβ with an extended nonsense C-terminus sequence and presented with failure to thrive, tachycardia uncontrolled by β-blockers and hypermetabolism. In such severe cases, one could consider short-term use of somatostatin to reduce TH levels or cautious administration of anti-thyroid drugs or triiodothyroacetic acide (TRIAC) [26].

In the majority of individuals with RTHβ, the increase in serum TH levels will compensate for the tissue resistance to TH. However, those with prior ablative treatment (with radioactive iodide or surgery) have limited thyroid reserve and should be treated with T_4. Adjusting the dosage in TH replacement in these patients is complicated and TSH concentration can be used to gauge T_4 requirements, which may be very high. Ablative treatment should be avoided, because management and follow up become very complicated.

TH is critical for mental development especially in the first years of life. Any sign of TH deficiency in an infant with diagnosed RTHβ (e.g., delayed bone maturation, failure to thrive, TSH elevation) deserves consideration for TH treatment. T_4 therapy is indicated in the rare patient homozygous for a *THRB* gene mutation, in whom several clinical findings suggestive of TH deficiency may be present. Factors such as linear growth, bone maturation, and mental development have to be considered when adjusting T_4 doses, especially in children. The growth and development patterns of older siblings who did not receive treatment may be used as a guideline. In addition, T_4 treatment should be considered in patients, in whom the degree of peripheral resistance cannot be overcome by the increase in endogenous TH production and secretion, and features of hypothyroidism are present. Dose adjustment should again be based on the individual's responses; biochemical indices to monitor include serum SHBG, CK, cholesterol and ferritin levels and basal metabolic rate.

The TH analogue TRIAC has been used in patients with RTHβ. TRIAC has a short half-life and some mutant TRβ proteins show greater affinity for TRIAC compared to T_3. These unique properties explain TRIAC's efficacy in reducing TSH and TH levels and improving the hypermetabolic state by acting more on the pituitary than on peripheral tissues [27, 28]. However, its efficacy has not been widely validated and its use remains limited.

In families with a known *THRB* gene mutation, genetic counseling is valuable and prenatal diagnosis of RTH can be made on DNA obtained by amniocentesis or chorionic villous sampling and testing for the presence or absence of maternal mutation in the fetus' sample. Of note, pregnant women with RTHβ carrying unaffected fetuses have a high rate of abortion and surviving unaffected infants have a significantly lower birth weight and suppressed TSH levels at birth, possibly due to an intrauterine environment with high TH levels incongruent to the fetal TH requirements [29]. In these cases, judicious use of antithyroid medication to maintain maternal fT_4 levels not higher than 20% above the upper limit of normal seems to prevent the otherwise expected low birth weight and suppressed postnatal TSH [30, 31]. There is no need for treatment of pregnant women carrying affected fetuses, as the intrauterine environment is congruent with the metabolic needs of the fetus harboring the same *THRB* gene mutation.

3 Conclusion

Resistance to thyroid hormone beta is a syndrome of reduced sensitivity of peripheral tissues to TH. The most common clinical manifestations are goiter, tachycardia, and hyperactivity. The biochemical hallmark of the disorder is the combination of high free serum T_4 level and non-suppressed TSH. Measurement of T_4 levels as part of the neonatal screening could lead to the early identification of the syndrome, whereas with the common measurement of only TSH most cases are missed. Genetic analysis and identification of a germline *THRB* mutation confirms the diagnosis of RTHβ and allows counseling of other family members. Clinical care should be individualized to treat the patient's symptoms rather than normalize TH levels. Ablative treatments with radioactive iodide or surgery should be avoided as they complicate follow-up and T_4 dose adjustment. Prenatal diagnosis of RTHβ can be useful in guiding the management of pregnant women with RTHβ in order to prevent adverse postnatal outcomes in unaffected infants, such as low birth weight and suppressed TSH.

References

1. Refetoff S, DeWind LT, DeGroot LJ (1967) Familial syndrome combining deaf-mutism, stuppled epiphyses, goiter and abnormally high PBI: possible target organ refractoriness to thyroid hormone. J Clin Endocrinol Metab 27(2):279–294.https://doi.org/10.1210/jcem-27-2-279
2. Sakurai A, Takeda K, Ain K, Ceccarelli P, Nakai A, Seino S, Bell GI, Refetoff S, DeGroot LJ (1989) Generalized resistance to thyroid hormone associated with a mutation in the ligand-binding domain of the human thyroid hormone receptor beta. Proc Natl Acad Sci U S A 86(22):8977–8981
3. Flamant F, Cheng SY, Hollenberg AN, Moeller LC, Samarut J, Wondisford FE, Yen PM, Refetoff S (2017) Thyroid hormone signaling pathways. Time for a more precise nomenclature. Endocrinology 158(7):2052–2057. https://doi.org/10.1210/en.2017-00250
4. Refetoff S, Weiss RE, Usala SJ (1993) The syndromes of resistance to thyroid hormone. Endocr Rev 14(3):348–399. https://doi.org/10.1210/edrv-14-3-348
5. Lazar MA, Chin WW (1990) Nuclear thyroid hormone receptors. J Clin Invest 86(6):1777–1782. https://doi.org/10.1172/JCI114906
6. Lafranchi SH, Snyder DB, Sesser DE, Skeels MR, Singh N, Brent GA, Nelson JC (2003) Follow-up of newborns with elevated screening T4 concentrations. J Pediatr 143(3):296–301.

https://doi.org/10.1067/S0022-3476(03)00184-7

7. Takeda K, Sakurai A, DeGroot LJ, Refetoff S (1992) Recessive inheritance of thyroid hormone resistance caused by complete deletion of the protein-coding region of the thyroid hormone receptor-beta gene. J Clin Endocrinol Metab 74(1):49–55. https://doi.org/10.1210/jcem.74.1.1727829
8. Weiss RE, Weinberg M, Refetoff S (1993) Identical mutations in unrelated families with generalized resistance to thyroid hormone occur in cytosine-guanine-rich areas of the thyroid hormone receptor beta gene. Analysis of 15 families. J Clin Invest 91(6):2408–2415. https://doi.org/10.1172/JCI116474
9. Adams M, Matthews C, Collingwood TN, Tone Y, Beck-Peccoz P, Chatterjee KK (1994) Genetic analysis of 29 kindreds with generalized and pituitary resistance to thyroid hormone. Identification of thirteen novel mutations in the thyroid hormone receptor beta gene. J Clin Invest 94(2):506–515. https://doi.org/10.1172/JCI117362
10. Yoh SM, Chatterjee VK, Privalsky ML (1997) Thyroid hormone resistance syndrome manifests as an aberrant interaction between mutant T3 receptors and transcriptional corepressors. Mol Endocrinol 11(4):470–480. https://doi.org/10.1210/mend.11.4.9914
11. Collingwood TN, Wagner R, Matthews CH, Clifton-Bligh RJ, Gurnell M, Rajanayagam O, Agostini M, Fletterick RJ, Beck-Peccoz P, Reinhardt W, Binder G, Ranke MB, Hermus A, Hesch RD, Lazarus J, Newrick P, Parfitt V, Raggatt P, de Zegher F, Chatterjee VK (1998) A role for helix 3 of the TRbeta ligand-binding domain in coactivator recruitment identified by characterization of a third cluster of mutations in resistance to thyroid hormone. EMBO J 17(16):4760–4770. https://doi.org/10.1093/emboj/17.16.4760
12. Ando S, Sarlis NJ, Oldfield EH, Yen PM (2001) Somatic mutation of TRbeta can cause a defect in negative regulation of TSH in a TSH-secreting pituitary tumor. J Clin Endocrinol Metab 86(11):5572–5576. https://doi.org/10.1210/jcem.86.11.7984
13. Sakurai A, Miyamoto T, Refetoff S, DeGroot LJ (1990) Dominant negative transcriptional regulation by a mutant thyroid hormone receptor-beta in a family with generalized resistance to thyroid hormone. Mol Endocrinol 4(12):1988–1994. https://doi.org/10.1210/mend-4-12-1988
14. Ono S, Schwartz ID, Mueller OT, Root AW, Usala SJ, Bercu BB (1991) Homozygosity for a dominant negative thyroid hormone receptor gene responsible for generalized resistance to thyroid hormone. J Clin Endocrinol Metab 73(5):990–994. https://doi.org/10.1210/jcem-73-5-990
15. Frank-Raue K, Lorenz A, Haag C, Hoppner W, Boll HU, Knorr D, Hentze S, Raue F (2004) Severe form of thyroid hormone resistance in a patient with homozygous/hemizygous mutation of T3 receptor gene. Eur J Endocrinol 150(6):819–823
16. Ferrara AM, Onigata K, Ercan O, Woodhead H, Weiss RE, Refetoff S (2012) Homozygous thyroid hormone receptor beta-gene mutations in resistance to thyroid hormone: three new cases and review of the literature. J Clin Endocrinol Metab 97(4):1328–1336. https://doi.org/10.1210/jc.2011-2642
17. Bochukova E, Schoenmakers N, Agostini M, Schoenmakers E, Rajanayagam O, Keogh JM, Henning E, Reinemund J, Gevers E, Sarri M, Downes K, Offiah A, Albanese A, Halsall D, Schwabe JW, Bain M, Lindley K, Muntoni F, Vargha-Khadem F, Dattani M, Farooqi IS, Gurnell M, Chatterjee K (2012) A mutation in the thyroid hormone receptor alpha gene. N Engl J Med 366(3):243–249. https://doi.org/10.1056/NEJMoa1110296
18. Brucker-Davis F, Skarulis MC, Grace MB, Benichou J, Hauser P, Wiggs E, Weintraub BD (1995) Genetic and clinical features of 42 kindreds with resistance to thyroid hormone. The National Institutes of Health prospective study. Ann Intern Med 123(8):572–583
19. Barkoff MS, Kocherginsky M, Anselmo J, Weiss RE, Refetoff S (2010) Autoimmunity in patients with resistance to thyroid hormone. J Clin Endocrinol Metab 95(7):3189–3193. https://doi.org/10.1210/jc.2009-2179
20. Campi I, Cammarata G, Bianchi Marzoli S, Beck-Peccoz P, Santarsiero D, Dazzi D, Bottari de Castello A, Taroni EG, Viola F, Mian C, Watuntantrige-Fernando S, Pelusi C, Muzza M, Maffini MA, Persani L (2017) Retinal photoreceptor functions are compromised in patients with resistance to thyroid hormone syndrome (RTHbeta). J Clin Endocrinol Metab 102(7):2620–2627. https://doi.org/10.1210/jc.2016-3671
21. Beck-Peccoz P, Chatterjee VK (1994) The variable clinical phenotype in thyroid hormone resistance syndrome. Thyroid 4(2):225–232. https://doi.org/10.1089/thy.1994.4.225
22. Weiss RE, Marcocci C, Bruno-Bossio G, Refetoff S (1993) Multiple genetic factors in the heterogeneity of thyroid hormone resistance. J Clin Endocrinol Metab 76(1):257–259. https://doi.org/10.1210/jcem.76.1.8421095

23. Persani L, Borgato S, Romoli R, Asteria C, Pizzocaro A, Beck-Peccoz P (1998) Changes in the degree of sialylation of carbohydrate chains modify the biological properties of circulating thyrotropin isoforms in various physiological and pathological states. J Clin Endocrinol Metab 83(7):2486–2492. https://doi.org/10.1210/jcem.83.7.4970
24. Ercan-Fang S, Schwartz HL, Mariash CN, Oppenheimer JH (2000) Quantitative assessment of pituitary resistance to thyroid hormone from plots of the logarithm of thyrotropin versus serum free thyroxine index. J Clin Endocrinol Metab 85(6):2299–2303. https://doi.org/10.1210/jcem.85.6.6625
25. Anselmo J, Refetoff S (2004) Regression of a large goiter in a patient with resistance to thyroid hormone by every other day treatment with triiodothyronine. Thyroid 14(1):71–74. https://doi.org/10.1089/105072504322783876
26. Wu SY, Cohen RN, Simsek E, Senses DA, Yar NE, Grasberger H, Noel J, Refetoff S, Weiss RE (2006) A novel thyroid hormone receptor-beta mutation that fails to bind nuclear receptor corepressor in a patient as an apparent cause of severe, predominantly pituitary resistance to thyroid hormone. J Clin Endocrinol Metab 91(5):1887–1895. https://doi.org/10.1210/jc.2005-2428
27. Radetti G, Persani L, Molinaro G, Mannavola D, Cortelazzi D, Chatterjee VK, Beck-Peccoz P (1997) Clinical and hormonal outcome after two years of triiodothyroacetic acid treatment in a child with thyroid hormone resistance. Thyroid 7(5):775–778. https://doi.org/10.1089/thy.1997.7.775
28. Takeda T, Suzuki S, Liu RT, DeGroot LJ (1995) Triiodothyroacetic acid has unique potential for therapy of resistance to thyroid hormone. J Clin Endocrinol Metab 80(7):2033–2040. https://doi.org/10.1210/jcem.80.7.7608251
29. Anselmo J, Cao D, Karrison T, Weiss RE, Refetoff S (2004) Fetal loss associated with excess thyroid hormone exposure. JAMA 292(6):691–695. https://doi.org/10.1001/jama.292.6.691
30. Weiss RE, Dumitrescu A, Refetoff S (2010) Approach to the patient with resistance to thyroid hormone and pregnancy. J Clin Endocrinol Metab 95(7):3094–3102. https://doi.org/10.1210/jc.2010-0409
31. Pappa T, Anselmo J, Mamanasiri S, Dumitrescu AM, Weiss RE, Refetoff S (2017) Prenatal Diagnosis of Resistance to Thyroid Hormone and Its Clinical Implications. J Clin Endocrinol Metab. 1;102(10):3775–3782. https://doi.org/10.1210/jc.2017-01251

Chapter 19

TRα Mutations in Human

C. Briet, Natacha Bouhours-Nouet, F. Illouz, D. Prunier-Mirebeau, and P. Rodien

Abstract

Resistance to thyroid hormone alpha is an emerging syndrome, with up to now a limited number of published cases. Some features are common to most of the patients, but there is still some work to provide a comprehensive description of the full spectrum of the syndrome. A survey of the strategy to screen for and characterize the mutations in TR α gene is given.

Key words Thyroid hormone receptor mutation, Resistance to thyroid hormone, *THRA* gene

1 Introduction

Since the cloning of Thyroid hormone receptors α and β, overlapping as well as specific functions of their respective isoforms, have been identified thanks to murine KO and KI models.

For many years, after the demonstration of the role of mutation of TR β in the syndrome of resistance to thyroid hormones (RTH) an unsolved question remained which was the phenotype the patients with mutations of TR α would exhibit.

The absence of identified patients with such mutations was a matter of intense speculation. Was lethality a trivial explanation or was the quest of such mutations focusing on the wrong phenotype? In other words, was the screening, based on abnormalities in the thyroid hormones concentration, fated to fail?

Based on the phenotyping of TR α KO and KI n animals, it was postulated that patients with an inactivating mutation of TR α, would be lean, suffering of hypothermia or at least would be chilly, and bradycardic and anxious or depressive [1].

No dramatic thyroid hormone or TSH changes were expected. Finally, it was proposed that such patients, would they exist, would putatively see their symptomatology improved by T4 or T3 treatment.

Michelina Plateroti and Jacques Samarut (eds.), *Thyroid Hormone Nuclear Receptor: Methods and Protocols*,
Methods in Molecular Biology, vol. 1801, https://doi.org/10.1007/978-1-4939-7902-8_19,
© Springer Science+Business Media, LLC, part of Springer Nature 2018

Nevertheless, the search for lean hypothermic bradycardic patients in the population of depressive patients was not successful.

2 An Emerging Syndrome

2.1 The First Patient

A 6 years old girl, was identified through whole exome sequencing, after elimination of genetic variants shared with her parents, selection of a restricted number of variants that would make sense with the phenotypical traits she exhibited [2].

The patient was examined and investigated because of severe constipation. She was short, especially with short legs, had a moderate retardation in neurologic development, was bradycardic. She had on several occasions, a low-normal to low FT4 concentration and a high-normal to high fT3 and TT3, with a normal TSH concentration. Reverse T3 was low. The ratio FT4/FT3 was low when compared to children of the same age.

When the mutation of TR α gene was identified, and the hormonal phenotype was picked out, it appeared that some of the symptoms the patient was affected by could be related to a kind of peripheral tissue hypothyroidism.

Several additional cases were reported in the following years, sharing the low FT4/FT3 ratio, variable growth retardation, and variable mental retardation [3–6].

Anemia appeared to be a quite common trait, some additional features were reported such as an increased head circumference, skin tags, seizures in some cases.

One patient investigated, because of a short stature with unequilibrated length of legs compared to body height, appeared to have a very severe bone phenotype with syndactilies, facial dysmorphia, clavicular agenesis and surprisingly no mental retardation, no bradycardia, a diarrhea, and inappropriate PTH secretion. In addition she was affected by congenital dyserythropoieis.

She was the first case identified with a mutation affecting both TR α1 and TR α2, and the discrepant phenotype was initially thought to be due to this specific mutation location, which retrospectively was probably not true.

In fact, shortly after, a family was reported with a mutation also involving the two α isoforms, and no obvious difference with the henceforth classical phenotype [4].

This family was however very informative in that the patients had been treated during childhood by thyroid hormones because of central hypothyroidism. This allowed the demonstration of some possible rescue with the treatment by thyroid hormones, as the growth was restored as well as the mental development. One feature was still present, despite the maskerading effect of the hormonal treatment, which was the increased head circumference. Also, the voice and the articulation were reported to be abnormal.

Based on this association of symptoms and despite extensive information in the pediatrician and genetician meetings, the search for new cases was poorly efficient.

However, one new case was identified through cautious reappraisal of the phenotype of a patient treated for congenital central hypothyroidism diagnosed during investigation of mental retardation. She still had a large head that prompted sequencing of the TR α gene, and identification of a mutation (unpublished). IT is expected that new cases will be identified through new generation sequencing techniques, in cases of unexplained mental retardation associated or not with other defects.

As pointed by Moran et al., the phenotype can be dramatically changed upon treatment by thyroid hormones and both untreated and treated presentations have to be kept in mind [6, 7].

2.2 Screening for New Cases?

The common features of patients include growth and mental retardation, bone maturation delay, increased head circumference, anemia, constipation. As mentioned already many of these traits can be modified by the hormonal treatment. The hormonal picture, low T4 and FT4, high T3 and fT3, and low T4/T3 ratio implies that valid reference values are available corresponding to the age and maturation of the patients. Thyroid hormone assays do not all have the same reliability, and screening based on these parameters is difficult. Age-dependent reference range in a reference method is eagerly awaited.

It should be kept in mind that isolated congenital thyrotropic insufficiency (low T4 and non-increased TSH) is an extremely rare condition, and that in such instance, it is worth looking for a mutation in TR α gene if no other cause (mutation in TSH beta gene, TRH receptor gene, IGSF1 gene in boys) has been identified.

3 Proving the Role of the TRα Mutation

3.1 In Silico Analysis

The identification of a mutation either by sequencing of the candidate gene, here TR α, or by exome sequencing, must lead to an analysis of the segregation in the family. Many cases reported up to now relate, however, to neomutations, which limits the power of segregation studies. In addition, several mutations have been found in anonymous databases which precludes any collection of information on the patients and their health status.

It is also necessary to characterize the mutation. This can proceed through complementary methods.

The first one may consist in an in silico analysis.

Several softwares are routinely used to predict the pathogenecity of a mutation in a gene, based on the conservation during evolution of the affected residue and on the physicochemical changes brought with the mutation. For example, changing a conserved negatively charged aminoacid into a positively charged one is supposed to be poorly tolerated and to affect the function of the protein.

As changes brought by a mutation may sometimes be much less dramatic, the impact of the mutation in a tridimensional model of the receptor or of part of it, may complete this in silico analysis. TR α and β are highly homologous, so that the alignment of the two sequences can reveal that a mutation in TR α has already been described and characterized in TR β, making it likely this is a pathogenic one. This may also point to very atypical location of a mutation, therefore partly explaining some phenotypical differences [2, 4–6].

3.2 In Vitro Characterization

In addition, the loss of function of the mutant receptor is usually assessed in vitro in eukaryotic cells transfected with a reporter gene and either the mutant or the wild-type receptor. COS-7 cells or HEK 293 cells are commonly used for this purpose. Different thyroid hormone response elements (palindromic, direct repeat, etc.) may be used.

Furthermore, since the mutations responsible for this new resistance syndrome are heterozygous, it is important to demonstrate that the mutant receptor is able to impair the normal function of the wild-type receptor, in cotransfection experiments, as it was demonstrated years ago for the TR β in the RTH β syndrome. This is the dominant negative effect [2–7, 9].

3.3 In Vivo Animal Models

New technologies such as CRISPR/Cas9, now allow for efficient and rapid production of KIn mice [8]. It may be useful to better understand the impact of the mutation in several aspects of the phenotype, but also to better understand the role of the wild type receptor, by using the mutant receptors as tools. It is also useful in cases of atypical phenotypes. Indeed, in cases of phenotypes diverging from the "classical" reported phenotype, and even when in vitro assays demonstrate the defects in the function of the receptors, KIn animals may be of value to confirm or dismiss the pathogenic role of the mutation. In complement to in vitro experiments, such an approach may help decipher the role a mutation may have during development.

3.4 Ex Vivo Assessment of the Impact of the Mutant

In vitro experiments with reporter genes, binding experiments, modelization may demonstrate the functional impact of a mutation in TR α. This is, however, based on very artificial conditions: overexpression of receptors and reported genes in cells that natively express no or low level of the receptor. In addition, the reporter genes are designed in order to amplify the signal, sometimes adding several copies of the hormone response element.

The in vivo demonstration of resistance to T3 may also be obtained by assessing the clinical and biological response to the administration of increasing dose of T3. This will provide a global assessment of tissue sensitivity to T3.

Specific target genes may be studied by ex vivo assessment of the response to T3 in cells endogenously expressing this gene. For example, the stimulation of the expression of the gene KLF9 has been studied ex vivo on peripheral mononuclear cells [9].

3.5 Prediction of the Effect of the Treatment by Thyroid Hormones?

Previously described in vitro and ex vivo experiments may be a way to try to predict the chance a treatment by T3 has to be efficient. Although a comparison between two mutations, with some correlation between in vitro and in vivo response to T3, has been reported [9], the real capacity to predict the efficiency of the treatment is uncertain. Especially, the age and particularly the developmental age of the patients will probably be a strong determinant of the impact of the treatment.

Nevertheless, these in vitro and ex vivo experiments provide tools to evaluate the impact of alternative treatment such as T3 versus T4, tiratricol versus T3, etc. in case T4 would be ineffective in vivo. This would avoid in vivo empirical trials.

References

1. Vennstrom B, Mittag J, Walis K (2008) Severe psychomotor and metabolic damages caused by a mutant thyroid hormone receptor alpha 1 in mice: can patients with a similar mutation be found and treated? Acta Paediatr 97:1605–1610
2. Bochukova E, Schoenmakers N, Agostini M, Schoenmakers E, Rajanayagam O, Keogh JM et al (2012) A mutation in the thyroid hormone receptor alpha gene. N Engl J Med 366:243–249
3. Van Mullem A, van Heerebeek R, Chrysis D, Visser E, Medici M, Andrikoula M et al (2012) Clinical phenotype and mutant TRα1. N Engl J Med 366:1451–1453
4. Moran C, Agostini M, Visser WE, Schoenmakers E, Schoenmakers N, Offiah AC et al (2014) Resistance to thyroid hormone caused by a mutation in thyroid hormone receptor (TR) alpha1 and TRalpha2: clinical, biochemical, and genetic analyses of three related patients. Lancet Diabetes Endocrinol 2:619–626
5. Espiard S, Savagner F, Flamant F, Vlaeminck-Guillem V, Guyot R, Munier M et al (2015) A novel mutation in THRA gene associated with an atypical phenotype of resistance to thyroid hormone. J Clin Endocrinol Metab 100:2841–2848
6. Moran C, Chatterjee K (2015) Resistance to thyroid hormone due to defective thyroid receptor alpha. Best Pract Res Clin Endocrinol Metab 29:647–657
7. Moran C, Chatterjee K (2016) Resistance to thyroid hormone α-emerging definition of a disorder of thyroid hormone action. J Clin Endocrinol Metab 101:2636–2639
8. Markossian S, Flamant F (2016) CRISPR/cas9: a breaktrough in generating mouse models for endocrinologists. J Mol Endcrinol 57:R81–R92
9. Moran C, Agostini M, McGowan A, Schoenmakers E, Fairall L, Lyons G et al (2017) Contrasting phenotypes in resistance to thyroid hormone alpha correlate with divergent properties of thyroid hormone receptor α1 mutant proteins. Thyroid 27:973–982

Chapter 20

Role of Thyroid Hormone Receptor in Amphibian Development

Liezhen Fu, Luan Wen, and Yun-Bo Shi

Abstract

The amphibian *Xenopus laevis* has long been used as a model for studying vertebrate cell and developmental biology largely due to the easiness to manipulate this system in vivo and in vitro. While most of the developmental studies have been on *Xenopus* embryogenesis, considerable efforts have been made to understand its metamorphosis, a process mimicking postembryonic development in mammals when many organs mature into their adult forms in the presence of high levels of thyroid hormone (T3). Amphibian metamorphosis is totally dependent on T3 and offers a number of advantages for experimental analyses compared to the late stage, uterus-enclosed mammalian embryos. Earlier studies on metamorphosis in *Xenopus laevis* have revealed dual functions of T3 receptors (TR) during premetamorphic development and metamorphosis as well as important roles of TR-interacting corepressors and coactivators during these two periods, respectively. The development of gene-editing technologies that functions in amphibians in recent years has made possible for the first time to study function of endogenous TRs, especially in the highly related diploid anuran species *Xenopus tropicalis*. Here, we first review the current mechanistic understanding of the regulation of metamorphosis by T3 and TR, and then describe a detailed method to use TALEN to knock out TRα for studying its role in gene regulation by T3 in vivo and *Xenopus* development.

Key words Xenopus laevis, Xenopus tropicalis, Metamorphosis, Gene-editing, Thyroid hormone receptor, Postembryonic development, TALEN

1 Introduction

Amphibian undergoes biphasic development with its embryogenesis producing a free-living larva, which later metamorphoses into the adult form [1]. Anuran amphibians undergo the most dramatic metamorphosis, with essentially every single organ/tissue changed during this process. This process bears strong similarity to the so-called postembryonic development in mammals, a period around birth when many organs/tissues are matured into the adult form, resembling the organ remodeling during anuran metamorphosis [1, 2]. Furthermore, both postembryonic development in

Michelina Plateroti and Jacques Samarut (eds.), *Thyroid Hormone Nuclear Receptor: Methods and Protocols*, Methods in Molecular Biology, vol. 1801, https://doi.org/10.1007/978-1-4939-7902-8_20,
© Springer Science+Business Media, LLC, part of Springer Nature 2018

mammals and anuran metamorphosis are characterized by peak levels of plasma thyroid hormone (T3) [1, 2]. Importantly, T3 is the causative agent of amphibian metamorphosis. Blocking synthesis of endogenous T3 leads to the formation of giant tadpoles that cannot metamorphose while exogenous T3 can induce premetamorphic tadpoles to undergo precocious metamorphosis prior to the synthesis of endogenous T3 [1, 2]. Furthermore, most, if not all, organs are genetically predetermined to undergo specific organ autonomous changes, making it possible to study anuran metamorphosis even in organ cultures. These properties together with the recent advancement in genetic tools for amphibian studies make anuran metamorphosis one of the best models to study the in vivo function of T3 and its underlying molecular mechanisms. Below, we summarize some of the studies on *Xenopus laevis* and *Xenopus tropicalis*, two highly related and best studied anuran species.

1.1 Dual Roles of Thyroid Hormone Receptors (TRs) in Xenopus Metamorphosis

T3 can exert its effect through both genomic and non-genomic pathways. It can regulate gene transcription through its nuclear receptors, TRs, while its non-genomic effects are believed to be mediated by cell surface and cytoplasmic binding proteins, including the small fractions of TRs present in the cytoplasm [3–12]. TRs are known to function mainly as heterodimers with 9-cis retinoic acid receptors (RXRs) on T3-inducible genes. TR/RXR heterodimers are nuclear proteins that bind constitutively to thyroid hormone response elements (TREs) in chromatin to repress or activate target gene expression in the absence or presence of T3, respectively.

Earlier studies have shown that zygotic transcription of TR, especially TRα, occurs well before the synthesis and secretion of endogenous T3. TRα and RXRα genes are expressed at high levels by stage 45 when tadpole feeding begins while plasma T3 is detectible only by stage 54, the onset of metamorphosis [13–16]. This and the ability of TR to repress or activate target gene in a T3-dependent manner have led to a dual function model for TR during frog development [17, 18]. That is, between stage 45, when tadpole feeding begins, and stage 54, the onset of metamorphosis, TRs are present mainly in the unliganded form due to only very low levels of T3, if any, in the tadpole plasma. The unliganded TR/RXR heterodimers repress T3 inducible genes to prevent precocious metamorphosis. After stage 54, the rising concentration of endogenous T3 leads to the binding of T3 to TR, and the liganded TR/RXR then activate the target genes to promote metamorphosis.

1.2 Analyses of TR Function via Overexpression during Xenopus laevis Development

Functional analysis of frog TRs initially came from microinjection studies. Overexpression of TR and RXR by microinjecting their mRNAs into fertilized eggs led to repression of endogenous target genes in the resulting embryos, and when the embryos were treated with exogenous T3, the target genes were activated, leading to embryonic defects and lethality [19]. RXR was found to be important for these effects, suggesting that TR functions as heterodimers with RXR in vivo during *Xenopus* development.

The development of the reliable sperm-mediated transgenic approach [20] made it possible to study the effect of overexpressed mutant TRs during *Xenopus* development. Several different laboratories analyzed the effect of overexpressing a dominant negative TR that had a C-terminal deletion to prevent T3 binding, leading to a constitutively unliganded TR that represses T3-inducible genes. Overexpression of such a dominant negative TR by either transgenesis in whole animals or in vivo gene transfer in the tail inhibited target gene induction by T3 and metamorphic transformation in *Xenopus laevis* [21–25]. Mechanistically, the transgenic dominant negative TR was found to compete against endogenous TR for binding to endogenous target genes, consequently preventing the regulation of the genes by T3 [25]. Thus, TR is required to mediate the effects of T3 on target gene induction and metamorphosis of different organs/tissues.

As indicated above, T3 has both genomic and non-genomic effects. To determine whether TR is sufficient to mediate the metamorphic effects of T3, a transgenic study with a dominant positive TR was carried out. This dominant positive TR consisted of the dominant negative TR above fused to the N-terminus a strong viral transcription activation protein, VP16 [26]. This mutant TR could not bind to T3 but functioned as a constitutive activator on T3 target genes due to the presence of the VP16 moiety. To prevent likely precocious induction of metamorphosis by the expression of the dominant positive TR, it was placed under the control of a heat shock-inducible promoter for transgenesis. When transgenic tadpoles and their wild-type siblings reached premetamorphic stages, they were subjected to heat shock treatment to induce the transgene expression. This enabled the transgenic but not the wild-type tadpoles to metamorphose just like premetamorphic tadpoles after T3 treatment [26]. Mechanistically, the dominant positive TR was shown to bind to endogenous target genes and specifically activate their expression [26]. Furthermore, when the intestine of such transgenic premetamorphic tadpoles was cultured in vitro, heat shock treatment also led to intestinal metamorphosis [27]. Thus, TR is sufficient to mediate the effects of T3 on individual organs during metamorphosis. These findings further demonstrate that the non-genomic effects of T3 are not necessary for amphibian metamorphosis.

1.3 Determining the Roles of Endogenous TR in *Xenopus tropicalis*

While molecular and transgenic studies have provided strong evidence to support the dual functions of TR during amphibian development, the roles of endogenous TRs had eluded researchers until recently. With the development of first TALEN and then CRISPR-nuclease mediated knockout/knockdown technologies that can effectively edit endogenous genes in both the pseudo-tetraploid *Xenopus laevis* and diploid *Xenopus tropicalis,* it becomes possible to study endogenous genes during *Xenopus* development [28–33]. Using the TALEN technology, we and the Buchholz's laboratory initially independently knocked down the endogenous TRα [34–37]. *Xenopus tropicalis* was chosen for such studies due to its diploid nature. Both groups microinjected mRNAs encoding a TALEN nuclease targeting TRα into fertilized eggs and studied the effect on the resulting animals, the F0 generation knockdown animals, with similar conclusions [34–37]. In our study, we generated a TALEN targeting exon 3, encoding part of the DNA binding domain of *Xenopus tropicalis* TRα (Fig. 1) [35]. Microinjecting the mRNA encoding the left and right arms of the TALEN into fertilized egg prior to the first cell division led to the expression of the TALEN in all cells of the early stage embryos and the resulting F0 tadpoles had genetic alterations at the TRα target site with about 90% efficiency by the time when feeding began (stage 45, 3 days of age) [35]. Such high efficiency suggested that we had nearly a complete knockout of the TRα gene and could study the phenotype in the F0 animals. Indeed, the findings in the F0 knockdown animals were subsequently confirmed by studies on complete TRα knockout animals derived from germ line breeding of the knockdown animals [38, 39].

Using either F0 knockdown or complete knockout animals, both the groups found that TRα knockdown/knockout led to precocious limb development (Fig. 2) [34–39]. Stage 54 is considered to be the onset of metamorphosis and is determined based on hindlimb morphology. Based on this, both groups observed that the TRα knockdown/knockout tadpoles initiated metamorphosis at younger age, with concurrent upregulation (derepression) of several well-known T3-inducible genes [35, 36, 38, 39]. In addition, the TRα knockdown/knockout tadpoles had reduced response to exogenous T3 treatment in terms of both gene regulation and morphological changes, indicating that TRα is important both for controlling the timing of the onset of metamorphosis and for mediating the response to T3. Consistently, when the knockdown/knockout animals were allowed to undergo natural metamorphosis, it was found that TRα knockdown/knockout animals took much longer time to develop from stage 54, the onset of metamorphosis, to stage 58, the early metamorphic climax. Furthermore, the studies on total TRα knockout animals also led to several novel observations [38, 39]. First, TRα knockout appears to have no gross effect on embryogenesis. Second and surprisingly, TRα knockout does not prevent the animal from completing metamorphosis or affect the total developmental time from fertilization to the end of metamorphosis as judged based on total resorption of the tail. Finally, TRα is important for temporal coordination of organ-specific transformations.

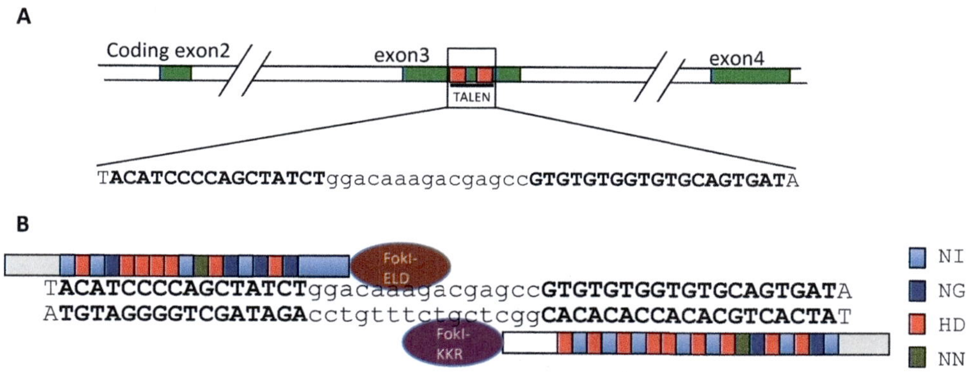

Fig. 1 Targeting TRα with TALEN in *Xenopus tropicalis* embryos. (**a**) Schematic diagram of the TRα gene showing the TALEN target site. The protein coding exons were shown as numbered, colored boxes. The TALEN-recognized sequences were shown as red boxes (top) or bold letters (below). (**b**) Schematic representation of the TRα TALEN and the targeted DNA sequence in the TRα gene. The four types of RVDs (repeat variable diresidues) recognizing nucleotide A, G, T, or C were depicted in different colors, respectively. The left arm contained the FokI-ELD nuclease and the right arm contained the FokI-KKR nuclease, which, when both arms bind to their respective binding sites, heterodimerize to form the functional nuclease to make a double-stranded break in the intervening sequence. See [35] for more details

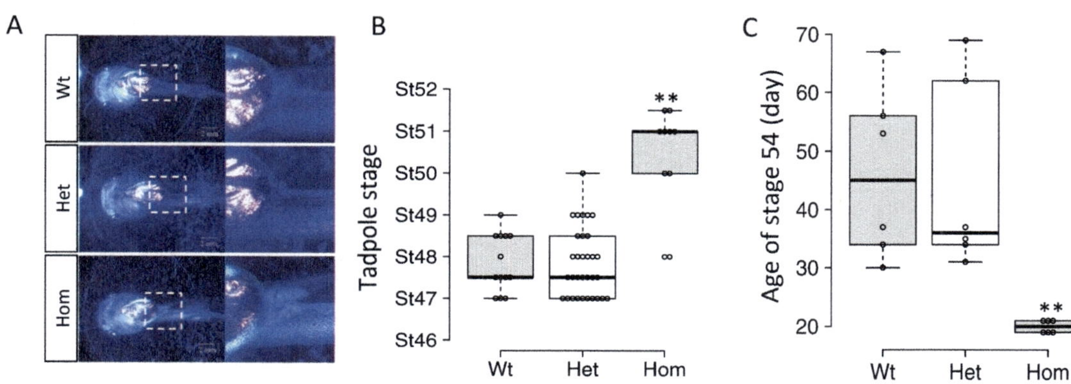

Fig. 2 TRα regulates metamorphic timing. (**a**) TRα knockout tadpoles develop faster. One representative animal is shown for each of the three genotypes: wild type, heterozygous, and homozygous TRα knockout siblings. Dashed boxes in the left panel were enlarged and shown in the right panel. Note that the homozygous TRα knockout tadpole had significantly larger and more advanced hindlimb buds. (**b**) TRα knockout tadpoles develop into more advanced stages. The stage of individual sibling tadpoles in the three genotypes was plotted with the median (darker solid line) for each genotype shown in the figure. Note that homozygous TRα knockout tadpole reached a median stage of 51 while the wild-type and TRα heterozygous knockout tadpoles reached a median stage of 47.5. (**c**) TRα knockout tadpoles initiate metamorphosis, i.e., reaching stage 54, at a younger age. The age in days for each tadpole to reach stage 54 was plotted with the median (darker solid line) for each genotype shown in the figure. "**" indicates significant difference between homozygous knockout and the other groups ($P < 0.01$). See [38] for more details

Thus, the gene-editing technologies make it possible to investigate TR function during postembryonic development. Furthermore, the ability to introduce TALEN into fertilized eggs and the high efficiency of TALEN-mediated mutagenesis also enable one to determine the function of TR in F0 generation animals within a few weeks after micro-injecting of the TALEN mRNAs, providing a fast and powerful tool to study gene function in vivo.

In addition, TALEN-mediated mutations in certain genes may lead to lethal phenotype during early embryogenesis. This makes it difficult to study the functions of such genes during postembryonic development. However, it is possible to microinject mRNAs/proteins into fertilized eggs to rescue the embryonic defect due to the gene alteration, leading to the formation of free-feeding tadpoles. As the mRNA or protein injected into the eggs only last a few days, one can use the resulting rescued animals to study the effect of the knockout on subsequent tadpole development and metamorphosis.

2 Materials

2.1 Animal

1. Adult *Xenopus tropicalis*, Nasco (Wisconsin).

2.2 Devices for Microinjections

1. Needle puller, World precision Instruments, PUL-1 (*see* **Note 1**).
2. Needles, Drummond Microcaps 1-000-0300 (*see* **Note 1**).

2.3 Plasmids and Constructs

1. Golden Gate TALEN and TAL Effector Kit (Addgene, Kit#1000000024) [40] (*see* **Note 2**).
2. pDP(RFP)HG construct ([41] and unpublished).
3. pDP(GFP)HG construct ([41] and unpublished).
4. pCS2+ vector-based TAL- KKR and TAL- ELD constructs [30].
5. pCR TOPO-TA vector (Invitrogen).

2.4 Enzymes

1. T4 DNA ligase (include 10× T4 DNA ligase buffer).
2. Restriction endonuclease BsaI (New England Biolabs).
3. Restriction endonuclease Esp31 (Fermentas or Fisher).
4. Restriction endonuclease AgeI (New England Biolabs).
5. Restriction endonuclease EcoRI (New England Biolabs).
6. Restriction endonuclease NotI (New England Biolabs).
7. PrimeSTAR HS DNA Polymerase (Clontech) (Note that any high fidelity Taq DNA polymerase would be fine).
8. GoTaq® Green Master Mix (Promega).

2.5 Chemicals and Reagents

1. 10 mM ATP.
2. SOC medium: 3.6 g/l glucose, 0.19 g/l potassium chloride, 0.58 g/l sodium chloride, 20 g/l tryptone, and 5 g/L yeast extract (Quality Biologicals, MD).
3. Human chorionic gonadotropin (HCG, Novarel).
4. 20 mg/ml X-gal in DMSO.
5. 10 mg/ml IPTG.
6. LB broth: 10 g/l sodium chloride, 10 g/l tryptone, and 5 g/l yeast extract (Quality Biologicals, MD).
7. 10 mg/ml tetracycline stock solution.
8. 50 mg/ml spectinomycin stock solution.
9. 100 mg/ml ampicillin stock solution.
10. 1× Marc's Modified Ringers (MMR): 100 mM NaCl, 2 mM KCl, 1 mM $MgCl_2$, 2 mM $CaCl_2$, 5 mM HEPES, pH 7.5.
11. 0.1× MMR.
12. 0.1× MMR/6% Ficoll.
13. Ficoll (Sigma).
14. Cysteine.
15. Agarose.
16. 1× TAE buffer for DNA electrophoresis.
17. dNTPs, individual solutions or mix of equal concentration for each, for PCR reactions.
18. Mineral oil (for microinjectors).

2.6 Primers

1. Primers for sequencing during TALEN assembly:

   pCR8_F1: 5′-TTGATGCCTGGCAGTTCCCT-3′

   pCR8_R1: 5′-CGAACCGAACAGGCTTATGT-3′

   TAL_F1: 5′-TTGGCGTCGGCAAACAGTGG-3′

   TAL_R2: 5′-GGCGACGAGGTGGTCGTTGG-3′

   TAL_Seq: 5′-CATCGCGCAATGCACTGAC-3′

   TAL_R3: 5′-GGCTCAGCTGGGCCACAATG-3′

2. Primers for modifying TALEN backbone constructs:

   Forward: 5′-CTTGGTCTCACCGGCACCATGGCTCCAAAGAAGAAG-3′ (AgeI-compatible end by BsaI digestion)

   Reverse: 5′- CTTGGTCTCCAATTCGTAATACGACTCACTATAGTTC-3′ (EcoRI-compatible end by BsaI digestion)

3. Primers for detecting TRα mutations due to TALEN expression [35]:

   Forward primer F: 5′-ATTGGGTTGGTTGTGGGTCG-3′

   Reverse primer R: 5′-TAAGGTGGGGGCAGTACAGG-3′

   Forward primer f: 5′-ATGTTCCATGAGACGCCCCT-3′

   Forward primer f1: 5′-CCAGCTATCTGGACAAAGAC-3′

2.7 Other Materials

1. RNA transcription kit, Ambion mMessage mMachine.
2. Plasmid DNA Miniprep kit (Qiagen).
3. PCR Purification kit (Qiagen).
4. DNA Gel Extraction kit (Qiagen).
5. pCR TOPO-TA Cloning kit (Invitrogen).
6. Chemically Competent cells TOP10 (Invitrogen).
7. LB plates with 10 mg/l tetracycline, 50 mg/l spectinomycin, or 50 mg/l ampicillin, respectively.

3 Methods

3.1 TALEN Assembly

1. Design of TALEN target

   Obtain *Xenopus tropicalis* TRα gene sequence and intron/exon structure from frog genome database (http://www.xenbase.org/entry/).

   Choose the sequence encompassing the exon 2 to exon 4 for TALEN target screening.

   Screen for TALEN targets via a web-based TALEN-targeter tool (https://tale-nt.cac.cornell.edu/node/add/talen).

   Select a pair of TALEN targets within the exon 3 [35] (*see* **Note 3**).

2. Modification on TALEN backbone constructs

   Golden Gate TALEN and TAL Effector Kit assemble both arms of a TALEN into a backbone vector which expresses TALEN containing the same FokI domain [40] (*see* **Note 2**). For assembling the two arms of a TALEN into a set of backbones such that one TALEN construct expresses TALEN containing a mutant FokI-ELD and the other TALEN construct expresses TALEN containing a mutant FokI-KKR in order to further increase the specificity [29, 30], the TALEN backbone constructs are modified to be compatible with the Golden Gate TALEN and TAL Effector Kit. The modified TALEN backbone constructs are based on two double-promoter transgenesis constructs, pDP(GFP)HG (with a green fluorescent protein gene driven by a γ-crystallin promoter) and the pDP(RFP)HG (with a red fluorescent protein gene driven by a γ-crystallin promoter), both of which can produce the TALEN mRNA from a SP6 promoter in vitro for microinjection or express the TALEN protein in transgenic animals from a heat shock-inducible promoter if desired [41] (*see* **Note 4**).

PCR-amplify the TAL- KKR and TAL- ELD coding regions from the corresponding pCS2+ vector-based constructs [29, 30] with the same set of primers (forward primer 5′-CTTGGTCTCACCGGCACCATGGCTCCAAAGAAGAAG-3′ and reverse primer 5′- CTTGGTCTCCAATTCGTAATACGACTCACTATAGTTC-3′).

Gel-purify the PCR fragments and digest with BsaI to produce AgeI and EcoRI compatible ends, respectively.

Digest the double-promoter transgenesis constructs pDP(GFP)HG (with a green fluorescent protein gene driven by a γ-crystallin promoter) and the pDP(RFP)HG (with a red fluorescent protein gene driven by a γ-crystallin promoter) with AgeI and EcoRI, respectively, and gel-purify the vector part.

Ligate the PCR amplicon of TAL-KKR containing AgeI and EcoRI-compatible ends with the pDP(RFP)HG vector bearing AgeI and EcoRI ends to make pDP(RFP)-TAL-KKR. Simultaneously, ligate the PCR amplicon of TAL-ELD containing AgeI and EcoRI-compatible ends with the pDP(GFP)HG vector bearing AgeI and EcoRI ends to make pDP(GFP)-TAL-ELD.

Transform the ligated products into competent *E. coli* TOP10 cells, plate the cells on LB plates containing 100 μg/ml ampicillin, and incubate at 37 °C overnight.

Pick up a couple of colonies each for bacteria growth in 5 ml of LB broth supplemental with ampicillin overnight and extract plasmid DNA from them.

Confirm the cloned TAL-KKR and TAL-ELD sequences by DNA sequencing.

3. TALEN assembly into pDP(GFP)-TAL-ELD and pDP(RFP)-TAL-KKR constructs

Assemble the TALEN arms essentially following the instructions for Golden Gate TALEN and TAL Effector Kit (https://www.addgene.org/taleffector/goldengatev2/) [40] except for the use of the TALEN backbone vectors.

Assemble the TALEN left arm (TALEN-L) and right arm (TALEN-R) into the double-promoter transgenesis constructs-based TALEN backbones pDP(GFP)-TAL-ELD and pDP(RFP)-TAL-KKR, respectively.

Confirm the TALEN-L and TALEN-R by DNA sequencing.

3.2 In Vitro Transcription

1. Linearize plasmid DNA with restriction enzyme NotI so that the SP6 RNA polymerase transcription would produce sense mRNA with a 3′-untranslated region followed by a poly(A)$_{30}$ tail suitable for TALEN translation (*see* **Note 5**). Prepare enough linearized DNA (e.g., digest about 20μg) for at least several transcription reactions at 1 μg per reaction.
2. Confirm linearization and DNA purity by running a small aliquot on an agarose gel (*see* **Note 6**).
3. Purify the linearized TALEN constructs following the manufacturer's instructions of a PCR Purification kit (*see* **Note 7**).
4. Transcribe RNA following directions of Ambion's mMessage mMachine kit (*see* **Note 8**). One μg of the DNA was transcribed in vitro with SP6 RNA polymerase.
5. Remove template DNA, purify RNA and quantitate it (*see* **Note 9**).
6. RNA can be aliquoted and stored at −80 °C for months.

3.3 In Vitro Fertilization of Xenopus Eggs

1. Xenopus *tropicalis* males and females are primed with 20 U of chorionic gonadotropin (HCG) the evening prior to the day that ovulation is desired (*see* **Note 10**).
2. The primed frogs are then boost with a 200 U of HCG injection early next morning. The females can be injected at different intervals of 1~2 h apart so they begin to ovulate at different times to ensure constant supply of eggs within the whole day.
3. A *Xenopus* male frog is sacrificed when the first batch of females begin to ovulate, the testes are removed, and keep in 1× MMR on ice.
4. Macerate 1 testis in 0.3 ml of 1 × MMR to make sperm solution, transfer into a 1.5 ml microcentrifuge tube, and keep on ice. The sperm solution can be used throughout the day.
5. Eggs are expelled from ovulating females onto petri dishes with gentle squeezing. Remove the water from the petri dish and add 0.1 ml of the sperm solution for about 200 eggs and 0.9 ml of ddH_2O.
6. After gentle splaying out of eggs into the sperm solution and 2–4 min incubation, the dishes are flooded with 0.1× MMR and incubated for an additional 15 min.
7. To remove the jelly coat, the fertilized eggs are washed with 3% cysteine, pH 8.0, until they started to touch each other (about 3–5 min), and then washed 4–6 times with 0.1× MMR before being transferred into fresh 0.1× MMR/6% Ficoll for microinjection (*see* **Note 10**).

3.4 Microinjection of TALEN mRNAs and Mutation Analysis

1. Place the fertilized eggs in an agar-coated plate and inject the mRNA into the cytoplasm of the fertilized eggs prior to the first division with a Drummond Nanojet II Auto-Nanoliter Microinjector.
2. For TALEN-targeted mutagenesis of TRα, equal amounts of TALEN-L and TALEN-R mRNA were mixed together and injected at 400 pg for each mRNA per egg (*see* **Note 11**).
3. The TALEN mRNA-injected eggs along with some uninjected eggs from the same batch of in vitro fertilization were incubated side by side in a 25 °C incubator overnight.
4. Examine the embryos for potential phenotypes following TALEN mRNA injection (*see* **Note 12**).
5. Collect embryos for mutation rates evaluation starting from day 3 after the TALEN mRNA injection (*see* **Note 13**). To do so, three or more embryos from the same group are pooled together and lysed with the lysis buffer (10 mM Tris–HCl, 100 mM NaCl, 10 mM EDTA, and 0.5% SDS, pH 8.0) containing 8 μg/ml proteinase K at 56 °C overnight.
6. Precipitate the genomic DNA with isopropanol, wash with 70% ethanol, dry in air for 10 min., and dissolve the DNA in 10 mM Tris–HCl, pH 8.0.
7. Subject the genomic DNA to PCR amplification for a DNA fragment encompassing the 2 TALEN-targeted regions by using primers F and R [35].
8. Clone the PCR fragments into pCR TOPO-TA vector (*see* **Note 14**), transform into E. coli competent cells, plate the cells on LB plates containing 100 μg/ml ampicillin, and incubate at 37 °C overnight.
9. Screen the colonies for mutations by colony PCR using semi-nested primers f or fl paired with R [35] (*see* **Note 15**).
10. Grow up the colonies containing mutant DNA as identified by PCR above for plasmid DNA purification and sequencing to determine the exact mutations (*see* **Note 16**)
11. If the mutation rate is high, grow up the remaining injected embryos for phenotypic analysis of the F0 knockdown animals, or use the F0 animals to produce total knockout animals for analysis (see [35, 38].

4 Notes

1. Many instruments and approaches can be used to pull needles and inject RNA or DNA samples including simple hand pulled needles using a Bunsen Burner and forced air injectors using "house air" found in most laboratories, etc. The crucial criteria are to choose a needle and a method that allow

for a small needle diameter (about 10 μm) to be pulled and a way to calibrate or control injection volumes (about 5–30 nl per injection).

2. There are several TALEN assembly tools available from different resources. TALEN constructs can be also custom-designed and made through commercial companies on the market. Golden Gate TALEN and TAL Effector Kit assembles both arms of a TALEN into a backbone vector which expresses TALEN containing the same FokI domain [40]. However, the two arms of a TALEN can be assembled into two separate backbones such that one TALEN construct expresses TALEN containing the mutant FokI-ELD and the other TALEN construct expresses TALEN containing the mutant FokI-KKR to further increase the specificity [29, 30]. FokI-ELD and FokI-KKR can form heterodimer to become active enzyme, but they do not form homodimer as the wild-type FokI itself [42].
3. *Xenopus tropicalis* TRα gene expresses two alternative splicing isoforms (TRα1 and TRα2) and both contain the DNA binding domain (DBD). The exon 3 encodes the start region of the DBD, thus mutation within this targeted region from TALEN-mediated mutagenesis would produce truncated protein lacking DBD and the ligand binding domain (LBD) or a TR with mutations in the DBD. In principle, TALEN targeted region should be as close to the translational start site of the gene as possible to disrupt protein function without worrying about potential side effects from a long truncated protein. Alternative splicing on mRNA or alternative translational start sites in the proximity of the TALEN target should also be taken into consideration when choosing an appropriate target region for TALEN.
4. TALEN-mediated mutations in certain genes may lead to lethal phenotype during early embryogenesis. Using inducible promoter to drive TALEN expression in transgenic animals may be an alternative approach to knock down a target gene at later developmental stages to investigate gene function. Our TALEN arms are also under the control of a heat shock-inducible promoter in the vectors. If they are used to generate transgenic tadpoles, the arms should not be expressed without heat shock in the transgenic animals, thus allowing the target gene to function normally. However, when the animals are treated with heat shocks at desired stages to induce TALEN expression, the target gene would be mutated in some cells, thus affecting development. As the eye-specific γ-crystallin promoter drives GFP expression from one TALEN construct for one of the arm and RFP expression from the TALEN construct for the other arm, one can easily identify animals express-

ing both arms for such conditional knockdown studies. On the other hand, while TALENs assembled in these backbone vectors work effectively in mRNA microinjection experiments [35, 43], their efficiency in transgenic studies remains to be determined.

5. When linearizing plasmids, avoid excess DNA sequence following the gene of interest (such as long multiple cloning sites found in some plasmids). This is because long untranslated sequences may form RNA secondary structures that inhibit translation. In addition, avoid linearizing DNA with restriction enzymes which leave 3′ overhangs (such as Pst I) as that may prime transcription from the inappropriate strand [44].
6. Use gel electrophoresis to confirm that the DNA is pure and completely linearized. Circular plasmid templates will generate extremely long, heterogeneous RNA transcripts because RNA polymerases are very processive.
7. The linearized DNA should be purified for high yield RNA production. Commercial DNA purification kits are effective and usually produce reproducible results in the following RNA production. The linearized DNA can be also purified through phenol/chloroform extraction and isopropanol precipitation; however, care must be taken to ensure that the process does not carry any inhibitory chemical residues to affect RNA synthesis efficiency from the RNA polymerase.
8. Though a commercial RNA transcription kit is optimized for producing translatable RNA, cost may prohibit the use of such a kit. Transcription can easily be performed with transcription components purchased individually. RNA polymerases, their appropriate buffers, nucleotides, m7G(5′)ppp(5′)G cap analogs, DNase, and other reagents are available from numerous companies. Care must be taken, however, to ensure that when these reagents are used together, the conditions are optimized for efficient transcription. This may take some trials and errors.
9. Once the transcription reaction is complete (following the removal of the DNA template with DNase), the RNA must be purified prior to its injection. LiCl precipitation is a good and quick method for removing unincorporated nucleotides and proteins. Phenol/chloroform extraction followed by ethanol precipitation is a more stringent method of purification, though care must be taken to remove all traces of phenol and chloroform prior to injection.
10. The in vitro fertilization for *Xenopus tropicalis* is essentially identical to that for *Xenopus laevis*. For details, see *Xenopus* fertilization protocols by [45].

11. Once injected, RNA stability (half-life) is dependent on a number of factors including transcript size, secondary structure, 5′ and 3′ UTR lengths and sequences, polyA tail lengths, and others. Most injected RNA (such as TR or RXR) can still be readily detectable after 10 h, though Northern or other analysis should be used to determine the half-life of the transcript used if needed. TALEN mRNA stability has not been measured, but the mutation rate is a useful endpoint for validating the TALEN mRNA injection. Injecting higher doses of RNA may produce a higher mutation rate but may also be toxic to the embryos. Typical nucleic acid toxicity phenotypes include gastrulation defects (ring embryos, sometimes called the spina bifida phenotype), kinked embryos (abnormal somitogenesis), anterio-dorsal deficiencies (small heads or eyes, lack of pigment cells) and death. Nucleic acids also become toxic at much lower concentrations if too big a volume is injected [44].
12. TRα knockdown may produce distinct phenotypes depending on the absence or presence of T3. Scoring (classification) of embryos must be achieved with readily visible criteria. Thus, axis length, body width, presence/size of eye or cement gland, and other finer details such as number and organization of somites can all be used to correlate with a treatment. Subsequently, more detailed molecular or biochemical analyses can be carried out. RNA can be isolated from embryos and Northern blot or RT-PCR analyses can be used to examine changes of expression in downstream genes. Soluble proteins can be isolated from embryos for Western analysis of various proteins of interest, or in gel mobility shift assays to examine the binding level of TR/RXR heterodimer to TREs. Finally, whole mount in situ hybridization analysis can be used to investigate gene expression, etc.
13. Upon TALEN mRNA injection, TALEN proteins would be produced and bind to the target sequences to cleave the DNA in between the two TALEN binding sites. Mutations could be introduced when the cells repair the DNA damage. The DNA break/repair cycle can continue as long as the TALEN binding sites are intact and the distance between the two TALEN binding sites is still suitable for the TALEN pair to form active FokI enzyme. The mutations would be mosaic in the F0 animals, and the mutation spectrum and rate should be evaluated for deciding if the animals are worthwhile to keep for germline breeding for the target gene knockout lines or phenotype analysis on the F0 animals. TALEN pairs that produce very low mutation rates, e.g., 10%, should be discarded.
14. If high-fidelity DNA polymerases are used for PCR amplification of the target DNA fragment, the fragment should be

cloned into pCR Blunt II TOPO vector (Invitrogen) or similar cloning vectors for blunt-ended PCR fragments. If possible, high-fidelity DNA polymerase should be used for this step of PCR amplification to avoid potential random mutations introduced by Taq DNA polymerase during PCR amplification.

15. For colony-PCR, PCR amplification system based on high-fidelity DNA polymerase should be avoided to ensure no correction occurs on the primer in order to distinguish the wild-type target sequences and the mutant DNA sequences within the target.
16. A number of methods can be used to facilitate the identification of mutations caused by designer nucleases such as TALEN. The enzyme mismatch cleavage assays (MCAs) such as Surveyor [46] or T7 endonuclease I (T7E1) [47] mismatch cleavage assay are widely adapted to quantify mutation rates as a primary evaluation of targeted mutagenesis by designer nucleases. Other methods, such as High Resolution Melt Assay (HRMA) [48], droplet digital PCR (ddPCR) [49], have also been applied with varied benefits and disadvantages. PCR cloning and sequencing of the DNA sequences encompassing the TALEN targets would directly identify the mutation types and determine the mutation rates. For the purpose of gene knockdown or knockout, only the mutations that introduce out-of-frame mutations in the coding sequences of the target gene are desirable. For this purpose, the recently developed LacZ recovery/disruption assay [50] and direct visualization of out-of-frame mutation assay based on mCherry-GFP or LacZα-GFP fusion protein expression [51] would be valuable tools.

Acknowledgments

This work in the laboratory was supported by the Intramural Research Program of National Institute of Child Health and Human Development, National Institutes of Health.

References

1. Shi Y-B (1999) Amphibian metamorphosis: from morphology to molecular biology. John Wiley & Sons, Inc., New York
2. Tata JR (1993) Gene expression during metamorphosis: an ideal model for post-embryonic development. BioEssays 15(4):239–248
3. Lazar MA (1993) Thyroid hormone receptors: multiple forms, multiple possibilities. Endocr Rev 14(2):184–193
4. Evans RM (1988) The steroid and thyroid hormone receptor superfamily. Science 240:889–895
5. Yen PM (2001) Physiological and molecular basis of thyroid hormone action. Physiol Rev 81(3):1097–1142
6. Davis PJ, Davis FB (1996) Nongenomic actions of thyroid hormone. Thyroid 6:497–504

7. Tsai MJ, O'Malley BW (1994) Molecular mechanisms of action of steroid/thyroid receptor superfamily members. Annu Rev Biochem 63:451–486
8. Buchholz DR, Paul BD, Fu L, Shi YB (2006) Molecular and developmental analyses of thyroid hormone receptor function in Xenopus laevis, the African clawed frog. Gen Comp Endocrinol 145(1):1–19
9. Davis PJ, Davis FB, Cody V (2005) Membrane receptors mediating thyroid hormone action. Trends Endocrinol Metab 16(9):429–435
10. Shi YB, Matsuura K, Fujimoto K, Wen L, Fu L (2012) Thyroid hormone receptor actions on transcription in amphibia: the roles of histone modification and chromatin disruption. Cell Biosci 2(1):42. https://doi.org/10.1186/2045-3701-2-42. 2045-3701-2-42 [pii]
11. Guigon CJ, Cheng SY (2009) Novel nongenomic signaling of thyroid hormone receptors in thyroid carcinogenesis. Mol Cell Endocrinol 308(1–2):63–69
12. Hiroi Y, Kim HH, Ying H, Furuya F, Huang Z, Simoncini T, Noma K, Ueki K, Nguyen NH, Scanlan TS, Moskowitz MA, Cheng SY, Liao JK (2006) Rapid nongenomic actions of thyroid hormone. Proc Natl Acad Sci U S A 103(38):14104–14109
13. Yaoita Y, Brown DD (1990) A correlation of thyroid hormone receptor gene expression with amphibian metamorphosis. Genes Dev 4(11):1917–1924
14. Wong J, Shi Y-B (1995) Coordinated regulation of and transcriptional activation by Xenopus thyroid hormone and retinoid X receptors. J Biol Chem 270:18479–18483
15. Wang X, Matsuda H, Shi Y-B (2008) Developmental regulation and function of thyroid hormone receptors and 9-cis retinoic acid receptors during Xenopus tropicalis metamorphosis. Endocrinology 149:5610–5618
16. Leloup J, Buscaglia M (1977) La triiodothyronine: hormone de la métamorphose des amphibiens. CR Acad Sci 284:2261–2263
17. Shi Y-B, Wong J, Puzianowska-Kuznicka M, Stolow M (1996) Tadpole competence and tissue-specific temporal regulation of amphibian metamorphosis: roles of thyroid hormone and its receptors. BioEssays 18:391–399
18. Sachs LM, Damjanovski S, Jones PL, Li Q, Amano T, Ueda S, Shi YB, Ishizuya-Oka A (2000) Dual functions of thyroid hormone receptors during Xenopus development. Comp Biochem Physiol B Biochem Mol Biol 126(2):199–211
19. Puzianowska-Kuznicka M, Damjanovski S, Shi Y-B (1997) Both thyroid hormone and 9-cis retinoic acid receptors are required to efficiently mediate the effects of thyroid hormone on embryonic development and specific gene regulation in xenopus laevis. Mol Cell Biol 17:4738–4749
20. Amaya E, Kroll KL (1999) A method for generating transgenic frog embryos. Methods Mol Biol 97:393–414
21. Nakajima K, Yaoita Y (2003) Dual mechanisms governing muscle cell death in tadpole tail during amphibian metamorphosis. Dev Dyn 227:246–255
22. Schreiber AM, Brown DD (2003) Tadpole skin dies autonomously in response to thyroid hormone at metamorphosis. Proc Natl Acad Sci U S A 100:1769–1774
23. Das B, Schreiber AM, Huang H, Brown DD (2002) Multiple thyroid hormone-induced muscle growth and death programs during metamorphosis in Xenopus laevis. Proc Natl Acad Sci U S A 99:12230–12235
24. Schreiber AM, Das B, Huang H, Marsh-Armstrong N, Brown DD (2001) Diverse developmental programs of Xenopus laevis metamorphosis are inhibited by a dominant negative thyroid hormone receptor. Proc Natl Acad Sci U S A 98:10739–10744
25. Buchholz DR, Hsia VS-C, Fu L, Shi Y-B (2003) A dominant negative thyroid hormone receptor blocks amphibian metamorphosis by retaining corepressors at target genes. Mol Cell Biol 23:6750–6758
26. Buchholz DR, Tomita A, Fu L, Paul BD, Shi Y-B (2004) Transgenic analysis reveals that thyroid hormone receptor is sufficient to mediate the thyroid hormone signal in frog metamorphosis. Mol Cell Biol 24:9026–9037
27. Hasebe T, Buchholz DR, Shi YB, Ishizuya-Oka A (2011) Epithelial-connective tissue interactions induced by thyroid hormone receptor are essential for adult stem cell development in the Xenopus laevis intestine. Stem Cells 29(1):154–161. https://doi.org/10.1002/stem.560
28. Guo X, Zhang T, Hu Z, Zhang Y, Shi Z, Wang Q, Cui Y, Wang F, Zhao H, Chen Y (2014) Efficient RNA/Cas9-mediated genome editing in Xenopus tropicalis. Development 141(3):707–714. https://doi.org/10.1242/dev.099853
29. Lei Y, Guo X, Deng Y, Chen Y, Zhao H (2013) Generation of gene disruptions by transcription activator-like effector nucleases (TALENs) in Xenopus tropicalis embryos. Cell Biosci 3(1):21. https://doi.org/10.1186/2045-3701-3-21. 2045-3701-3-21 [pii]
30. Lei Y, Guo X, Liu Y, Cao Y, Deng Y, Chen X, Cheng CH, Dawid IB, Chen Y, Zhao H (2012)

Efficient targeted gene disruption in Xenopus embryos using engineered transcription activator-like effector nucleases (TALENs). Proc Natl Acad Sci U S A 109:17484–17489. https://doi.org/10.1073/pnas.1215421109. 1215421109 [pii]
31. Nakayama T, Fish MB, Fisher M, Oomen-Hajagos J, Thomsen GH, Grainger RM (2013) Simple and efficient CRISPR/Cas9-mediated targeted mutagenesis in Xenopus tropicalis. Genesis 51(12):835–843. https://doi.org/10.1002/dvg.22720
32. Blitz IL, Biesinger J, Xie X, Cho KW (2013) Biallelic genome modification in F(0) Xenopus tropicalis embryos using the CRISPR/Cas system. Genesis 51(12):827–834. https://doi.org/10.1002/dvg.22719
33. Wang F, Shi Z, Cui Y, Guo X, Shi YB, Chen Y (2015) Targeted gene disruption in Xenopus laevis using CRISPR/Cas9. Cell Biosci 5:15. https://doi.org/10.1186/s13578-015-0006-1
34. Yen PM (2015) Unliganded TRs regulate growth and developmental timing during early embryogenesis: evidence for a dual function mechanism of TR action. Cell Biosci 5:8. https://doi.org/10.1186/2045-3701-5-8
35. Wen L, Shi YB (2015) Unliganded thyroid hormone receptor alpha controls developmental timing in Xenopus tropicalis. Endocrinology 156:721–734. https://doi.org/10.1210/en.2014-1439
36. Choi J, Suzuki KI, Sakuma T, Shewade L, Yamamoto T, Buchholz DR (2015) Unliganded thyroid hormone receptor alpha regulates developmental timing via gene repression as revealed by gene disruption in Xenopus tropicalis. Endocrinology 156:735–744. https://doi.org/10.1210/en.2014-1554
37. Sachs LM (2015) Unliganded thyroid hormone receptor function: amphibian metamorphosis got TALENs. Endocrinology 156(2):409–410. https://doi.org/10.1210/en.2014-2016
38. Wen L, Shibata Y, Su D, Fu L, Luu N, Shi Y-B (2017) Thyroid hormone receptor α controls developmental timing and regulates the rate and coordination of tissue specific metamorphosis in Xenopus tropicalis. Endocrinology 158(6):1985–1998
39. Choi J, Ishizuya-Oka A, Buchholz DR (2017) Growth, development, and intestinal remodeling occurs in the absence of thyroid hormone receptor alpha in tadpoles of Xenopus tropicalis. Endocrinology 158(6):1623–1633
40. Cermak T, Doyle EL, Christian M, Wang L, Zhang Y, Schmidt C, Baller JA, Somia NV, Bogdanove AJ, Voytas DF (2011) Efficient design and assembly of custom TALEN and other TAL effector-based constructs for DNA targeting. Nucleic Acids Res 39:e82
41. Fu L, Buchholz D, Shi YB (2002) Novel double promoter approach for identification of transgenic animals: a tool for in vivo analysis of gene function and development of gene-based therapies. Mol Reprod Dev 62(4):470–476
42. Doyon Y, Vo TD, Mendel MC, Greenberg SG, Wang J, Xia DF, Miller JC, Urnov FD, Gregory PD, Holmes MC (2011) Enhancing zinc-finger-nuclease activity with improved obligate heterodimeric architectures. Nat Methods 8(1):74–79. https://doi.org/10.1038/nmeth.1539
43. Wen L, Fu L, Guo X, Chen Y, Shi YB (2015) Histone methyltransferase Dot1L plays a role in postembryonic development in Xenopus tropicalis. FASEB J 29:385–393. https://doi.org/10.1096/fj.14-252171
44. Vize PD, Melton DA, hemmati-Brivanlou A, Harland RM (1991) Assays for gene function in developing Xenopus embryos. Methods Cell Biol 36:367–387
45. Wu M, Gerhart J (1991) Raising Xenopus in the laboratory. Methods Cell Biol 36:3–18
46. Qiu P, Shandilya H, D'Alessio JM, O'Connor K, Durocher J, Gerard GF (2004) Mutation detection using surveyor nuclease. BioTechniques 36(4):702–707
47. Vouillot L, Thelie A, Pollet N (2015) Comparison of T7E1 and surveyor mismatch cleavage assays to detect mutations triggered by engineered nucleases. G3 (Bethesda) 5(3):407–415. https://doi.org/10.1534/g3.114.015834
48. Dahlem TJ, Hoshijima K, Jurynec MJ, Gunther D, Starker CG, Locke AS, Weis AM, Voytas DF, Grunwald DJ (2012) Simple methods for generating and detecting locus-specific mutations induced with TALENs in the zebrafish genome. PLoS Genet 8(8):e1002861. https://doi.org/10.1371/journal.pgen.1002861
49. Sedlak RH, Liang S, Niyonzima N, De Silva Feelixge HS, Roychoudhury P, Greninger AL, Weber ND, Boissel S, Scharenberg AM, Cheng A, Magaret A, Bumgarner R, Stone D, Jerome KR (2016) Digital detection of endonuclease mediated gene disruption in the HIV provirus. Sci Rep 6:20064. https://doi.org/10.1038/srep20064
50. Hisano Y, Ota S, Arakawa K, Muraki M, Kono N, Oshita K, Sakuma T, Tomita M, Yamamoto T, Okada Y, Kawahara A (2013) Quantitative assay for TALEN activity at endogenous genomic loci. Biol Open 2(4):363–367. https://doi.org/10.1242/bio.20133871
51. Fu L, Wen L, Luu N, Shi YB (2016) A simple and efficient method to visualize and quantify the efficiency of chromosomal mutations from genome editing. Sci Rep 6:35488. https://doi.org/10.1038/srep35488

Chapter 21

De Novo Transcriptomic Approach to Study Thyroid Hormone Receptor Action in Non-mammalian Models

Nicolas Buisine, Gwenneg Kerdivel, and Laurent M. Sachs

Abstract

Thyroid hormones are pleiotropic hormones involved in chordates physiology. Understanding their functions and mechanisms is also instrumental to diagnose dys-regulations and get a predictive power that can applied to medicine, ecology, etc. Today, high-throughput sequencing technologies offer the opportunity to address this issue not only in model organisms but also in non-model organisms. Here, we describe a method that makes use of RNA-seq to address differential expression analysis in non-model organism.

Key words Thyroid hormones, Transcriptome assembly, Transcriptome annotation, Differential expression analysis

1 Introduction

Thyroid hormones (THs) are phylogenetically conserved molecules that affect many aspects of development, growth, and metabolism of chordates [1]. THs were originally known to promote metamorphosis, the maturation of juvenile animals and maintain the basal metabolic rate. In fact, the roles of THs are more diverse, with an action during developmental and adult neuro-plasticity [2, 3], animal dispersal [4], hibernation [5], sex differentiation [6, 7], reproduction [8], lactation [5], stress and inflammatory responses [9]. Identifying the molecular and cellular mechanisms that govern the diversity of these effects is critical for understanding physiology because perturbations of TH signaling can have very diverse outcome affecting health, life quality, biodiversity, and ecosystems. Unfortunately, these effects are rarely probed in non-model organisms, thus restricting our knowledge of the key genes involved. In addition, even in model organisms, the molecular dissection of physiological processes has been mostly driven by candidate gene approaches and is thus based on an a priori. Today, the availability of next-generation sequencing (NGS) offers the opportunity to

Michelina Plateroti and Jacques Samarut (eds.), *Thyroid Hormone Nuclear Receptor: Methods and Protocols*, Methods in Molecular Biology, vol. 1801, https://doi.org/10.1007/978-1-4939-7902-8_21,
© Springer Science+Business Media, LLC, part of Springer Nature 2018

access naively the complete spectrum of genes involved in a particular biological process, and focus on species and environment-specific regulations of gene expression [10]. However, scientific communities working on non-model organisms frequently suffer from limited budget and it is not always easy to choose the best approach given a specific biological question and challenging sample material.

In this article, we describe a cost effective and reasonably fast method to measure gene expression by RNA-Seq, for species where no genome sequence is available. This is a two-step process: assemble de novo a set of coding sequences from paired-end RNA-Seq that will be used as a proxy for genome sequences, and measure gene expression by conventional (single-end) RNA-Seq. Although less resolutive, this nonetheless bypass the need for whole genome sequencing, assembly and annotation, which are notoriously difficult, time consuming, costly and often stay beyond reach of the small scientific communities centered around non-model organisms.

2 Materials

2.1 Computational Requirements

1. A 64-bit computer or server running any flavor of Unix or the GNU Linux operating system, for routine work.
2. Access to a server with multiple cores and large amount (>100 GB) of RAM for assembly. The TRINITY assembler typically requires around 1 GB of RAM per million of paired-end reads. Other assemblers may require more RAM. The amount of RAM is dependent on the total number of reads to assemble.
3. Access to a farm of computing cores, for annotation. For sequence alignment of assembled sequences against databases, such as NCBI's nr, the emphasis should be put on the number of computing cores rather than the total amount of RAM available, since sequence comparison can be easily parallelized on multiple CPUs and is not memory intense. Although not essential per se, GPU servers (with up to 4096 computing units) scales up computing time almost linearly with the number of computing cores, which makes them attractive solutions.

2.2 Data

1. Protein sequence data bases: NCBI's non redundant (nr) protein sequence database and any protein sequences database from a reference species closely related to the studied species.
2. Assembly: RNA should be sequenced with paired-end protocols, in FASTQ format (or FASTA format, together with

QUAL files). At least 100–150 million raw reads are needed (>50 million high quality reads).

3. Measure of gene expression: conventional, single end RNA sequencing with biological replicates, with around 50–60 million raw reads per sample.

2.3 Softwares

1. Quality control and read pre-processing: reads quality can be assessed with the FASTQC toolkit (http://www.bioinformatics.babraham.ac.uk/projects/fastqc/). The pre-processing of reads can be performed using any software designed to remove low-quality bases and trim residual adapter sequences (*see* **Note 1**).
2. Assembly: many de novo assemblers may be used depending on the available computational resources. In our lab, we use the Trinity assembler [11, 12] but other assembler, such as SOAPdenovo [13], AbySS [14], Velvet [15], or Oases [16], may also be used.
3. Blast2 [17] (ftp://ftp.ncbi.nlm.nih.gov/blast/executables/blast2/) or Blast+ [18] (ftp://ftp.ncbi.nlm.nih.gov/blast/executables/blast+/).
4. Bowtie [19] or Bowtie2 [20] (http://bowtie-bio.sourceforge.net).
5. DESeq [21] or DESeq2 [22] (http://bioconductor.org/packages/release/bioc/html/DESeq.html).
6. Bedtools [23] (https://code.google.com/archive/p/bedtools/).
7. The R environment (http://www.r-project.org).
8. Assembly statistics: Assemblathon tools [24].

3 Methods

3.1 General Workflow

This review section focuses on the bioinformatic part of the process. Although not described here, animal treatments, tissue/RNA extraction, and high-throughput sequencing should be produced with the maximum quality possible. Importantly, transcriptome assembly and measure of gene expression each require a specific protocol and specific type of NGS data.

The conceptual steps for the detection of differential expression after de novo transcriptome assembly are the following:

- Clean up the reads to remove low quality reads or trim the 5′ and/or 3′ end (*see* **Note 1**).
- Assemble the transcriptome (*see* **Note 2**).
- Clusterize contigs, so that multiple contigs are coalesced into single mRNA species.

- Annotate clustered contigs, by sequence comparison against databases of known sequences (*see* **Note 3**).
- Identify differentially expressed genes, after mapping of conventional RNA-Seq reads onto reference sequences.

3.2 A Word of Warning

Bioinformatic processing of large datasets typically produce hundreds (if not thousands) of files, which may be hard to store and keep track of. The success of the project thus relies on rigorous good laboratory practices that minimize human errors (*see* **Notes 4** and **5**). Furthermore, we should emphasize the power of the versatile toolkits readily available in the Unix/Linux environments and we strongly suggest making heavy use of them (*see* **Notes 6–9**). In terms of project management, three points are important: a well-planned project with a realistic budget, producing large sets of high-quality raw sequences, set regular checkpoints during the project and impose stringent quality controls (*see* **Notes 10–12**).

3.3 Quality Control and Read Pre-Processing

1. Assess the quality of paired-end-reads using FASTQC toolkit. For each file, run the following command from a bash terminal:

```
fastqc <read file>
```

2. If necessary (average per base sequence quality lower than 28, over-representation of adaptator sequences), use any cleaning software to correct for any biases revealed by the quality control (*see* **Notes 1** and **13**). AlienTrimmer can be run with the following command:

```
AlienTrimmer -if <forward read file> -ir
<reverse read file> -c alien.txt
```

   Example:

```
AlienTrimmer -if 500892_0035_AHVL_1.fastq -ir
500892_0035_AHVL_2.fastq -c alien.txt
```

3. This creates three output files, containing trimmed forwards and reverse reads files, plus an additional file containing singletons, respectively:

   `500892_0035_AHVL_1.fastq.at.fq`,
   `500892_0035_AHVL_2.fastq.at.fq` and
   `500892_0035_AHVL_1.fastq.at.sgl.fq`

   The file `alien.txt` is a text file containing the sequences of adaptors and of all putative contaminating sequences (one per line). Other options may be considered, such as `-q` that allows us to set a threshold for quality score and `-l` that define the minimum length of trimmed reads that will be kept.

4. Re-assess the quality of paired-end-reads using FASTQC toolkit and verify that low quality reads have been filtered out and reads trimmed appropriately. It is not uncommon to observe an inefficient biases correction. In this case, consider using other read cleaning software (*see* **Note 1**).

3.4 de novo Transcriptome Assembly and Evaluation of the Assembly Procedure

1. Merge the FASTQ files corresponding to the forward and reverse reads file, separately. Merge datasets from different tissues/conditions to increase the total number of reads if needed (*see* **Note 14**). On bash-like environment, this is easily carried out with the `cat` command (that outputs the content of the files pointed to) and redirection (with the > sign) into a new file:

```
cat <pair1_1.fastq> <pair1_2.fastq> … >
pair1_all.fastq
```

```
cat <pair2_1.fastq> <pair2_2.fastq> … >
pair2_all.fastq
```

Example:

```
cat 500892_0035_AHVL_1.trim.fastq
500882_0033_AHVL_1.trim.fastq > 5008-92-
82_00-35-33_AHVL_1.trim.fastq
```

```
cat 500892_0035_AHVL_2.trim.fastq
500882_0033_AHVL_2.trim.fastq > 5008-92-
82_00-35-33_AHVL_2.trim.fastq
```

2. If needed, copy files to the server running TRINITY with the bash command `scp`, iterating over a loop. The following command copies all the FASTQ files in the current folder (`*.fastq`) and copies them in the `/datasets/` folder of the `my_server.my_institution` server, under the `me` account:

```
for CURRENT_FILE in *.fastq; do
scp $CURRENT_FILE
me@my_server.my_institution:/datasets/ ;
done
```

3. de novo assembly can be performed with numerous tools, many of which are freely available (*see* Subheading 2). In this method, we will focus on the TRINITY assembler. According to TRINITY's manual, it is recommended to use a computer with around 1 GB of RAM per million of paired-end reads used for the assembly. Several options may be used to perform de novo assembly with TRINITY (*see* **Note 15**).

   Calling TRINITY by the command line has the following synopsis:

```
Trinity.pl [options] --left <left FASTQ
paired file> --right <right FASTQ paired file>
```

Example:

```
Trinity.pl --seqType fq --max_memory 50G --left
5008-92-82_00-35-33_AHVL_1.trim.fastq --right
5008-92-82_00-35-33_AHVL_2.trim.fastq --CPU 16
--PasaFly 2> 20180707.trinity.error.log >
20180707.trinity.log
```

The `--PasaFly` option specifies the use of the alternative PASA-like algorithm, which is more conservative and thus detects fewer isoforms. The number of CPU and the maximum amount of RAM to use are set by the `--CPU` and `--max_memory` options. These two options should match the maximum resources allocated on the server. As discussed above, they condition the total number of reads to use for assembly. Error messages (after `2>`) are stored in the `20180707.trinity.error.log` file and standard output (after `>`) in `20180707.trinity.log`. TRINITY output is located in the `trinity_out_dir` folder. The file `Trinity.fasta` contains the contig sequences, in FASTA format:

```
>TR1|c0_g1_i1 len=285 path=[263:0-284] [-1,
263, -2]

AGGTCCAATAGTGCGCTTTCTTGCGTATTTGTGCTTTTATGGTG
GTGTGCCACTGCACGA

CCATTGGAGGTACCAATTTTCCTAACCCTTTATTCAGTCTGCTT
TTATTTTTGGGTCCCT

CTGCACTATGGAGCGCCTTTCTGTTTAACTTTTGCAATCGCTTG
AATGATACTGAGTTGC

TCTCAGGCTATACATGAGATCACAGGCCTGAGAATAGACCACAG
TGCGCAATAGAGCTCC

GAAGCCTGTGCAGTCAGCTGATTGGCAGGGATCCCAAGTTTTG
TG

>TR2|c0_g1_i1 len=317 path=[397:0-25 398:
26-26 399:27-316] [-1, 397, 398, 399, -2]

AATGCATGGGTGAAATGCATGTTGCGGGTGCTGAATGGTCTGGA
TGCCGCCTATTACCAG

ATTACTATGCCACTGGCTTATGATGCTCTTCTGGTTTATATATA
TATGCATCCCTTGGAT

ACTCTGCACTTTAGTGCTATATGTTCTACATATCTGTCCTATAT
ACCTTGATCTCTTGTG
```

Notice that contig names provide a number of information, including their length. It is good practice to rename of output file:

```
mv trinity_out_dir/Trinity.fasta./
20180707.Trinity_jaccPasa.fas
```

4. Collect statistics describing the total number of contigs. To this end, the name of each contig contains its size. This is easily collected with a combination of `awk` and `sed` commands, under Unix/Linux environment:

```
awk '{print $2;}'20180707.Trinity_jaccPasa.
fas | sed 's/len=//' > 20180707.Trinity_
jaccPasa.fas.contig_size
```

5. The contig length cumulative density distribution is then plotted from `R` or any spreadsheet program (Fig. 1a–d).
6. Transcript assembly generally generates a large number of short sized contigs, corresponding mostly to assembly artifacts and/or very partially assembled transcripts (typically, a pair of overlapping forward and reverse reads). It is good practice to discard all contigs shorter than some threshold, established from the statistics above (e.g., 200 bp). This can be carried out with a moderately advanced `awk` script:

```
BEGIN {RS = ">" ; ORS = "\n"; FS="\n";
THRESHOLD=MIN_SIZE;}
{sq="";
for (i = 2; i <= NF; i++){sq=sq "" $i;}
 if (length(sq)>=THRESHOLD){print ">"$1;for(
j=0;j<=length(sq)/60;j+=1){print substr(sq,1
+(j*60),60);}}
```

Save the script into a text file (called `filter_fasta.awk`, for example) and call it from the command line:

```
awk -f filter_fasta.awk -v MIN_SIZE=1000
20180707.Trinity_jaccPasa.fas > 20180707.
Trinity_jaccPasa.fas.length_mt_2000.fas
```

The option `-v MIN_SIZE = 1000` sets the threshold of the minimum contig length allowed (1000 bp in this case). Notice the name of the output file.

3.5 Clustering of Contigs and Annotation of de novo Assembled Transcriptomes

1. Clustering of contigs into a set of nonredundant reference sequences. This may be carried out with CD-HIT-EST [5], which was initially designed to clusterize ESTs. Option `-T` specifies the number of CPUs to be used. If set to 0, all CPUs will be used. Option `-n` specifies word length (between 8 and 10 for similarity threshold between 90% and 100%), `-r 1` sets to compare both strands, and `-o` specifies the output file.

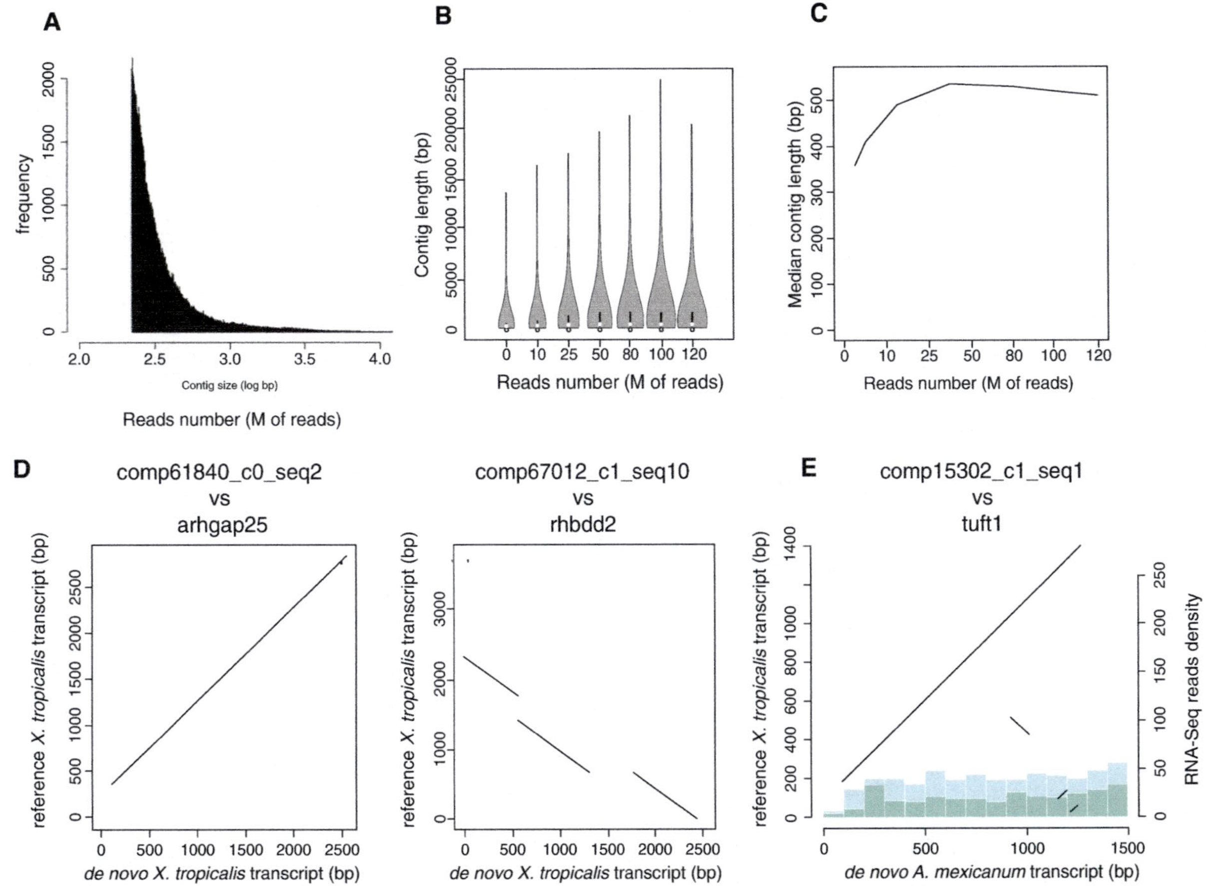

Fig. 1 (**a**) Size distribution of the contigs produced by TRINITY, before filtering. Example from real dataset. Notice that the vast majority of contigs are very short sized. (**b**, **c**) Violin plot and median value of the contigs length produced, as a function of the total number of reads used for assembly. (**d**) Dot-plot between assembled transcripts and expected sequence, from *Xenopus tropicalis*. (**e**) Dot-plot between the assembled tuft1 transcript in *Ambystoma mexicanum* and the orthologous sequence from *Xenopus tropicalis*. Histograms represent RNA-Seq reads density, in control experiment and after hormone treatment. Example from real dataset

```
cd-hit-est -r <strand> -T <CPUs> -i <Fasta
file of redundant sequences> -o <output file>
-c <similarity threshold> -n <word length>
```

Example:

```
cd-hit-est -r 1 -T 0 -i 20180707.Trinity_
jaccPasa.fas.length_mt_2000.fas -o 20180707.
Trinity_jaccPasa.fas.length_mt_2000.
clusterc95.fas -c 0.95 -n 8
```

The clustering details are stored in an additional `clstr` file (`20180707.Trinity_jaccPasa.fas.length_mt_2000.clusterc95.fas.clstr`).

2. Annotation of the references sequences. This corresponds to defining orthologous relationships with known protein sequences. This is typically performed with the nonredundant (nr) protein sequence database (NCBI) and/or, if available, to the RefSeq protein sequences of a closely related species. This can be carried out with the BLASTX (*see* **Note 16**), although this typically requires the use of custom made python or perl script to parse and filter out the results. Alternatively, these services are provided by CRB-BLAST [25], which runs BLAST, detects the Best-Reciprocal-Blast-Hits, and sets orthologous relationships. CAUTION: this step can be computationally intensive, and the use of dedicated hardware with multiple cores can significantly accelerate the process (*see* above). CRB-BLAST is simple to use:

```
crb-blast --query <reference sequences>
--target <known proteins> --threads <CPUs>
--output <Output file>
```

Example:

```
crb-blast --query 20180707.Trinity_
jaccPasa.fas.length_mt_2000.clusterc95.
fas --target nr.fa --threads 256 --output
20180707.Trinity_jaccPasa.fas.length_
mt_2000.clusterc95.annotation.tsv
```

The output is a tabulated file where the first two fields correspond to the query name (name of the unannotated sequence) and the target name (name of the putative ortholog). Other information (% similarity, alignment length, blast e_value and bit_score, etc.) are provided, and it may be worth exploring them to adjust the stringency further if needed. This can easily be carried out with a simple `awk` command or script, that would allow us to set various thresholds based on alignment length, bit_score, etc. In case known transcripts are available, they can be used to tweak the stringency and better control the specificity and sensitivity of the search.

A simple python script can update the name of references sequences. In principle, a hash table (*a.k.a.* dictionary) is built from the annotation.tsv file and is used to update individual sequence names. The following script (rename.py) is a straightforward example:

```
#! /usr/bin/env python
import sys
ANNOTATIONFN=sys.argv[1]
SEQUENCE_FILE_NAME=sys.argv[2]
# Build dictionary
annotation={}
inputfile=file(ANNOTATIONFN,'r')
for records in inputfile:
 old,new=records.strip().split()[:2]
 annotation[old]=new
inputfile.close()
# Parse sequence file
sequence_file=file(SEQUENCE_FILE_NAME,'r')
for cl in sequence_file:
 # Change sequence name if needed
 if cl[0]=='>' and cl.split()[0] in annota-
tion.keys():
  print ">%s"%annotation[cl.split()[0]]
else:
  print cl,
```

Annotated sequences are printed to the standard output. Invoke the rename.py script with the following command:

```
python ./rename.py <annotation file> <se-
quence file> > <output file>
```

where the annotation file is:

```
20180707.Trinity_jaccPasa.fas.length_
mt_2000.clusterc95.annotation.tsv
```

and the sequence file:

```
20180707.Trinity_jaccPasa.fas.length_
mt_2000.clusterc95.fas.
```

In order to be consistent, the output file name could be named:

```
20180707.Trinity_jaccPasa.fas.length_
mt_2000.clusterc95.annotation.fas.
```

Notice the redirection symbol ('>') at the end of the command line, so that the output is properly redirected into the file.

3.6 Differential Analysis of Gene Expression

Let us consider a typical RNA-Seq experiment with two datasets (control and treated with THs), composed of three biological replicated where RNA have been sequenced with the simple Illumina

50 bp single-end protocol (`RFTK.00356.28`, `RFTL.00356.29` and `RFTM.00356.30` for control, and `XTD.024.72`, `XTD.025.73` and `XTD.026.74` after TH treatment). The point of this section is to detect the genes (more specifically clustered contigs, used as proxies for genes) which are differentially expressed between the two experimental conditions. The workflow is straightforward: for each RNA-Seq library, reads are mapped to the reference sequences produced above and a count table is derived (i.e., the number of reads per contig). Differential analysis is run with the `R` package DESeq. Although not detailed here, quality controls (*see* above), clipping, and adaptors removal also apply here as a prerequisite.

1. Reads mapping. For each condition, align RNA-Seq reads to the set of reference sequences using a read such as BOWTIE. BOWTIE2 is more appropriate for mapping reads longer than 50 bp. Mapping parameters are largely dependent on reads length and quality, and although a large palette of parameters may be suitable, it is important to highlight that mapping specificity is function of the seed length (`-l` parameter), the number of mismatches allowed in the seed (`-n`) and the maximum number of alignments tolerated (`-m`). For best results, one should consider using stringent parameters (large `-m`, low `-n`) and keeping only uniquely mapped reads (`-m 1`). Also, it is important to remove the PCR amplification biases that may be generated during the sequencing library construction by collapsing all the reads mapping at the same location. This is carried out with the Unix/Linux tools `sort` and `uniq`. At the same time, BOWTIE output is transformed into a standard format (BED).

   Preparing the reference transcriptome for mapping and building the bowtie has the form:

```
bowtie-build <reference sequences> <name of
bowtie index>
```

Example:

```
bowtie-build 20180707.Trinity_jaccPasa.fas.
length_mt_2000.clusterc95.annotation.fas
species.transcriptome
```

For each RNA-Seq library, mapping single end reads requires a single call to BOWTIE piped to `awk` and `uniq`:

```
bowtie -l <seed length> -n <mismatches>
-m <max alignments> <index> <reads file>
```

Example:

```
bowtie -l 45 -n 1 -m 1 species.transcriptome
RFTK.00356.28.fastq | awk 'BEGIN{OFS="\t"}
```

```
{print $4,$5,$5+length($6),".",0,$3;}' |
sort -k 1,1 -k 2,2n -k 6,6 | uniq >
RFTK.00356.28.fastq.bow.sorted.uniq.bed
```

2. Derive an annotation of the features to detect. In our case, there is a single feature per reference, starting with the first base pair (counted as position 0) and as long as the reference. This can be produced with the help of the following awk script (contigs2bed.awk) acting on the reference sequences file:

```
 BEGIN {RS = ">" ; ORS = "\n"; FS="\
n";OFS="\t"; }
 {sq="";
  for (i = 2; i <= NF; i++){sq=sq "" $i;}
  {print $1,0,length(sq);}}
```

This script is called:

```
awk -f contigs2bed.awk <reference sequences>
> <annotated features>
```

Example:

```
awk -f contigs2bed.awk annotation.fas > an-
notation.bed
```

For the sake of keeping commands onto a single line, annotation.fas stands for 20180707.Trinity_jaccPasa.fas.length_mt_2000.clusterc95.annotation.fas and annotation.bed for 20180707.Trinity_jaccPasa.fas.length_mt_2000.clusterc95.annotation.bed.

3. Derive individual read count tables. INTERSECTBED, from the BEDTOOLS package, is a very popular tool to count bed file entries (reads) overlapping with annotated features (*see* above). This should be applied onto all the .bed files. The call to INTERSECTBED is straightforward:

```
intersectBed -c -a <annotated references> -b
<reads> > <individual read count>
```

Example:

```
 intersectBed -c -a annotation.bed -b
RFTK.00356.28.fastq.bow.sorted.uniq.bed >
RFTK.00356.28.fastq.bow.sorted.uniq.bed.cnt
```

In fact, this process can be automated with a simple loop:

```
for CURRENT_FILE in *.bed ; do intersect-
Bed -c -a annotation.bed -b $CURRENT_FILE >
$CURRENT_FILE.cnt ; done
```

4. Combine reads count onto a single table. This requires two steps. First, the number of reads per reference sequence corresponds to the fourth column of the individual .cnt files. Thus, for each .cnt file, this column will be extracted and

stored in a temporary file with the `cut` command. All these columns will then be added one after the other, ***column wise***, with the `paste` command. The resulting table can be called `annotation.treatments.cnts`. The first line creates a header in the output file, and the last line redirects the output of the `paste` command after the header (notice the >>, instead of the > symbol):

```
echo -e "gene\tRFTK.00356.28\tRFTL.00356.29\
tRFTM.00356.30\tXTD.024.72\tXTD.025.73\
tXTD.026.74" > annotation.treatments.cnts
cut -f 4 RFTK.00356.28.fastq.bow.sorted.
uniq.bed.cnt > 28.col4
cut -f 4 RFTL.00356.29.fastq.bow.sorted.
uniq.bed.cnt > 29.col4
cut -f 4 RFTM.00356.30.fastq.bow.sorted.
uniq.bed.cnt > 30.col4
cut -f 4 XTD.024.72.fastq.bow.sorted.uniq.
bed.cnt > 72.col4
cut -f 4 XTD.025.73.fastq.bow.sorted.uniq.
bed.cnt > 73.col4
cut -f 4 XTD.026.74.fastq.bow.sorted.uniq.
bed.cnt > 74.col4
paste annotation.bed 28.col4 29.col4 30.col4
72.col4 73.col4 74.col4 >> annotation.treat-
ments.cnts
```

5. Filter out reference sequences with low reads counts. Low reads count can strongly affect differential analysis, since stochastic variations of a few reads may artifactually overestimate the biological response. It is therefore good practice to filter them out. Although this threshold is set empirically, reference sequences associated with a total number of reads (across all conditions) less than 50 will be discarded. One might want to adjust this threshold according to the sequencing depth or other variables. Again, this is carried out with a simple `awk` script:

```
awk 'NR==1{print $0;}NR>1{if
($2+$3+$4+$5+$6+$7>=50){print $0;}}'
annotation.treatments.cnts > annotation.
treatments.filtered.cnts
```

The resulting file, `annotation.treatments.filtered.cnts`, will be used for differential analysis.

6. Differential analysis. DESeq [21] models variance with a generalization of the Poisson distribution (the negative binomial distribution). The following `R` script, based on the DESeq abundant documentation, reads the `annotation.treatments.filtered.cnts` file, creates a DESeq object, normal-

izes sequencing depth, estimates dispersion, and performs the statistical test. It also generates a number of diagnostic graphs (MAplot, PCA, etc.). Importantly, the `nbinomTest` parameters should match those indicated in the condition vector (`condition <- c("cntrl", "cntrl", "cntrl","treat","treat","treat")`). Results are saved in the `.csv` file. The following `R` commands can be typed directly from within the `R` environment:

```
library(DESeq)
input_file_name<-'annotation.treatments.
filtered.cnts'
d<-read.table(input_file_name,sep="\t",row.
names=1,header=TRUE)

# The '#' symbol marks a comment ignored by
the R interpretor.
# This is useful to annotate a script

# prepare labels/designs
condition <- c( "cntrl", "cntrl",
"cntrl","treat","treat","treat" )

# Simple comparison
## Let's make an object DEseq can manipulate
cds<-newCountDataSet(d,condition)

## Normalization
cds<-estimateSizeFactors(cds)

## Estimate dispersion
cds<-estimateDispersions(cds,method="pooled"
,sharingMode="fit-only",fitType="local")
# PCA on stabilized variance
vsdFull = varianceStabilizingTransformation
(cds)

png(paste(outfileBaseName,"PCA","png",s
ep="."))
plotPCA(vsdFull)
dev.off()

# binomial test
res = nbinomTest(cds,"cntrl","treat")
outfileBaseName=paste(input_file_
name,"DE",sep='.')

# output MAplot
png(paste(outfileBaseName,"MAplot","png",s
ep="."))
plotMA(res)
dev.off()
```

```
# output histogram pValues
png(paste(outfileBaseName,"histPval","png",s
ep="."))
hist(res$pval,breaks=100, col="skyblue",bord
er="slateblue", main="")
dev.off()

# output results
out=paste(outfileBaseName,'csv',sep=".")
write.table(res,file=out,sep="\
t",quote=FALSE,row.names=FALSE)
```

Alternatively, it would be more efficient to save this set of commands in a text file (with an explicit name like `DiffAnalysis.R`) and run through the `R` interpretor from the `bash` command line (*see* `R` documentation for details):

```
Rscript ./DiffAnalysis.R
```

4 Notes

1. This list of reads pre-processing tools is not exhaustive: Trimmomatic (http://www.usadellab.org/cms/?page=trimmomatic), AlienTrimmer (ftp://ftp.pasteur.fr/pub/gensoft/projects/AlienTrimmer/), AdapterRemoval (https://github.com/MikkelSchubert/adapterremoval), fastx_clipper (http://hannonlab.cshl.edu/fastx_toolkit/), Btrim (http://graphics.med.yale.edu/trim/).
2. For assembly, the stress is on the computing resources available. If not available on site, computing time can be paid for at cloud service providers. The cost should be planned from the start of the project.
3. For transcripts annotation, the emphasis of computing power should be put on the number of computing cores rather than the total amount of RAM available, since sequence comparison can be easily parallelized on multiple CPUs and is not memory intense. Computing time scales up almost linearly with the number of computing cores, which makes GPU servers (with up to 4096 computing units) attractive solutions.
4. There are two main constraints: (1) storage space is limited. Keeping all files at intermediate processing levels will consume storage space quickly. On the other hand, computing resources are limited and re-running processing steps multiple times may not be practical. It is just a matter of finding the good balance that optimizes both storage and computing time. (2) It is often necessary to explore a number of alternative softwares and parameter sets, which increases dramatically the number of files

produced in the course of the project. Although it is not a scientific issue per se, the large number of files is a clear source of human error, amplified by any lack of clear and rigorous file naming convention. We therefore strongly advise to establish a strict file naming convention, in addition to regular laboratory good practices.

5. A possible naming convention would rely on the fact that file names suffer from little length limitations (in fact, <257 characters on most modern file systems) and accept any number of dots under Unix/Linux. In this case, a processed file would inherit the name of the parent input files together with a suffix related to the processing step, suffixes being separated by a dot. The whole point is to make the file name explicit relative to its content. Suppose a folder contains the following files:

    1 - `20180707.Trinity_jacc.fas`

    2 - `20150707.Trinity_jaccPasa.fas`

    3 - `20180707.Trinity_jaccPasa.fas.stats`

    4 - `20180707.Trinity_jaccPasa.fas`

    5 - `20180707.Trinity_jaccPasa.cd-hit-est_clusterc95.fas.gz`

    6 - `20180707.Trinity_jaccPasa.cd-hit-est_clusterc95.annotated.fas.gz`

    7 - `20180707.Trinity_jaccPasa.fas`

    8 - `20180707.Trinity_jaccPasa.length_mt_2000.cd-hit-est_clusterc95.fas.gz`

    9 - `20180707.Trinity_jaccPasa.length_mt_2000.cd-hit-est_clusterc95.annotated.fas.gz`

    10 - `20180707.Trinity_jacc.fas`

    11 - `20180707.Trinity_jacc.cd-hit-est_clusterc95.fas.gz`

    12 - `20180707.Trinity_jacc.cd-hit-est_clusterc95.annotated.fas.gz`

    13 - `20180707.Trinity_jacc.fas`

    14 - `20180707.Trinity_jacc.length_mt_2000.cd-hit-est_clusterc95.fas.gz`

    15 - `20180707.Trinity_jacc.length_mt_2000.cd-hit-est_clusterc95.annotated.fas.gz`

    The name of file 1 can be decomposed into `20150707`, `Trinity_jacc` and `fas`, which corresponds to the raw sequences files `20180707` subjected to assembly with `Trinity` and `jacc` option. The file is a multiple FASTA file (`fas`). For file 8, `Trinity` was run with `jacc` and `Pasa` options, where only contigs longer than 2000 bp were kept (`length_mt_2000`), before clustering with the cd-hit-est software and `clusterc95` option. This file contains the sequence of clustered contigs (`fas`) and is compressed (`gz`). File 9 is similar except that contigs have been annotated. Looking at the name of a file is enough to know what is corresponded to.

6. Although the command line interface looks puzzling first, the galaxy of GNU tools (available on virtually all Unix systems and Linux) offer a la1rge tool set of everyday use in bioinformatics. These include softwares that can be used to combine files into a single file (`cat`), sort records by chromosome position (`sort`), remove identical entries (`uniq`). Also, most bioinformatic packages are available from public depots, which make software installation simple and fast.
7. At some point of the project, one will naturally want to develop his/her own tools or deploy some file processing facility (e.g., custom filters). The `awk` program provides a simple (but powerful) scripting interface for file processing services. Although remarkably efficient on tabulated text files, it proves also valuable on other file types as well (such as FASTA files). The syntax is simple and numerous tutorials (web pages and videos) are available on the Internet. In a nutshell, `awk` processes each line of a file, one after the other. For each line, columns are identified by a dollar sign and a number (the column number), so that the first column is `$1`, the second is `$2`, etc. Printing a line (`$0`, the entire line) depending on whether the value of one column (e.g., `$3`) is larger than or equal to a threshold is illustrated by the following (simple) statement:

   ```
   if ($3>=THRESHOLD) {print $0;}.
   ```

   Arithmetic (and a wealth of other) operations are also possible. *See* `awk` documentation for details. The initial investment will quickly become worthwhile as the learning curve is very steep.
8. Another scripting language worth considering is `python`. This language provides several programing paradigms (procedural, object oriented) and may be more appropriate for complex tasks. Nonetheless, `python` is very versatile, easy to learn, its syntax is simple and cookbook approaches work well. There are also numerous packages and tutorials available on the Internet, including in bioinformatics. Historically, the `perl` language was used but its syntax is less rigorous and it may be more difficult to analyze and reuse someone else's script. `Perl` interpreter is also slower than `python's`. For these reasons, nowadays, `perl` attracts less attention from researchers.

9. Finally, the power of Unix/Linux also comes with the command interpreter. Many exist but BASH and CSH/TCSH are found most of the time. In addition to interpret commands, they are also a rich scripting language with loops, control structures, and moderately complex datatypes (such as arrays). It is good practice to copy all commands in a text file (and thus make a bash script) for two reasons: (1) keep a trace of what has been done (good laboratory practice), and (2) if needed, it is possible to re-run all the commands in it by simply executing the script.
10. Sequencing is not the (only) limiting step, and it is useless to sequence without proper manpower and know-how to process and analyze the datasets. Given the large time stamp of the bioinformatic component of the project, manpower is much more expensive than a few sequencing runs. Other non-negligible sources of cost also include softwares, computing hardware, hosting facilities, etc. In some cases, it might be worth considering to rent computation resources from a cloud. All this should be budgeted as early as at the design phase of the project. Too often, emphasis is restricted just on sequencing and projects later suffer from limited bioinformatics resources.
11. The total number of (high-quality) reads directly impacts the assembly quality and resolution. Trying to assemble a transcriptome with single end reads, too little reads or poor quality reads will almost certainly fail, and efforts should focus on the generation of high-quality and abundant raw sequences. We found that from >50 million (high quality) reads (meaning >100–150 M raw reads), adding new reads does not significantly help in detecting novel transcripts (although contigs length still increases). Sequencing multiple libraries originating from different tissues would also capture more tissue specific transcripts, increase the diversity of the sequences assembled de novo, and help describe more fully the repertoire of expressed sequences. Two critical aspects of transcriptome assembly are sequencing depth and reads quality.

    Depending on the FASTQC quality controls, it may be worth clipping the low quality 5′ and/or 3′ end of reads. At the same time, reads length is an important parameter of sequence assembly and over-clipping might have a strong negative effect on assembly efficiency. It is all a matter of finding the good balance between reads quality and reads length and, although necessary, this puts stress on computing time and storage space. To this regard, a set of known transcripts (in databases or characterized by molecular biology and annotated "by hand") may prove to be very valuable resources to limit the parameter space search and quickly channel toward a set of "biologically sound" parameters, where the pipeline independently reconstructs known transcripts. In fact, creating a high-quality resource (a gold standard) is an integral part of the project.

12. During the course of the project, it is important to derive a number of quality controls at each milestone, in order to help decide whether or not to continue to the next step. There are numerous pitfalls: poorly curated raw sequence reads, the use of single end instead of paired-end reads, using non-optimal bioinformatics parameter sets, etc. In addition to experimental and bioinformatic reconstruction biases, biological diversity adds an extra level of complexity (shared exons between transcripts, transcripts originating from duplicated genes, repeated/transposable elements in coding sequences, etc.). As a result, chimerism is relatively common, UTRs are often missing... and stringent quality controls, together with experimental validation of a selected number of transcripts, are essential. One such quality control is shown in Fig. 1e. This represents the dot-plot of a randomly picked-up reference sequence to its cognate ortholog, showing high co-linearity between them, and thus high-quality assembly. The averaged density of mapped reads between the two experimental conditions is shown at the bottom, nicely showing that there is no chimerism and that the contrast of reads density can be attributed to the expected transcript.
13. For optimal representation of transcripts description in a specie, consider merging reads originating from different tissues and/or experimental conditions. The larger the number of reads, the higher sensitivity and specificity. Ultimately, the reconstruction of lowly expressed transcripts and the capture of a large diversity of transcripts is dependent on the total number of reads available and the processive power (number of CPUs, RAM) available on the server.
14. Among others, two options may be interesting to consider when using TRINITY. The first option, `--jaccard_clip`, is typically used with paired end reads to limit the artifactual assembly of fusion transcript [11, 12]. The `--PasaFly` option is typically used to provide a good balance between transcript fragmentation (optimized total count of assembled contigs) and the reconstruction of full-length transcripts. The choice between these parameters is not trivial and depends, in part, on the genome and genes structure and may be difficult to set a priori.
15. Transcriptome assembly generally generates hundreds thousands of sequence that can contain artifacts and chimeric tran-

scripts as well as various alternative transcripts for each gene. This can also include new genes/transcripts specific to the species under study. If the aim of the work is to compare different species, it may be useful not to consider these specific transcripts individually but to perform the analysis only at the gene level, after clustering of contigs.

16. When using BLAST in search for orthologs, two main parameters have to be looked at: word length (`-W`) and the substitution matrix. BLAST recodes sequences (query and subject) in an alphabet of `W` characters and then looks for identical subsequences. Thus, increasing `W` reduces computing speed but at the cost of sensitivity. Inversely, lowering `W` increases computing time and sensitivity, but may also increase significantly background noise. Again, using a gold standard may be instrumental in setting appropriate parameters. BLOSUM matrices are popular matrices and the rule of thumb is that BLOSUM45 should be used when the expected similarity level is about 45%, BLOSUM60 for similarity ~60%.

Acknowledgment

This work was supported by the "Centre National de la Recherche Scientifque" (PEPS ExoMod "Triton" to L.M.S.) and the "Muséum National d'Histoire Naturelle" (Action Transversale du Muséum "Formes possibles, Formes réalisées" and Action Transversale du Muséum "Cycles biologiques" to L.M.S., and the Scientifc council post-doctoral position to G.K.). We thank J. Pedraza and the PCIA services for providing intensive computation infrastructure.

References

1. Laudet V (2011) The origins and evolution of vertebrate metamorphosis. Curr Biol 21(18):R726–R737
2. Raymaekers SR, Darras VM (2017) Thyroid hormones and learning-associated neuroplasticity. Gene Comp Endocrinol 247:26–33
3. Rovet JF (2014) The rôle of thyroid hormones for brain development and cognitive function. Endocr Dev 26:26–43
4. Edeline E, Bardonnet A, Bolliet V et al (2005) Endocrine control of *Anguilla anguilla* glass eel dispersal: effect of thyroid hormones on locomotor activity and rheotactic behavior. Horm Behav 48(1):53–63
5. Wilsterman K, Buck CL, Barnes BM, Williams CT (2015) Energy regulation in context: free-living female artic ground squirrels modulate the relatioship between thyroid hormones and activity among life history stages. Horm Behav 75:111–119
6. Sharma P, Tang S, Mayer GD, Patiño R (2016) Effects of thyroid endocrine manipulation on sex-related gene expression and population sex ratios in Zebrafish. Gen Comp Endocrinol 235:38–47
7. Sun BJ, Li T, Mu Y et al (2016) Thyroid hormone modulates offspring sex ratio in a turtle with temperature-dependent sex determination. Proc Biol Sci 283(1841). pii:20161206
8. Nishiwaki-Ohkawa T, Yoshimura T (2016) Molecular basis for regulating seasonal reproduction in vertebrates. J Endocrinol 229(3):R117–R127

9. Geven EJ, Klaren PH (2017) The teleost head kidney: integrating thyroid and immune signalling. Dev Comp Immunol 66:73–83
10. da Fonseca RR, Albrechtsen A, Themudo GE et al (2016) Next-generation biology: sequencing and data analysis approaches for non-model organisms. Mar Genomics 30:3–13
11. Grabherr MG, Haas BJ, Yassour M et al (2011) Full-length transcriptome assembly from RNA-Seq data without a reference genome. Nat Biotechnol 29(7):644–652
12. Haas BJ, Papanicolaou A, Yassour M et al (2013) De novo transcript sequence reconstruction from RNA-Seq: reference generation and analysis with trinity. Nat Protoc 8(8):1494–1512
13. Luo R, Liu B, Xie Y et al (2012) SOAPdenovo2: an empirically improved memory-efficient short-read de novo assembler. Gigascience 1(1):18
14. Birol I, Jackman SD, Nielsen CB et al (2009) De novo transcriptome assembly with ABySS. Bioinformatics 25(21):2872–2877
15. Zerbino DR, Birney E (2008) Velvet: algorithms for de novo short read assembly using de Bruijn graphs. Genome Res 18(5):821–829
16. Schulz MH, Zerbino DR, Vingron M, Birney E (2012) Oases: robust de novo RNA-seq assembly across the dynamic range of expression levels. Bioinformatics 28(8):1086–1092
17. Altschul SF, Madden TL, Schaffer AA et al (1997) Gapped BLAST and PSI-BLAST: a new generation of protein database search programs. Nucleic Acids Res 25:3389–3402
18. Camacho C, Coulouris G, Avagyan V et al (2009) BLAST+: architecture and applications. BMC Bioinformatics 10:421
19. Langmead B, Trapnell C, Pop M, Salzberg SL (2009) Ultrafast and memory-efficient alignment of short DNA sequences to the human genome. Genome Biol 10(3):R25
20. Langmead B, Salzberg SL (2012) Fast gapped-read alignment with Bowtie 2. Nat Methods 9(4):357–359
21. Anders S, Huber W (2010) Differential expression analysis for sequence count data. Genome Biol 11(10):R106
22. Love MI, Huber W, Anders S (2014) Moderated estimation of fold change and dispersion for RNA-seq data with DESeq2. Genome Biol 15(12):550
23. Quinlan AR, Hall AM (2010) BEDTools: a flexible suite of utilities for comparing genomic features. Bioinformatics 26(6):841–842
24. Bradnam KR, Fass JN, Alexandrov A et al (2013) Assemblathon 2: evaluating de novo methods of genome assembly in three vertebrate species. Gigascience 2(1):10
25. Aubry S, Kelly S, Kümpers BMC et al (2016) Deep evolutionary comparison of gene expression identifies parallel recruitment of *Trans*-factors in two independent origins of C_4 photosynthesis. PLoS Genet 12(5):e1006087

Chapter 22

Role of TRs in Zebrafish Development

Federica Marelli and Luca Persani

Abstract

In recent years, the zebrafish has become a powerful model not only for the developmental biology studies, but also for genetic analyses and drug screenings, mostly thanks to the ease with which its embryos can be manipulated and to its translucent body, which allows in vivo imaging. In this chapter, we will provide an overview of the current knowledge about the role of thyroid hormone receptors during zebrafish embryonic development. Moreover, we will explore the methodologies applied to zebrafish biology to knock down a gene of interest and to analyze in vivo the molecular mechanisms of the mutated receptors.

Key words Thyroid hormone, Thyroid hormone receptor, THRA, THRB, Zebrafish

1 Introduction

Thyroid hormones (TH) play fundamental roles in regulating development, differentiation, and metabolism of all vertebrates [1]. Although TH action has been studied mainly in mammals and amphibians, several observations indicate that TH are important in fish development and that almost all of the components of the thyroid physiology have been identified, with their structure and function closely resembling those found in higher vertebrates [2–5].

High concentrations of TH of maternal origin are stored in the yolk of eggs and embryos that provide TH action before the development of a functional thyroid gland from the 72 h post-fertilization (hpf). TH was shown to be a critical regulator of zebrafish embryonic growth and larval transition [6–8], as well as neural development [9, 10]. Exogenous TH can induce premature embryo-to-larva transition and differentiation of the pectoral fins (analogous to the forelimbs of tetrapods), while goitrogens have negative and opposite effects on zebrafish development. In particular, they also inhibit the development of scales and pigment pattern, and impair the growth of both pectoral and pelvic fins [6].

Michelina Plateroti and Jacques Samarut (eds.), *Thyroid Hormone Nuclear Receptor: Methods and Protocols*, Methods in Molecular Biology, vol. 1801, https://doi.org/10.1007/978-1-4939-7902-8_22,
© Springer Science+Business Media, LLC, part of Springer Nature 2018

1.1 TH Receptors in Zebrafish: Nomenclature and Structure

As the other vertebrates, TH act via nuclear thyroid receptors (TRs) to regulate gene expression, typically by binding thyroid response elements (TREs) in their promoter region [11, 12]. Thanks to the introduction of molecular biology techniques, it was possible to clone the genes encoding for TRs in zebrafish [6, 11, 13]. Both TRα and TRβ mRNAs were found to be present in zebrafish embryos and to respond to exogenous TH [4–8]. In contrast to those observed in mammals and chicken in which two genes, one encoding TRα and the other TRβ, have been identified, due to genome duplication, the zebrafish genome encodes two TRα genes (*thraa* and *thrab*), together with a single TRβ gene (*thrb*). Zebrafish *thraa* gene generates two TRα1 isoforms that showed high similarity with TRαs from other vertebrates [11]. A revision of the nomenclature of these isoforms has recently been proposed by us in order to avoid confusion with the classic nomenclature of mammalian TRs [12] (Fig. 1). The short TRα1 (TRα1S, previously reported as TRα1A1) isoform has high homology with human TRα1, whereas the long TRα1 isoform (TRα1L, previously reported as TRα1A1-2) incorporates an additional 14-aminoacid peptide extending its carboxy-terminal "F-domain" that are not found in any other known TR. The so-called F-domain extension does not alter the overall structure of TRα1S, but it reduces the transcriptional activity of the receptor by changing its affinity for the zebrafish coactivator NCoA2 [14]. The sequence of the ligand binding domain of TRαB, the transcript encoded by the *thrab* gene, is very similar to the one of TRα1L, but based on the predicted sequence, it probably has a shorter and not-functional amino-terminal domain and it is lost along evolution [11, 14]. The TRβ isoforms, transcribed from the zebrafish *thrb* locus, are all able to bind DNA and retain the T3-dependent transactivation activity. The zebrafish TRβ1 (or TRβ1S) has the typical structure of mammalian TRβs including the short N-terminal domain. A second TRβ1 isoform (TRβ1L) was described that had a 9-amino acid insert in the hinge region between the DNA-(DBD) and the ligand binding domain (LBD), a feature found in several teleost TRβs but not in other vertebrate classes [15, 16]. Interestingly, the TRβ1L has a diminished affinity for T3 in comparison with TRβ1S [12, 17]. Finally, similar to mammals, birds, and amphibians, zebrafish has the TRβ2 isoform, with a longer N-terminal, which has been reported so far to play important functions in the zebrafish retina [15, 18]. The schematic representation of zebrafish TRs (zTRs) is reported in Fig. 1.

Each TR consists of an amino-terminal regulatory domain, a central DNA binding domain, and a carboxy-terminal hormone binding domain. The high level of conservation between mammals, birds, and fish (sequence homology is as high as 90–95% in the LBD or DBD) indicates that the TRs of all these vertebrates bind T3 and recognize similar TREs, and should therefore undergo

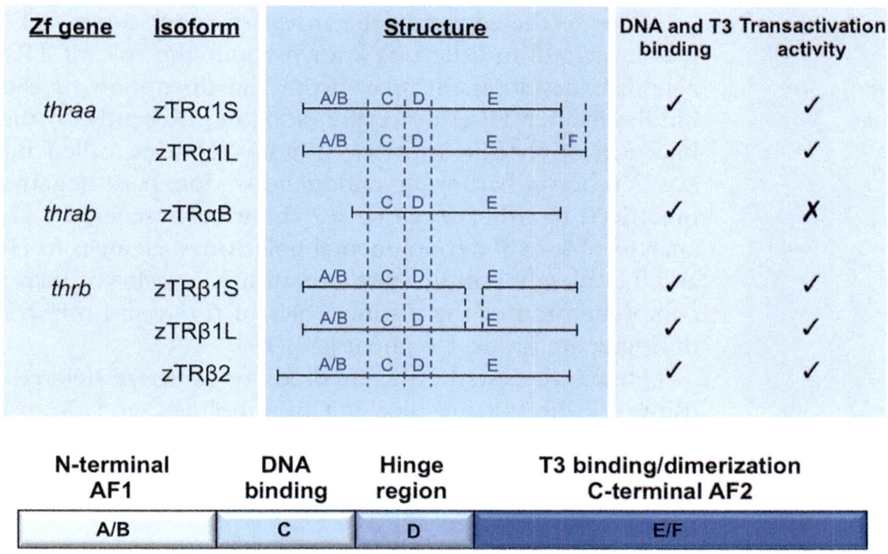

Fig. 1 Representation of the zebrafish TR isoforms (modified from Ref. 12)

dimerization and regulate gene transcription in a similar fashion. This view was recently confirmed by showing that zebrafish and human TRs are functionally interchangeable [12, 17].

1.2 Patterns of Expression of Zebrafish TRs

Confirming the data obtained from other groups using Northern blot and semiquantitative PCR analysis, the quantification of TRs expression by Real-time PCR (qRT-PCR) revealed that both transcripts had maternal and zygotic expression since they were highly expressed from the first hours of development with a predominant *thraa* expression over *thrb* [12]. At the beginning of gastrulation and up to the end of the segmentation period, the expression of *thraa* and *thrb* decreased and became undetectable. After 1 day post-fertilization (dpf), both mRNAs increased significantly and persisted until the larval stage (5 dpf). The expression data obtained using qRT-PCR and whole-mount in situ hybridization (WISH) techniques in embryos and adult tissues confirm that the zebrafish TRs are expressed in a spatio-temporal divergent pattern and that *thraa* is more widely distributed [12, 19]. Similarly to chick [11], the *thraa* is predominantly expressed in brain, heart, pronephros, and gastrointestinal tract, whereas *thrb* is specifically expressed in the pituitary, eyes, and otic vesicles [12]. The expression of *thraa* has been also detected in mature primary neuromast cells that are derived from ectodermal placode and account for the anterior and posterior lateral line organs [20], the mechanic-receptive systems for vibration and pressure that are typical of fish and disappear in the land vertebrates. By WISH experiment, it has been also described that both zTRs are co-expressed at 5 dpf in pectoral fins, swim bladder, and jaw cartilages [12, 19].

Prior to the advent of the antisense knock-down (KD) strategies in zebrafish, little was known about the role of TRs during zebrafish development apart from the disruption of the rostral hindbrain after TRα1 overexpression [21]. Recently, by the microinjection of specific antisense oligonucleotides, called morpholinos, embryos harboring endogenous dominant-negative (DN) mutations in either *thraa* or *thrb* have been generated [12]. The analysis of several developmental milestones, changes in HPT axis, and TH metabolism in these zebrafish KD embryos increased the comprehension of the distinct roles of *thraa* and *thrb* transcripts during zebrafish development [12, 19].

Here, we describe the methods to generate deleted TRs by using specific splicing-blocking morpholinos, and the procedure required to test the pathological mechanism of the mutated receptors using a new in vivo luciferase assay.

2 Materials

1. 30× Danieu Buffer: Reagents to prepare 1 L of 30× stock solution: 101.7 g of NaCl, 1.56 g of KCl, 2.96 g of $MgSO_4 \cdot 7H_2O$, 4.25 g of $Ca(NO_3)_2$ and 35.75 g of HEPES buffer. Add dH_2O to 1 L and stir until dissolved. Check the pH (7.6) and store at 4 °C. Dilute the working solution at 1× in dH_2O.
2. Embryo Medium: Reagents to prepare 5 L of EM: 0.5 g of NaCl, 0.5 g of Istant Ocean, 0.95 g of $CaCl_2$, 2.5 ml of 10× methylene blue. Add dH_2O to 1 L and stir until dissolved. Store at room temperature.
3. Tricaine (3-amino benzoic acid ethyl ester also called ethyl 3-aminobenzoate) comes in a powdered form from Sigma (Cat. # A-5040). It is also available as Finquel (Part No. C-FINQ-UE) from Argent Chemical Laboratories, Inc. Make solution for anesthetizing fish by combining in a glass bottle with a screw cap 400 mg tricaine powder, 97.9 ml dH_2O and 2.1 ml 1 M Tris (pH 9). Adjust pH to ~7. Store this solution in the freezer. To use tricaine as an anesthetic combine in a 250 ml beaker 4.2 ml tricaine and 100 ml clean tank water (0.016%).
4. Protein Extraction Buffer (RIPA): 10 mM Tris (pH 7.4), 2% Triton-X 100, 1 mM PMSF, 1 mM aprotinin, 1 mM leupeptin, and 1 mM trypsin inhibitor. Add 100 ml of ddH_2O. Store at −20 °C.
5. *thraa*- and *thrb* splice-blocking morpholinos (MOs) to knock down the function of *thraa* and *thrb* genes. To generate a functional mRNA with an in-frame deletion follows these tips: (a) the MOs should cover an exon-intron boundary, (b) the exon

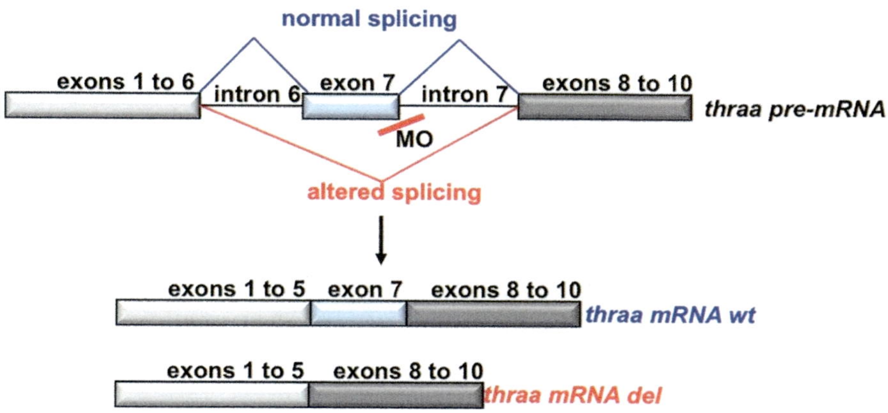

Fig. 2 Morpholino target regions spanning exon-intron boundaries of *thraa* pre-mRNA. The block of the splice donor causes the skipping of the exon 7 during the maturation of the RNAs (aberrant splicing). After MO injection both wild-type and deleted mRNA are obtained

recognized by the MO must contain a number of nucleotides multiple of three, and (c) the MO should bind regions that are not implicated in dimerization or DNA binding (e.g., N-term domain or LBD). To avoid off-target effects, control MOs with five mismatches distributed along the target sequences should be designed. An example of outcomes of splice-blocking MO is reported in Fig. 2. MOs are manufactured and delivered pre-quantitated, freeze-dried, sterile, salt free and lyophilized solid in an amber glass vial by Gene Tools Inc. Solubilize the oligos in double distilled water at a final concentration of 1 mM (stock solution). Heat for 5–10 min at 65 °C for dissociating solid aggregates and briefly vortex (cool them to room temperature prior to use). Maintain stock solutions sterile, tightly capped and in the dark at room temperature.

3 Methods

3.1 Microinjection of MOs in Zebrafish Embryos

1. Once the stock solution is made, MOs are diluted in phenol red and 1× Danieu Buffer. To determine the optimal working concentration dilute each MO to the final concentration of 0.3, 0.6, 0.8, and 1 pmol/embryo (*see* **Note 1**). At higher amounts (above 1.5–2 pmol/embryo, depending on the MO) morpholinos often cause nonspecific effects, such as brain or general cell death (*see* **Note 2**).
2. Obtain eggs from natural spawning of zebrafish adults. Collect the eggs at 1–2 cells stage and rinse them in embryo medium (EM). Arrange the collected embryo in a petri dish with a slide.

Remove nearly all the EM medium for injection that provides surface tension that prevents the embryo/chorion from sticking to the needle when trying to remove it after injection.

3. Hold the needle with the MO mixtures and mount it in the microinjector. Set the pressure and gently break the needle to calibrate the volume to inject by measuring the droplet size. Carefully extend the needle through the chorion and into the yolk and inject. If the injection is successful, a red spot in the yolk or cytoplasm is clearly visible.
4. Rinse the injected embryos in EM and grow them in an incubator at 28 °C.

3.2 Analysis of MO Efficiency

1. Select the injected embryos at the prim-25 developmental stage [22] and resuspend 50 embryos in 200 μl of Trizol Reagent (Life Technologies) for RNA extraction (*see* **Notes 3** and **4**).
2. Disaggregate embryos pipetting and vortexing and add 800 μl of Trizol. Add 200 μl of chloroform and invert 4–5 times. Centrifuge at 13,000 rpm (15,871 × *g*) for 10 min at 4 °C. Collect the aqueous phase in a sterile tube and add 500 μl of cold isopropanol. Invert 50 times and centrifuge at 13,000 rpm (15,871 × *g*) for 10 min at 4 °C. Discard the isopropanol and wash the pellet with 500 μl of cold 75% EtOH. Centrifuge at 13,000 rpm (15,871 × *g*) for 5 min at 4 °C. Discard the EtOH and add 20 μl of nuclease-free water.
3. Retro-transcribe 1 μg/μl of RNA from each condition using the ImProm-II Reverse Transcription System (Promega) and Random primers according to the manufacturer's instructions. Perform a PCR of *β-actin* to check cDNA quality and possible genomic contamination.
4. To analyze the relative expression of the wild-type or deleted TRs by qRT-PCR, design different sets of primers that were used to amplify the different zTRs in which the reverse primers discriminate the wild-type and morphant exon-exon boundaries (Fig. 3).
5. Perform a classic qRT-PCR using SYBRGreen MasterMix (Life Technology). The PCR reaction comprises an initial denaturation step at 95 °C for 30 s, followed by 40 cycles at 95 °C for 5 s, 60 °C for 15 s, and 72 °C for 45 s. At the end of the amplification phase, perform the melting temperature-determining dissociation step at 95 °C for 30 s, 60 °C for 30 s, and 72 °C for 60 s. Analyze the dissociation curve to check the specificity of PCR products. Measure the relative mRNA expression normalizing the raw data against the housekeeping gene *eef1α* using the $2^{-\Delta\Delta Ct}$ method. In our study, we select MO doses, which allow the expression of nearly 1:1 ratio of the zTR_wt/zTR_del (Fig. 3).

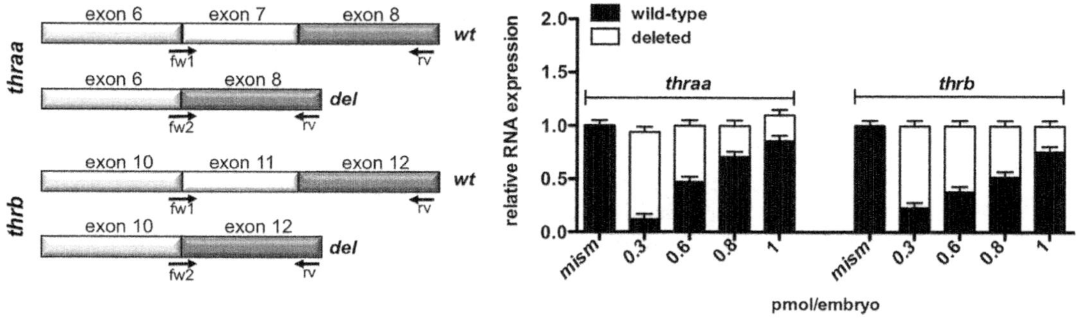

Fig. 3 Scheme of forward and reverse primer targets that allow the discrimination between the wild-type and deleted transcripts. Relative expression of wt and del TRs after MOs injection, analyzed by qRT-PCR

3.3 Cloning the Wild-Type and Deleted TRs

1. Microinject 1–2 cells stage embryos with *thraa* and *thrb* MOs at the selected doses. At 1, 2, and 3 dpf anesthetize the embryos with 0.016% tricaine (Sigma) in EM. For *thraa*, remove the eyes with a needle, and cut the head-trunk region of about 100 embryos and collect them in 200 μl of Trizol. For *thrb*, collect the eyes of about 200 injected embryos in Trizol. Extract and retro-transcribe the RNA as described above (*see* **Note 5**).
2. Design different sets of primers to amplify the entire coding-sequence of wild-type and deleted receptors (zTRα1S, zTRα1L, zTRβ1S, zTRβ1L, and zTRβ2) adding the EcoRI and XhoI restriction sites at the 5′ and 3′, respectively. Analyze the PCR products by gel electrophoresis and cut the bands (*see* **Note 6**).
3. Clone the purified bands into the pCS2+ plasmid. Digest both PCR and plasmid with EcoRI and XhoI for 2 h at 37 °C. Purify the digestion with the Qiagen's enzymatic reaction cleanup kit (Qiagen) and perform the ligation adding 1 μl of T4 ligase and incubating overnight at 16 °C. Transform 100 μl of chemical competent *E. coli* cells with 5 μl of plasmids and incubate for 5 min in ice. Heat-shock the cells for 30 s at 42 °C and immediately transfer the tubes on ice. Add 250 μl of S.O.C. medium and shake at 37 °C for 1 h. Spread 50 μl from each transformation on a pre-warmed LB plate with 50 μg/ml of ampicillin and incubate overnight at 37 °C.
4. Pick 24 colonies from each plate and use PCR to directly analyze positive transformants. For PCR primers, use the T7 Forward and a reverse primer that hybridizes with your insert. Analyze the PCR by gel electrophoresis and confirm by automatic sequencing (*see* **Note 7**).
5. In vitro transcription of the RNA from the selected clones using the mMessage mMachine T7/SP6 Kit (Ambion): linearize the constructs with a restriction enzyme downstream the insert to be transcribed. Add 1/20th volume 0.5 M EDTA,

1/10th volume of 3 M Na acetate, and 2 volumes of EtOH. Mix well and chill at −20 °C for 30 min. Pellet the DNA for 15 min at 13,000 rpm (15,871 × *g*). Resuspend in dH_2O at a final concentration of 1 μg/μl. Assemble the transcription reaction as follows: 10 μl 2× NTP/CAP, 2 μl of 10× Reaction Buffer, 2 μl T7 polymerase, 1 μg/μl of linear template DNA, and nuclease-free water to 20 μl. Mix gently and incubate for 2 h at 37 °C. Add 1 μl of TURBO DNase and incubate for 15 min at 37 °C. Purify the RNAs using the MEGAclear Kit (Ambion) following the manufacturer's instructions.

6. Quantify the RNAs by UV-spectrometry and dilute them at the final concentrations 250 and 125 ng/μl in Danieu Buffer (*see* **Note 8**).

3.4 In Vivo Luciferase Assay

1. Prepare two different mixes for the injection. The first mix (basal control) contains Phenol Red and Danieu Buffer 1×, the pAL*tk*-luciferase reporter (0.25 μg/μl) and the constitutive expressed CMV-driven-β galactosidase reporter (0.25 μg/μl). The second mix is made as the basal control plus 1 pmol/embryo of *thraa* and *thrb* ATG-MOs each (atg-mix) (*see* **Note 9**).
2. Microinject at least embryos at 1–2 cells stage with 4 nl of the control mix and the atg-mix.
3. Depending on the experiments (*see* Subheading 3.5), proceed with a second microinjection of the atg-injected embryos with the diluted RNAs (100 embryos for each transcript).
4. After 6 h post-injection transfer batches of 50 embryos into 3 ml embryo medium that contains either vehicle (DMSO) or the appropriate concentration of T3 (5 μM) and incubate for 24 h at 28 °C.
5. Dechorionate the embryos and transfer them into 1.5 ml tubes (15 embryos/tube) with 1 ml of RIPA buffer. Proceed with the freeze-thaw cycle in liquid nitrogen and vortexing and clear the lysates by centrifugation (15,871 × *g*, 15 min).
6. Measurement of the β-galactosidase and luciferase activity using the Dual Reporter Luciferase Assay (Promega): the activities of firefly and Renilla luciferases are measured sequentially from a single sample. The firefly luciferase reporter is measured first by adding Luciferase Assay Reagent II (LAR II, Promega) to generate a stabilized luminescent signal. After quantifying the firefly luminescence, this reaction is quenched, and the Renilla luciferase reaction is simultaneously initiated by adding Stop & Glo Reagent (Promega) to the same tube that produces a stabilized signal, which is used to normalize the luciferase activity. The transcriptional activity of the zebrafish wild-type and deleted receptors is given in relative luciferase units (RLU × 1000). The schematic representation of this technique is reported in Fig. 4.

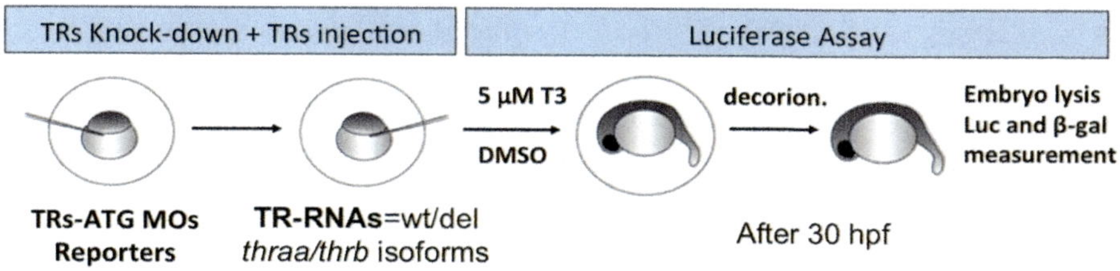

Fig. 4 Representation of methods used for the in vivo luciferase assay

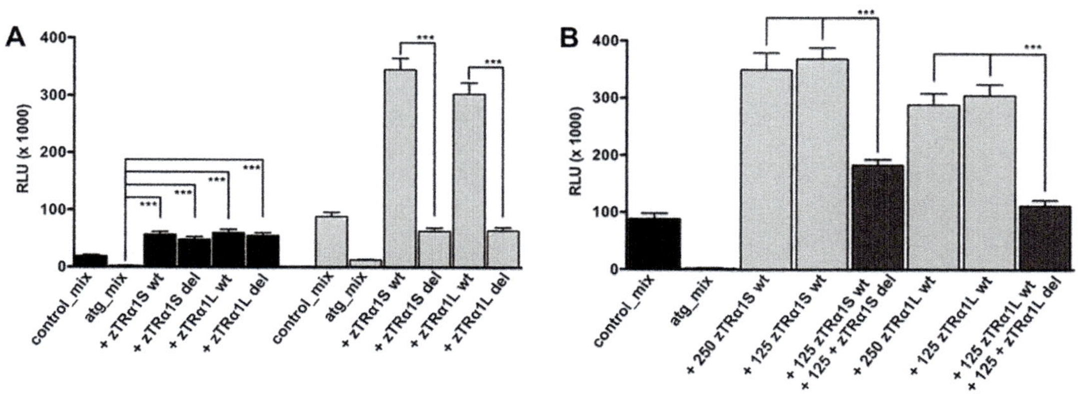

Fig. 5 (**a**) Activity of endogenous and exogenous zTRα1 isoforms in zebrafish embryos in the absence (black bars) or presence (gray bars) of exogenous T3 added into the harvesting water. (**b**) Analysis of the DN mechanism of the deleted zTRα1, in the presence of 5 μM T3. ***$p < 0.001$ (modified from Ref. 12)

3.5 In Vivo Analysis of the Transcriptional Activity of TRs

1. Analysis of the T3-dependent activity of the deleted receptors. Microinject the wild-type and deleted *thraa* (zTRα1S and zTRα1L) and *thrb* (zTRβ1S, zTRβ1L and zTRβ2) mRNAs at 250 ng/μl in embryos previously injected with the atg-mix. One can observe that the wild-type TRs are sensitive to T3 stimulation by the enhanced luciferase activity compared with the vehicle-treated condition (DMSO). In contrast, the deleted TRs present only a negligible increase of luciferase activity (an example is reported in Fig. 5a). These results are consistent with the impaired T3 binding activity of mutant TRs lacking a portion of the LBD [12].
2. DN assay. Co-inject equal amounts of wild-type and deleted TRs transcripts diluted at 125 ng/μl in embryos previously injected with the atg-mix. One can observe a significant fall of the T3-dependent activity compared with the wild-type TRs alone (inject the full or half dose, 250 and 125 ng/μl) (*see* **Note 10**). The results reported in Fig. 5b confirm that the deleted TRs, which are characterized by the loss of T3 binding activity, maintain the ability to bind the DNA and exert a dominant negative effect on the wild-type counterpart.

4 Notes

1. To prepare 5 μl of your working MO solution follow these guidelines: a drop injected in a single embryo corresponds to 4 nl. With 5 μl of solution it is possible to inject 1250 embryos. If you want to inject 0.8 pmol/embryo (0.8 pmol × 1250 embryos = 1 nmol), dilute 1 μl of 1 nM MO with 3 μl of Danieu Buffer and 1 μl of Phenol Red.
2. Common nonspecific effects of MOs are a general delay in development and widespread cell death. Even low amounts of MO delay development, so it is important to stage-match control embryos, rather than age-match them. A second non-specific effect is cell death, which can be low and transient, frequently restricted to the brain, or widespread causing early lethality. By using a p53 mutant or coinjection of a p53 MO, however, cell death can be suppressed [23].
3. After MO injection, a series of morphological defects are visible in your embryos. The amount of MO in each single drop of injection can vary resulting in different degrees of pathological phenotypes. At 1 dpf, screen your injected embryos and discard embryos that present severe defects (brain necrosis, haemorrhages, severe heart malformations, aberrant body axes formation). If the injection is successful, at least 70–80% of the injected embryos present similar phenotype.
4. RNAses are everywhere! Carefully treat the working superficies and your instruments with RNAzap solution (Sigma) or similar.
5. The *thraa* isoforms are mainly localized in brain and heart, whereas the *thrb* isoforms are preferentially detectable in eyes of embryos at 1–3 dpf. Removing the eyes from the *thraa*-injecting embryos (or picking them in the *thrb*-embryos) is necessary to concentrate the transcript of interest avoiding contaminations.
6. Different sets of primers need to be designed to amplify the entire coding sequence of the zTRα1S (fw:ATGGAAA ACACAGAGCAGGA; rv:ACCTAAATTCATACCTCCTG) and zTRα1L (fw:ATGGAAAACACAGAGCAGGA; rv:TCA CCTTAAGCAGGAACCG) isoforms. The same primers pair can be used for the amplification of the zTRβ1S and zTRβ1L isoforms (fw:ATGATGGTTAAGGCCTAATA; rv:ATCAGT CTTCAAACACTTCC), since they differ for 27 bp-insert into the LBD. For zTRβ2 use the following primers: fw:AT GAATTACTGCATGTCAGA; rv:ATCAGTCTTCAAACAC TTCC [12].
7. Using the primer pairs described in **Note 6**, you obtain PCR bands of: 1354 bp for zTRα1S; 1821 for zTRα1L; 2020 and

2011 bp for zTRβ1S and zTRβ1L, respectively; 2135 for zTRβ2. Screen a high number of positive clones to find your targets, especially for the zTRβ1 isoforms that are amplified together.

8. If the RNA appears degraded (e.g., smeared), remove residual RNase from the DNA template preparation before in vitro transcription. Do this by digesting the DNA prep with proteinase K (100–200 μg/ml) in the presence of 0.5% SDS for 30 min at 50 °C, follow this with phenol/chloroform extraction. The RNase Inhibitor that is present in the transcription reaction can only inactivate trace RNase contamination. Large amounts of RNase contamination will compromise the size and amount of transcription products.
9. The ATG MOs against the *thraa* and *thrb* mRNAs are designed between the 5′ UTR and the transcription starting sites [14]. Thus the MO will inhibit the injected RNA as well, and the injected RNA may simply titrate out the MO resulting ineffective silencing of the wild-type gene product. To circumvent this problem, each embryo is injected twice, first with the ATG MOs and then with the RNA. In this way, one can be certain that the amount of MO injected is at the appropriate level to cause the silencing of the endogenous zTRs and that the mRNA is truly expressed. The morpholino is injected first. Then remove all the MO injected embryos. Place 1/2 back in the trough and inject them with RNA.
10. To perform a dominant-negative assay is important to use two different doses of WT (125 and 250 ng/μl in our case) to discriminate if changes in luciferase expression after co-injection of 125 ng/μl of WT + 125 ng/μl of delTR are caused by the DN activity of the mutant receptor or by the half WT amount.

References

1. Brown DD (1997) The role of thyroid hormone in zebrafish and axolotl development. Proc Natl Acad Sci U S A 94(24): 13011–13016
2. Porazzi P, Calebiro D, Benato F, Tiso N, Persani L (2009) Thyroid gland development and function in the zebrafish model. Mol Cell Endocrinol 312(1–2):14–23. https://doi.org/10.1016/j.mce.2009.05.011
3. Opitz R, Maquet E, Zoenen M, Dadhich R, Costagliola S (2011) TSH receptor function is required for normal thyroid differentiation in zebrafish. Mol Endocrinol 25(9):1579–1599. https://doi.org/10.1210/me.2011-0046
4. Porazzi P, Marelli F, Benato F, de Filippis T, Calebiro D, Argenton F, Tiso N, Persani L (2012) Disruptions of global and JAGGED1-mediated notch signaling affect thyroid morphogenesis in the zebrafish. Endocrinology 153(11):5645–5658. https://doi.org/10.1210/en.2011-1888
5. Heijlen M, Houbrechts AM, Darras VM (2013) Zebrafish as a model to study peripheral thyroid hormone metabolism in vertebrate development. Gen Comp Endocrinol 188:289–296. https://doi.org/10.1016/j.ygcen.2013.04.004
6. Liu YW, Chan WK (2002) Thyroid hormones are important for embryonic to larval transitory phase in zebrafish. Differentiation 70(1):36–45. https://doi.org/10.1046/j.1432-0436.2002.700104.x
7. Walpita CN, Van der Geyten S, Rurangwa E, Darras VM (2007) The effect of 3,5,3′-triiodothyronine supplementation on zebrafish (*Danio rerio*) embryonic development and expression

of iodothyronine deiodinases and thyroid hormone receptors. Gen Comp Endocrinol 152(2–3):206–214. https://doi.org/10.1016/j.ygcen.2007.02.020

8. Walpita CN, Crawford AD, Janssens ED, Van der Geyten S, Darras VM (2009) Type 2 iodothyronine deiodinase is essential for thyroid hormone-dependent embryonic development and pigmentation in zebrafish. Endocrinology 150(1):530–539. https://doi.org/10.1210/en.2008-0457
9. Campinho MA, Saraiva J, Florindo C, Power DM (2014) Maternal thyroid hormones are essential for neural development in zebrafish. Mol Endocrinol 28(7):1136–1149. https://doi.org/10.1210/me.2014-1032
10. Zada D, Tovin A, Lerer-Goldshtein T, Vatine GD, Appelbaum L (2014) Altered behavioral performance and live imaging of circuit-specific neural deficiencies in a zebrafish model for psychomotor retardation. PLoS Genet 10(9):e1004615. https://doi.org/10.1371/journal.pgen.1004615
11. Darras VM, Van Herck SL, Heijlen M, De Groef B (2011) Thyroid hormone receptors in two model species for vertebrate embryonic development: chicken and zebrafish. J Thyroid Res 2011:402320. https://doi.org/10.4061/2011/402320
12. Marelli F, Carra S, Agostini M, Cotelli F, Peeters R, Chatterjee K, Persani L (2016) Patterns of thyroid hormone receptor expression in zebrafish and generation of a novel model of resistance to thyroid hormone action. Mol Cell Endocrinol 424:102–117. https://doi.org/10.1016/j.mce.2016.01.020
13. Essner JJ, Breuer JJ, Essner RD, Fahrenkrug SC, Hackett PB Jr (1997) The zebrafish thyroid hormone receptor alpha 1 is expressed during early embryogenesis and can function in transcriptional repression. Differentiation 62(3):107–117. https://doi.org/10.1046/j.1432-0436.1997.6230107.x
14. Takayama S, Hostick U, Haendel M, Eisen J, Darimont B (2008) An F-domain introduced by alternative splicing regulates activity of the zebrafish thyroid hormone receptor alpha. Gen Comp Endocrinol 155(1):176–189. https://doi.org/10.1016/j.ygcen.2007.04.012
15. Jones I, Srinivas M, Ng L, Forrest D (2003) The thyroid hormone receptor beta gene: structure and functions in the brain and sensory systems. Thyroid 13(11):1057–1068. https://doi.org/10.1089/105072503770867228
16. Marchand O, Safi R, Escriva H, Van Rompaey E, Prunet P, Laudet V (2001) Molecular cloning and characterization of thyroid hormone receptors in teleost fish. J Mol Endocrinol 26(1):51–65
17. Marelli F, Carra S, Rurale G, Cotelli F, Persani L (2016) In vivo functional consequences of human THRA variants expressed in the zebrafish. Thyroid. https://doi.org/10.1089/thy.2016.0373
18. Suzuki SC, Bleckert A, Williams PR, Takechi M, Kawamura S, Wong RO (2013) Cone photoreceptor types in zebrafish are generated by symmetric terminal divisions of dedicated precursors. Proc Natl Acad Sci U S A 110(37):15109–15114. https://doi.org/10.1073/pnas.1303551110
19. Bertrand S, Thisse B, Tavares R, Sachs L, Chaumot A, Bardet PL, Escriva H, Duffraisse M, Marchand O, Safi R, Thisse C, Laudet V (2007) Unexpected novel relational links uncovered by extensive developmental profiling of nuclear receptor expression. PLoS Genet 3(11):e188. https://doi.org/10.1371/journal.pgen.0030188
20. Tingaud-Sequeira A, Andre M, Forgue J, Barthe C, Babin PJ (2004) Expression patterns of three estrogen receptor genes during zebrafish (*Danio rerio*) development: evidence for high expression in neuromasts. Gene Expr Patterns 4(5):561–568. https://doi.org/10.1016/j.modgep.2004.02.002
21. Essner JJ, Johnson RG, Hackett PB Jr (1999) Overexpression of thyroid hormone receptor alpha 1 during zebrafish embryogenesis disrupts hindbrain patterning and implicates retinoic acid receptors in the control of hox gene expression. Differentiation 65(1):1–11. https://doi.org/10.1046/j.1432-0436.1999.6510001.x
22. Kimmel CB, Ballard WW, Kimmel SR, Ullmann B, Schilling TF (1995) Stages of embryonic development of the zebrafish. Dev Dyn 203(3):253–310. https://doi.org/10.1002/aja.1002030302
23. Robu ME, Larson JD, Nasevicius A, Beiraghi S, Brenner C, Farber SA, Ekker SC (2007) p53 activation by knockdown technologies. PLoS Genet 3(5):e78. https://doi.org/10.1371/journal.pgen.0030078

INDEX

Michelina Plateroti and Jacques Samarut (eds.), *Thyroid Hormone Nuclear Receptor: Methods and Protocols*,
Methods in Molecular Biology, vol. 1801, https://doi.org/10.1007/978-1-4939-7902-8,

© Springer Science+Business Media, LLC, part of Springer Nature 2018

S

T

V

X

Z

Printed by Printforce, the Netherlands